p 111
energy intensity
p 86 - graph

John Mironowski
ch. 3 pp. 68ff

Agriculture as a Producer and Consumer of Energy

*This book is dedicated in remembrance of
Barry D. McNutt,
senior policy analyst, US Department of Energy.
A superb civil servant, a wonderful human being and a great friend.
We miss his enthusiasm.*

Agriculture as a Producer and Consumer of Energy

Edited by

Joe L. Outlaw

Agricultural and Food Policy Center
Texas A&M University
College Station, Texas
USA

Keith J. Collins

US Department of Agriculture
Washington, DC
USA

James A. Duffield

US Department of Agricutlure
Washington, DC
USA

CABI Publishing

CABI Publishing is a division of CAB International

CABI Publishing
CAB International
Wallingford
Oxfordshire OX10 8DE
UK

CABI Publishing
875 Massachusetts Avenue
7th Floor
Cambridge, MA 02139
USA

Tel: +44 (0)1491 832111
Fax: +44 (0)1491 833508
E-mail: cabi@cabi.org
Web site: www.cabi-publishing.org

Tel: +1 617 395 4056
Fax: +1 617 354 6875
E-mail: cabi-nao@cabi.org

©CAB International 2005. All rights reserved. No part of this publication may be reproduced in any form or by any means, electronically, mechanically, by photocopying, recording or otherwise, without the prior permission of the copyright owners.

A catalogue record for this book is available from the British Library, London, UK

Library of Congress Cataloging-in-Publication Data

Agriculture as a producer and consumer of energy / edited by Joe L. Outlaw, Keith J. Collins, James A. Duffield.
   p. cm.
Based on papers from a conference held June 24-25, 2004 in Arlington, Va.
Includes bibliographical references and index.
ISBN 0-85199-018-5 (alk. paper)
   1. Alcohol fuel industry--United States--Congresses. 2. Agriculture and energy--United States--Congresses. 3. Agricultural wastes--Recycling--United States--Congresses. 4. Renewable energy sources--United States--Congresses. 1. Outlaw, Joe L. II. Collins, Keith J. III. Duffield, James A. IV. Title.

HD9502.5A433A47 2005
33.79'4--dc22

2005003337

ISBN 0 85199 018 5

Printed and bound in the UK by Cromwell Press, Trowbridge, from copy supplied by the editors.

Contents

Foreword	viii
Contributors	x
Unit Conversion Table	xiii
Currency Conversion Table	xiv
Acknowledgements	xv

Part I: Survey of Current Knowledge

1. **Energy and Agriculture at the Crossroads of a New Future** — 1
   Keith J. Collins and James A. Duffield

2. **Agriculture as a Producer of Energy** — 30
   Vernon R. Eidman

3. **Energy Consumption in US Agriculture** — 68
   John A. Miranowski

4. **Energy Systems Integration: Fitting Biomass Energy from Agriculture into US Energy Systems** — 112
   Otto C. Doering III

5. **US Oil and Gas Markets: A Scenario for Future Strong Inter-fuel Competition** — 131
   Kevin J. Lindemer

Part II: Current Research about Agriculture and Energy

Section 1: The Economics of Ethanol and Biodiesel from Grain

6. **Dry-mill Ethanol Plant Economics and Sensitivity** — 139
   Douglas G. Tiffany and Vernon R. Eidman

7. **An Econometric Analysis of the Impact of the Expansion in the US Production of Ethanol from Maize and Biodiesel from Soybeans on Major Agricultural Variables, 2005-2015** — 156
   John N. (Jake) Ferris and Satish V. Joshi

8. **Ethanol Policies, Programmes and Production in Canada** — 168
   K.K. Klein, Robert Romain, Maria Olar and Nancy Bergeron

Section 2: The Economics of Ethanol from Lignocellulosic Sources

9. **Economic Analysis of Alternative Lignocellulosic Sources for Ethanol Production** — 181
   Brian K. Herbst, David P. Anderson, Michael H. Lau, Joe L. Outlaw, Steven L. Klose and Mark T. Holtzapple

10. **The Supply of Maize Stover in the Midwestern United States** — 195
    Richard G. Nelson, Marie E. Walsh and John J. Sheehan

11. **Economic Modelling of a Lignocellulosic Biomass Biorefining Industry** — 205
    Francis M. Epplin, Lawrence D. Mapemba and Gelson Tembo

12. **Economic Impacts of Ethanol Production from Maize Stover in Selected Midwestern States** — 218
    Burton C. English, R. Jamey Menard, Daniel G. De La Torre Ugarte and Marie E. Walsh

Section 3: Energy Conservation and Efficiency in Agriculture

13. **Livestock Watering with Renewable Energy Systems** — 232
    R. Nolan Clark and Brian D. Vick

14. **Trends in US Poultry Housing for Energy Conservation** — 243
    John W. Worley, Michael M. Czarick and Brian F. Fairchild

Section 4: New Methods and Technologies

15. **Experiences Co-firing Grasses in Existing Coal-fired Power Plants** — 254
    Doug M. Boylan, Jack Eastis, Kathy H. Russell, Steve M. Wilson and Billy R. Zemo

16. **Animal Waste as a Source of Renewable Energy** — 266
    Soyuz Priyadarsan, Kalyan Annamalai, Ben Thien, John Sweeten and Saqib Mukhtar

17. **Development of Genetically Engineered Stress Tolerant Ethanologenic Yeasts using Integrated Functional Genomics for Effective Biomass Conversion to Ethanol** — 283
    Z. Lewis Liu and Patricia J. Slininger

18. **Case Studies of Rural Electric Cooperatives' Experiences with Bioenergy** — 295
    Carol E. Whitman

Section 5: Environmental Impacts and Considerations

19. **Potential for Biofuel-based Greenhouse Gas Emission Mitigation: Rationale and Potential** 300
    Bruce A. McCarl, Dhazn Gillig, Heng-Chi Lee, Mahmoud El-Halwagi, Xiaoyun Qin and Gerald Cornforth

20. **Life Cycle Assessment of Integrated Biorefinery-Cropping Systems: All Biomass is Local** 317
    Seungdo Kim and Bruce E. Dale

Glossary 332
Index 338

Foreword

Steve A. Halbrook

As this volume is being completed in the autumn of 2004, crude oil prices have exceeded US$50/barrel, a record nominal price and the highest real price since the energy crisis of the late 1970s. Natural gas prices are at record levels and are expected to remain high for the indefinite future. Energy demand is growing rapidly in China and other parts of the developing world. Natural gas has become the clean fuel of choice for electrical generation in the US, tightening a traditionally soft market. While supplies of crude oil appear to be adequate to meet current demand, the unstable political environment in the oil producing regions of the Middle East and elsewhere has resulted in a 'terror' premium on crude oil prices. This new energy price environment has spurred renewed interest in energy from renewable sources such as agriculture.

This volume grew from a project designed to:

- Provide baseline material on energy use by agriculture to address questions about economic impacts of energy price or supply 'shocks';
- Provide comprehensive baseline information about agriculture as a producer of energy;
- Provide a 'technology roadmap' for agricultural energy consumption and production, including barriers to technological advances, research needs, investment opportunities and economic (price) relationships; and
- Provide the US Department of Agriculture (USDA) with information to assess energy issues for US agriculture and improve administration of its energy-related programmes.

Much of the research base on ethanol and other renewable sources of energy from agriculture was developed during the late 1970s. This project grew from the frustration of using 1970s data and technology to answer the policy questions of the 21st century.

This new energy price environment has generated activity by industry, the US Congress and non-governmental organizations. The *Farm Security and Rural Investment Act of 2002* was the first US 'farm bill' to include an energy title. The US Congress has proposed energy legislation that contains numerous provisions for renewable energy, including a renewable fuels standard. Environmental and conservation groups have proposed raising the Corporate Average Fuel Economy (CAFE) standards for the automobile industry. The petroleum industry is using more ethanol and biodiesel in transportation fuels. Electric power companies are experimenting with co-firing, using various sources of biomass and increasing the use of wind power.

Four principal papers were commissioned by the project planning committee in early 2003. A call for papers was issued in November 2003, with 30 papers selected for presentation. All commissioned and selected papers were presented at a conference June 24-25, 2004, in

Arlington, Virginia. Abstracts of all the papers presented at the conference and a conference executive summary can be found on the conference website http://www.farmfoundation.org/projects/03-35AgAsEnergyProducerAndConsumer.htm

The first part of this volume contains the four commissioned papers and a special introductory chapter by Keith Collins, chief economist, USDA and James Duffield, Office of Energy Policy and New Uses, USDA. Part two of this volume is a compendium of current research about agriculture and energy selected from the other papers presented at the conference. These papers focus on the economic viability of ethanol and biofuels, energy conservation and efficiency in agriculture, new methods and technologies and environmental impacts and considerations. The book also contains a comprehensive glossary and a chart for converting imperial and industry measures to metric measures. Please be aware that some totals in data tables may not sum exactly due to spreadsheet rounding.

Contributors

David P. Anderson, associate professor and extension economist, Department of Agricultural Economics, Texas A&M University, College Station, Texas.
Kalyan Annamalai, professor, Department of Mechanical Engineering, Texas A&M University, College Station, Texas.
Nancy Bergeron, research associate, Centre for Research in the Economics of Agrifood, Laval University, Ste. Foy, Quebec, Canada.
Doug M. Boylan, research engineer, Southern Company, Birmingham, Alabama.
R. Nolan Clark, laboratory director and research leader, Conservation and Production Research Laboratory, Agricultural Research Service, US Department of Agriculture, Bushland, Texas.
Keith J. Collins, chief economist, US Department of Agriculture, Washington, DC.
Gerald C. Cornforth, research scientist, Department of Agricultural Economics, Texas A&M University, College Station, Texas.
Michael M. Czarick, senior public service associate, Department of Biological and Agricultural Engineering, University of Georgia, Athens, Georgia.
Bruce E. Dale, professor, Department of Chemical Engineering and Materials Science, Michigan State University, East Lansing, Michigan.
Daniel G. De La Torre Ugarte, research associate professor, Department of Agricultural Economics, University of Tennessee, Knoxville, Tennessee.
Otto C. Doering III, professor, Department of Agricultural Economics, Purdue University, West Lafayette, Indiana.
James A. Duffield, agricultural economist, Office of Energy Policy and New Uses, Office of the Chief Economist, US Department of Agriculture, Washington, DC.
Jack Eastis, senior research specialist, Southern Company, Birmingham, Alabama.
Vernon R. Eidman, professor, Department of Applied Economics, University of Minnesota, St. Paul, Minnesota.
Mahmoud El-Halwagi, professor, Department of Chemical Engineering, Texas A&M University, College Station, Texas.
Burton C. English, professor, Department of Agricultural Economics, University of Tennessee, Knoxville, Tennessee.
Francis M. Epplin, professor, Department of Agricultural Economics, Oklahoma State University, Stillwater, Oklahoma.
David P. Ernstes, research associate, Agricultural and Food Policy Center, Texas A&M University, College Station, Texas.
Brian F. Fairchild, assistant professor, Department of Poultry Science, University of Georgia, Athens, Georgia.
John N. (Jake) Ferris, professor emeritus, Department of Agricultural Economics, Michigan State University, East Lansing, Michigan.
Dhazn Gillig, economist, American Express Corporation, Phoenix, Arizona.
Steve A. Halbrook, vice president, Farm Foundation, Oak Brook, Illinois.

Brian K. Herbst, research associate, Agricultural and Food Policy Center, Texas A&M University, College Station, Texas.
Mark T. Holtzapple, professor, Department of Chemical Engineering, Texas A&M University, College Station, Texas.
Satish V. Joshi, assistant professor, Department of Agricultural Economics, Michigan State University, East Lansing, Michigan.
Seungdo Kim, research associate, Department of Chemical Engineering and Materials Science, Michigan State University, East Lansing, Michigan.
K.K. Klein, professor and research chair, Department of Economics, University of Lethbridge, Lethbridge, Alberta, Canada.
Steven L. Klose, assistant professor and extension economist, Department of Agricultural Economics, Texas A&M University, College Station, Texas.
Michael H. Lau, research associate, Agricultural and Food Policy Center, Texas A&M University, College Station, Texas.
Heng-Chi Lee, assistant professor, Department of Economics, National Taiwan Ocean University, Keelung, Taiwan.
Kevin J. Lindemer, director, strategy and business, Irving Oil, Ltd., Portsmouth, New Hampshire.
Z. Lewis Liu, research molecular biologist, National Center for Agricultural Utilization Research, Agricultural Research Service, US Department of Agriculture, Peoria, Illinois.
Lawrence D. Mapemba, research assistant, Department of Agricultural Economics, Oklahoma State University, Stillwater, Oklahoma.
Bruce A. McCarl, regents professor, Department of Agricultural Economics, Texas A&M University, College Station, Texas.
R. Jamey Menard, research associate, Department of Agricultural Economics, University of Tennessee, Knoxville, Tennessee.
John A. Miranowski, professor, Department of Economics, Iowa State University, Ames, Iowa.
Saqib Mukhtar, associate professor and extension specialist, Department of Biological and Agricultural Engineering, Texas A&M University, College Station, Texas.
Richard G. Nelson, director and department head, Engineering Extension Programs, Kansas State University, Manhattan, Kansas
Maria Olar, research associate, Centre for Research in the Economics of Agrifood, Laval University, Ste. Foy, Quebec, Canada.
Joe L. Outlaw, associate professor, extension economist and co-director, Agricultural and Food Policy Center, Texas A&M University, College Station, Texas.
Soyuz Priyadarsan, graduate student, Department of Mechanical Engineering, Texas A&M University, College Station, Texas.
Xiaoyun Qin, graduate assistant-research, Department of Chemical Engineering, Texas A&M University, College Station, Texas.
Robert Romain, professor and director, Centre for Research in the Economics of Agrifood, Laval University, Ste. Foy, Quebec, Canada.
Kathy H. Russell, operations and maintenance manager, Plant Mitchell, Georgia Power Company, Albany, Georgia.
John J. Sheehan, senior researcher, National Renewable Energy Laboratory, Golden, Colorado.
Patricia J. Slininger, chemical engineer, National Center for Agricultural Utilization Research, Agricultural Research Service, US Department of Agriculture, Peoria, Illinois.

John M. Sweeten, professor and resident director, Texas A&M University Agricultural Research and Extension Center, Amarillo, Texas.

Gelson Tembo, research fellow, Michigan State University's Food Security Research Project, Lusaka, Zambia.

Ben Thien, research engineering associate, Energy Systems Laboratory, Texas A&M University, College Station, Texas.

Douglas G. Tiffany, research fellow, Department of Applied Economics, University of Minnesota, St. Paul, Minnesota.

Brian D. Vick, mechanical engineer, Conservation and Production Research Laboratory, Agricultural Research Service, US Department of Agriculture, Bushland, Texas.

Marie E. Walsh, adjunct associate professor, Department of Agricultural Economics, University of Tennessee, Knoxville, Tennessee.

Carol E. Whitman, legislative principal in environmental affairs, National Rural Electric Cooperative Association, Arlington, Virginia.

Steve M. Wilson, manager power technologies, Southern Company, Birmingham, Alabama.

John W. Worley, extension engineer and associate professor, Department of Biological and Agricultural Engineering, University of Georgia, Athens, Georgia.

Billy R. Zemo, senior engineer, Gadsden Steam Plant, Alabama Power Company, Gadsden, Alabama.

Unit Conversion Table

Energy industry units

Unit	Equivalent
1 barrel (bbl)	42.000 US gallons (gal)
1 million barrels (MMbbl)	136,370.000 metric tons (MT)
1 British thermal unit (Btu)	1,055.056 joules (J)
1 quadrillion Btu (quad)	1.055 exajoules (EJ)
1 kilowatt-hour (kWh)	3.600 megajoules (MJ)

Imperial units to metric equivalents

Imperial unit	Metric equivalent
1 foot (ft)	0.304 metres (m)
1 gallon (gal)	3.785 litres (L)
1 barrel (bbl)	158.987 litres (L)
1 pound (lb)	0.456 kilograms (kg)
1 ton	0.907 metric tons (MT)

Metric units to imperial equivalents

Metric unit	Imperial equivalent
1 metre (m)	3.281 feet (ft)
1 litre (L)	0.264 gallons (gal)
1 kilogram (kg)	2.205 pounds (lb)
1 metric ton (MT)	1.102 tons

Metric prefixes

Number		Prefix	Abbreviation
10^{18}	1,000,000,000,000,000,000	exa	E
10^{15}	1,000,000,000,000,000	peta	P
10^{12}	1,000,000,000,000	tera	T
10^{9}	1,000,000,000	giga	G
10^{6}	1,000,000	mega	M
10^{3}	1,000	kilo	k
10^{-2}	-100	centi	c
10^{-3}	-1,000	milli	m
10^{-6}	-1,000,000	micro	μ
10^{-9}	-1,000,000,000	nano	n
10^{-12}	-1,000,000,000,000	pico	p

Agricultural industry units

1 bushel (bu) of shelled maize is equal to 56 pounds (lb) or 25.4 kilograms (kg).
1 bushel (bu) of soybeans is equal to 60 pounds (lb) or 27.2 kilograms (kg).

References

Economic Research Service. (1992) *Weights, Measures, and Conversion Factors for Agricultural Commodities and Their Products*. Agricultural Handbook 697. US Department of Agricutlure, Economic Research Service, Washington, DC.

Society of Petroleum Engineers. (2004) *Unit Conversion Factors*. Society of Petroleum Engineers, Richardson, Texas. [Accessed 2004.] Available from http://www.spe.org/spe/jsp/basic/0,,1104_1732,00.html

Currency Conversion Table

Yearly average exchange rate of one US dollar in foreign currency.[a]

	Canada Dollar	Brazil Real	United Kingdom Pound	Germany Mark	European Union Euro[b]
1971	1.01	N/A	2.44	3.48	N/A
1972	0.99	N/A	2.50	3.19	N/A
1973	1.00	N/A	2.45	2.67	N/A
1974	0.98	N/A	2.34	2.59	N/A
1975	1.02	N/A	2.22	2.46	N/A
1976	0.99	N/A	1.81	2.52	N/A
1977	1.06	N/A	1.75	2.32	N/A
1978	1.14	N/A	1.92	2.01	N/A
1979	1.17	N/A	2.12	1.83	N/A
1980	1.17	N/A	2.33	1.82	N/A
1981	1.20	N/A	2.03	2.26	N/A
1982	1.23	N/A	1.75	2.43	N/A
1983	1.23	N/A	1.52	2.55	N/A
1984	1.30	N/A	1.34	2.85	N/A
1985	1.37	N/A	1.30	2.94	N/A
1986	1.39	N/A	1.47	2.17	N/A
1987	1.33	N/A	1.64	1.80	N/A
1988	1.13	N/A	1.63	1.61	N/A
1989	1.18	N/A	1.64	1.88	N/A
1990	1.17	N/A	1.78	1.62	N/A
1991	1.15	N/A	1.77	1.66	N/A
1992	1.21	N/A	1.76	1.56	N/A
1993	1.29	N/A	1.50	1.65	N/A
1994	1.37	N/A	1.53	1.62	N/A
1995	1.37	0.92	1.58	1.43	N/A
1996	1.36	1.01	1.56	1.50	N/A
1997	1.38	1.08	1.64	1.73	N/A
1998	1.48	1.16	1.66	1.76	N/A
1999	1.49	1.82	1.62	1.84	1.07
2000	1.49	1.83	1.52	2.12	0.92
2001	1.55	2.35	1.44	2.19	0.90
2002	1.57	2.92	1.50	N/A	0.95
2003	1.40	3.08	1.63	N/A	1.13
2004	1.30	2.93	1.83	N/A	1.24

N/A = not available or applicable.

[a] Averages of daily figures. Noon buying rates in New York City, New York, for cable transfers payable in foreign currencies.

[b] The euro is the currency of twelve European Union countries: Belgium, Germany, Greece, Spain, France, Ireland, Italy, Luxembourg, the Netherlands, Austria, Portugal and Finland.

Reference

Federal Reserve Bank of St. Louis. (2005) Exchange Rates. *Economic Research.* Federal Reserve Bank of St. Louis. St. Louis, Missouri. [Accessed 2005.] Available from http://research.stlouisfed.org/fred2/categories/15

Acknowledgements

Steve A. Halbrook

This book is the product of a partnership between Farm Foundation, Oak Brook, Illinois, and the Office of Energy Policy and New Uses (OEPNU), United States Department of Agriculture (USDA), to take a fresh look at agriculture as a producer and consumer of energy.

Partnerships and group efforts succeed when each party makes unique contributions to the effort. USDA provided the idea, financial resources and technical expertize. Farm Foundation contributed flexibility, management and contacts throughout the agricultural and academic communities to put the pieces together. This project was inspired by Keith Collins, chief economist, USDA, and made possible by Roger Conway, director, OEPNU, USDA.

The project planning committee was a model of efficiency. They set the agenda, did the work and met all deadlines. In addition to Keith and Roger, the committee included Otto Doering, Purdue University; James Duffield, OEPNU; Marvin Duncan, OEPNU; Vernon Eidman, University of Minnesota; James Fischer, Energy Efficiency and Renewable Energy, US Department of Energy; Steve Halbrook, Farm Foundation; John Miranowski, Iowa State University; Joe Outlaw, Texas A&M University; and Hosein Shapouri, OEPNU. Special recognition is due James Fischer, the only engineer in a group of economists.

Projects of this magnitude reach a successful conclusion when a few people step up and go the extra mile. James Duffield managed the project within USDA and coauthored Chapter 1. Joe Outlaw served as co-editor of this volume and managed the selected paper process. David Ernstes assisted Joe as technical editor. Gretchen Van Houten helped us identify and select a publisher. Mary Thompson coordinated all project communications, publications and press contacts, and Sandy Young managed conference registration and all the details I forgot.

Chapter 1

Energy and Agriculture at the Crossroads of a New Future

Keith J. Collins and James A. Duffield

Energy Risks and Agriculture

As recently as the early 1900s, energy sources around the world were mostly agriculturally derived and industrial products were primarily made from plant matter. Early transportation fuels also came from agriculture. The first petrol engine developed by Henry Ford was designed to run on ethanol and Rudolf Diesel's engine initially was fuelled by peanut oil. However, by 1920 petroleum increasingly became the fuel of choice for transportation, and petrochemicals began to replace starch, vegetable oil and cellulose, that once served as the feedstocks for many early industrial products (Morris and Ahmed, 1992). It was not until the energy crisis of the 1970s and the risks that volatile energy markets presented to the national economy that renewed interest in producing energy from agricultural products was initiated.

The Evolution of Today's Energy Risks

By the early 1970s, domestic oil reserves were shrinking rapidly and the USA was transforming from a major oil producer to a nation dependent on foreign oil. Rapid growth in world energy demand began to out pace world supply, and after years of abundance, oil shortages began to appear in the industrialized world (Yergin, 1991). Taking advantage of tight supplies, exporting countries increased their prices, causing the market price for crude oil to double between 1970 and 1973 (Fig. 1.1). The early 1970s was also a period of increasing political tension in the Middle East. In October 1973, Egypt and Syria made a surprise attack on Israel. Fearing its ally would be defeated, the USA airlifted critically needed munitions to Israel. In retaliation, the Arab oil ministers cut oil production and embargoed the USA, causing a worldwide energy crisis.

The Arab oil embargo symbolized a new era, demonstrating that the USA was no longer the dominant leader in world oil (Yergin, 1991). The embargo quickly caused petrol shortages, gas lines appeared throughout the USA and retail prices climbed 40%. Shortages continued into the early months of 1974, but with oil supply controls beginning to lose their effectiveness, diplomatic efforts finally succeeded in lifting the embargo on March 18. Oil supplies quickly returned to normal; however, retail petrol prices never retreated to pre-embargo levels (Energy Information Administration, 1991). By the mid-1970s, US oil companies had lost their grip on foreign oil and the Organization of the Petroleum Exporting Countries (OPEC) had transformed itself into a powerful cartel. Its members controlled vast financial resources and

© CAB International 2005. *Agriculture as a Producer and Consumer of Energy* (eds J.L. Outlaw, K.J. Collins and J.A. Duffield)

the embargo showed the world that controlling oil supplies could also be an effective political weapon.

The next energy shock took place in 1978 when political unrest in Iran, the second largest oil exporter, resulted in labour strikes, seriously curtailing production. In December 1978, the Iranian oil industry shut down and exports were halted. Revolution in Iran disrupted world oil supplies causing uncertainty in oil markets and triggering panic buying that sent prices on an upward spiral (Yergin, 1991). US refineries that were dependent on Iranian oil experienced shortages and petrol inventories fell in some parts of the country. Petrol lines once again formed around the country and retail prices increased 30% from December 1978 to June 1979 (Energy Information Administration, 1991).

About the time the Iranian oil industry began to recover and oil markets stabilized, war broke out between Iran and Iraq in September 1980. Exports from Iran and Iraq soon dissipated and world oil production dropped significantly, causing prices to rise sharply. However, market conditions had changed since the two previous energy shocks and high prices were not sustained. This potential third energy shock was avoided due to declining demand from a worldwide recession and increasing oil supplies from non-OPEC countries. Higher oil prices since 1973 had spurred investment in new oil sources from Alaska, Mexico, Norway, Russia and other non-OPEC countries (Yergin, 1991).

The fact that the Iran-Iraq war did not cause a third energy shock in 1980 indicated the capacity for oil markets to adjust over time. After 1973, the private and government sectors developed strategies to respond to both supply and demand issues. On the supply side, oil companies increased exploration activity in non-OPEC countries and the supply of oil and oil reserves increased significantly. Natural gas development increased in North America and became the fuel of choice for power generation. Over time, the USA has reduced oil imports from the Arab OPEC countries from 46% of total imports in 1978 to 26% in 2003 (Energy Information Administration, 2004f). Industries switched their energy sources from liquid fuels to coal, natural gas and electricity. The US Congress funded the Trans Alaska Pipeline and created the Strategic Petroleum Reserve. Policymakers began to look to agriculture as a source of energy supply and legislation was passed to encourage renewable fuel production and fund research on developing ethanol, biodiesel and solar power.

Americans became more conservation minded and industries increased their energy efficiency. US farmers decreased their energy use significantly following the 1978 energy shock and liquid fuel consumption on farms dropped 30% by 1985 (US Department of Agriculture, 1997). The US Congress set fuel efficiency standards for the automobile industry. The US government adopted building energy-efficiency standards and required government motor fleets to purchase alternative fuelled vehicles. Federal and state tax credits were adopted to encourage renewable fuel production and power generation (North Carolina Solar Center, 2004). Supply and demand adjustments helped reverse the trend of rising oil prices of the 1970s. OPEC's market power lost its potency and lower prices prevailed throughout the 1980s and most of the early 1990s. Prices did spike briefly beginning in August 1990 when Iraq invaded and occupied Kuwait (Fig. 1.1). This conflict led to the Persian Gulf War that began in January 1991 when an international coalition of military forces led by the USA began a massive air strike on Iraqi targets. The war ended quickly when Iraq surrendered to coalition forces on March 3. With tensions easing in the Middle East, oil prices quickly returned to pre-Gulf War levels, remaining low throughout most of the 1990s (Energy Information Administration, 2004e).

By the end of the 1990s, increasing world energy demand once again began to exert upward pressure on prices and energy markets became increasingly volatile. Crude oil prices in the USA exceeded US$30/barrel (bbl) in January 2000 and remained above US$25/bbl throughout most of 2001 and 2002 (Energy Information Administration, 2004e). The 2003 Iraq war and its aftermath, including continuing terrorist actions, increased tensions in the world oil market, helping to push daily oil prices above US$30/bbl throughout much of the year. Combined with strong global economic growth and oil demand, daily oil prices reached the US$40/bbl range in mid-2004, and exceeded US$55/bbl by the end of October.

Recent supply disruptions in the natural gas industry have also caused major price shocks. In the winter of 2000/2001 natural gas prices rose sharply and moved well above historic levels (Energy Information Administration, 2004d). Retail prices of natural gas peaked in February 2001 at US$9.80 per million cubic feet (Mcf), a 65% increase from February 2000. As the supply of natural gas recovered, prices fell back to normal levels and remained stable throughout 2002. However, prices spiked again in early 2003, reaching US$8.96/Mcf in March, not as high as the 2001 peak price, but 42% higher than March 2002. Moreover, prices did not fall back to the historical trend, as they did after the 2001 price shock. Estimates for natural gas reserves in North America were adjusted downward during the first half of 2004 and industry forecasters projected the 2005 annual spot price to almost double the level of the 2002 price (Cambridge Energy Research Associates, 2004). Higher natural gas prices have caused a major restructuring of the US fertilizer industry that is currently struggling with record high input costs. Prices of nitrogen fertilizer and fuel approached historic highs as farmers began the 2004 growing season (National Agricultural Statistics Service, 2004a).

Understanding the Challenge of Energy Risk for US Agriculture

Energy security has taken on a new dimension since the September 11, 2001, terrorist attack on the Pentagon and the World Trade Center. Policymakers, more than ever, are concerned with energy shortages and escalating prices. During the 1970s energy crisis, the Carter

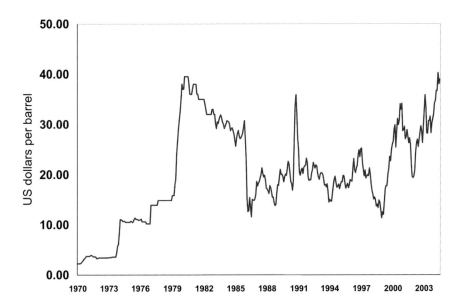

Source: Commodity Research Bureau (2004).

Figure 1.1. Monthly average petroleum prices, West Texas intermediate crude, 1970-2004.

administration created USDA's Office of Energy to develop energy polices and strategies for US agriculture. Twenty-five years later, the Office of Energy, now called the Office of Energy Policy and New Uses (OEPNU), continues to develop information on agriculture's consumption and production of energy and assists in identifying long-term solutions to US energy problems.

In cooperation with the Farm Foundation, OEPNU organized the 'Agriculture as a Producer and Consumer of Energy' conference on June 24-25, 2004, that brought together industry experts, economists, government leaders and others to gather baseline information and identify agriculture's role in formulating sound energy policies. The goals of the conference were to provide comprehensive baseline material on energy use by agriculture; to provide comprehensive baseline information about agriculture as a producer of energy; to generate information and analysis that could help private businesses and federal and state governments develop and administer energy-related activities and programmes; and to foster better relationships and exchange of information between USDA, other federal and state agencies, the private sector, environmental and other non-profit organizations and academia.

The conference provided much of the research, analysis and discussion presented in this book. This chapter continues with an overview of agriculture as an energy producer and consumer, closing with a discussion generated from the conference. The next four chapters provide a survey of current knowledge, obtained through commissioned papers from leading economists and energy experts. Vernon Eidman examines the economics, resource base and potential growth of bioenergy in Chapter 2. John Miranowski presents energy use on the farm and discusses rural energy security in Chapter 3. In Chapter 4, Otto Doering identifies the challenges of integrating biomass energy into current US energy systems. Chapter 5 provides a global view of the current energy situation and a long-run energy outlook by Kevin Lindermer. The remaining chapters include material from eight selected paper sessions organized for the conference. The papers analyse the economic and environmental effects of bioenergy production; evaluate various bioenergy technologies; assess the effectiveness of government policy on bioenergy development; estimate farm energy expenditures; and show the benefits of adopting energy-saving technologies and alternative energy sources on crop and livestock farms. The conference papers represent the current state of knowledge, provide baseline data and identify research needs.

Interaction between Agriculture and Energy Markets

Agriculture and energy markets have much in common. The US consumer wants affordable, healthy and abundant food and energy. Over the last several decades, this goal periodically has been both elusive and achievable, as supplies and prices of food and energy have often fluctuated widely. The US Department of Commerce reports the Consumer Price Index (CPI) each month and the core CPI – the CPI excluding food and energy – is watched as a crucial indicator of inflation. But food and energy are omitted from the core CPI because their high price variability relative to other goods may often reflect supply shocks, rather than general inflation in the economy. While the up-and-down necessities of food and energy may be linked in the public's mind due to their effects on household budgets, the two are much more fundamentally linked economically and that linkage is evolving in new ways.

Farm Economic Cycles and Productivity Gains Point to Energy Markets

The long-term trends in the US farm economy have been powerful forces for generating strong interest in energy markets among the farm economy's participants. The tendency for agricultural markets to generate surpluses and low market prices has focused attention on the role of purchased production inputs that are important contributors to farm production costs, such as diesel fuel or fertilizer, and what can be done to lessen their impacts on farm profitability. Low-price farm commodity markets also have stimulated a search for new markets, such as ethanol, biodiesel and other energy and industrial products, to generate increased demand for farm commodities.

Agricultural markets are cyclical, responding to the strength of the global economy, weather around the world and policy changes in trading countries. Since the oil crisis of the early 1970s, the US farm economy has experienced a series of boom and bust periods. The variability in US net cash farm income (gross cash income minus gross cash expenses for all US farming operations) earned from the marketplace illustrates these periods (Fig. 1.2). Excluding government payments to producers, US farm income shows the influence of market forces unobscured by government financial support to producers. In the mid-1970s, a surge in foreign demand for US grain and oilseeds, led by the entry of the Soviet Union into world grain markets, depleted US commodity stocks and farm prices soared. Farmers responded to rising market returns by expanding acreage, urged on by policy officials calling for planting from 'fence row to fence row'.

By the late 1970s and into the early 1980s, the prosperity was turning to economic stress as US agricultural production increased and farm prices declined. The problem was then exacerbated by a slowing global economy, rising foreign production and a strengthening value of the dollar. Reduced exports and farm prices and rising interest rates led to a 'credit crisis' in the mid-1980s, where farmers who had expanded and bid up land prices sharply

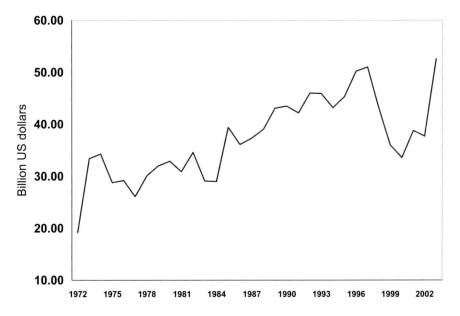

Source: Economic Research Service (2004b).

Figure 1.2. US net cash farm income excluding goverment payments, 1972-2003.

could no longer service their debts. Agricultural markets generally recovered in the late 1980s and prospered into the 1990s, reaching a zenith in the mid-1990s as the global economy grew sharply and foreign agricultural production declined. That boom came to an abrupt end in 1997 and 1998 when the currencies of a number of countries collapsed and a global economic slowdown reduced US farm sales. By 2003, markets had recovered, with farm income (including government payments) reaching a record high in 2003, and US farm prices setting new highs in the spring of 2004.

US government farm price and income support programmes have buffered the periodic declines in farm incomes by providing mechanisms to increase prices, such as production control programmes in existence prior to 1996, and direct payments. Many of the direct payment programmes have been and continue to be related to market prices, so that declines in market prices increase federal spending on price and income support programmes. The desire to reduce farm programme spending, particularly when federal budget deficits prevail, has been another force behind the interest in finding new markets for farm commodities.

Although weather, the strength of the global economy and other factors periodically contribute to declining farm prices, a persistent factor affecting long-term price trends has been the strong productivity growth of US agriculture. The annual average growth in US agricultural productivity from 1948 through 1999 was 1.9%, one of the highest rates of growth among sectors of the US economy (Heimlich, 2003). This growth reduces the resource cost of producing a unit of output, which in competitive markets, results in declining real prices of farm commodities for buyers. Figure 1.3 illustrates that the price of maize, adjusted for inflation, has been trending down for decades. The impact of productivity growth, while salutary for food consumers and the rate of growth of the overall economy, means an ever-decreasing number of farmers can supply what the market demands. The upshot is that marginal producers exit production agriculture and remaining producers consolidate (Paarlberg and Paarlberg, 2000).

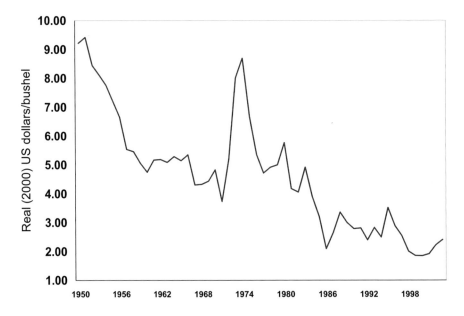

Source: Data adjusted for inflation from Economic Research Service (2004a).

Figure 1.3. Real maize prices (2000 US dollars), 1950-2003.

Efforts to mitigate the adverse effects of productivity gains on some producers have spurred the search for expanding markets in several ways, such as finding new buyers, producing and marketing new farm products and developing new uses for farm products. A primary effort to find new buyers has been to expand exports to faster growing markets overseas, particularly where food consumption responds more strongly to income growth than in the USA. While total US agricultural exports over the past two decades have grown, most of this growth has primarily come from increased sales of value-added and high value products – horticultural, livestock, semi-processed and consumer-ready products.

Export growth for bulk commodities (wheat, rice, coarse grains, soybeans, cotton and tobacco) has been flat over the past two decades (Fig. 1.4). Area planted by US producers to bulk commodities was 250 million acres in 2004, accounting for 77% of total area planted to crops, with hay accounting for most of the remaining share of acreage (National Agricultural Statistics Service, 2004b). Bulk commodities used for animal feed have benefited through exports of livestock products, which represent indirect exports of animal feeds. Bulk commodities, such as wheat and rice, face slow growth in demand in many countries because as incomes rise, consumption patterns shift to processed products, meats and horticultural products, slowing the increase in consumption for wheat and rice. US exports of bulk commodities also face strong competition in the world market. In addition to traditional competition from large exporters, such as Australia, Argentina, Canada and the European Union, another group of countries has been rapidly expanding bulk exports in recent years, including India, Russia, Ukraine and Brazil. Brazil, for example, has increased area planted to soybeans by nearly 21 million acres between 1999 and 2003. Thus, the dominance of bulk crops on the US landscape, the limited growth in bulk exports, and the emerging global competition provide a strong impetus to find new markets for US bulk commodities.

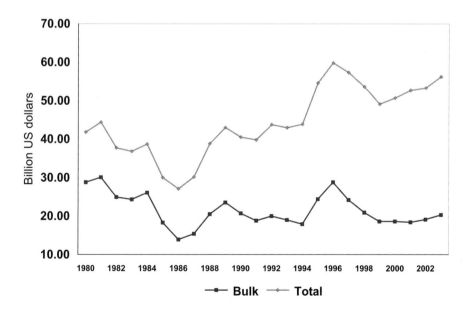

Source: Foreign Agricultural Service (2004).

Figure 1.4. US agricultural exports, 1980-2003.

Energy Market Trends Signal New Opportunities for Agriculture

Energy markets, like agricultural markets, reflect the interaction of variable global production and demand, giving rise to sharp fluctuations in prices. For agriculture, as a consumer of energy, spikes in oil and natural gas prices have raised considerable economic concern. The rise of crude oil prices since the late 1990s led to sharply higher prices of petrol and diesel fuel. Petrol is used in older and smaller farm equipment and in small trucks. Diesel, the most widely used farm fuel, is used for trucks, tractors, combines and other equipment. Irrigation pumps, used on 38 million acres of US farmland, are powered by liquid fuels, natural gas and electricity. Natural gas is also important in crop drying, building temperature control and fertilizer production. When natural gas prices rose from less than US$2/Mcf in the late 1990s to about US$10/Mcf during the winter of 2000-2001 and remained at more than double the levels of the late 1990s, the prices of nitrogen fertilizers increased sharply. Nitrogen fertilizer production costs are 75% to 90% natural gas. The energy price increases of 2000-2001 were especially alarming to agricultural producers as fertilizer plants began shutting down, and producers became concerned about availability of fertilizer supplies for crops planted in spring 2001 (Gullickson, 2001). Natural gas prices spiked again in 2003, staying at levels well above the 1990s. A return to the earlier levels is not expected (Energy Information Administration, 2004g).

The realignment of energy prices that occurred in the early 2000s, once again has spurred interest in the development of alternative energy sources and energy conservation. President George W. Bush created a National Energy Policy Development Group (NEPDG) in early 2001 to '...promote dependable, affordable, and environmentally sound production and distribution of energy for the future' (National Energy Policy Development Group, 2001, p. viii). The NEPDG concluded that the US energy policy should encourage a diverse and clean portfolio of domestic energy supplies and that renewable energy supplies help diversify the US energy portfolio. They argued, 'These nondepletable sources of energy are domestically abundant and often have less impact on the environment than conventional sources. They can provide a reliable source of energy at a stable price, and they can also generate income for farmers, landowners, and others who harness them' (National Energy Policy Development Group, 2001, p. 6-1).

The economic and national security incentives for developing technologies that enable integration of agricultural-based renewable energy into the national energy portfolio are largely provided by the changing supply, demand and price environments of bulk agricultural commodities and fossil energy. After the initial jump in crude oil prices during the oil crisis of the early 1970s, oil prices stabilized in the late 1970s, but then pushed up to record high levels in the early 1980s. Real oil prices (annual average for the contiguous 48 states) reached their record high of US$56/bbl in 1981 (2002 dollars) before dropping to US$19/bbl in 1989. The 1990s saw substantial variability as well, with real oil prices ranging from about US$16.50/bbl to US$28.50/bbl.

The US Department of Energy's (US DOE) reference case projects oil reserves in the lower 48 states steadily declining through 2025, and reserves declining after 2008, even under a high price scenario (Energy Information Administration, 2004a). Crude oil production in the reference case declines after 2008. Alternative outcomes are possible depending on technology development and crude oil prices; however, under all scenarios US oil consumption rises while US oil production, including Alaska, trends down through 2025.

These projections have two important implications for US agriculture. First, US dependency on imported oil is expected to increase. US imports of crude oil accounted for 53% of US consumption in 2002, and rise to 70% in the US DOE reference case. Under these projections, national security and vulnerability to supply disruptions would continue to be paramount concerns for the US public. Second, the real price of crude oil is expected to trend up through 2025. The rising dependency on petroleum imports and increasing real prices are expected to intensify efforts to slow the rate of growth of imports by increasing energy derived from renewable sources.

Environmental Issues Further Suggest Increased Role for Renewable Energy

The use of fossil energy is a major contributor to air pollution. The primary sources are electricity generation, other industrial processes and vehicles. Key pollutants include sulphur dioxide (SO_2), nitrogen oxides (NO_x), mercury, carbon monoxide (CO) and volatile organic compounds (VOC) (National Energy Policy Development Group, 2001). These pollutants cause cancer, heart disease, respiratory problems and damage ecosystems. Regulatory action and technological advances have reduced emissions of the key pollutants, despite the growth in the US economy and more vehicle miles driven. A range of strategies are in development to continue reducing pollutants, such as new pollution control technologies that reduce emissions from combustion as well as changes in fuels that prevent emissions. Shifting to low emission energy sources, such as natural gas, nuclear, or renewable energy, is a growing choice for avoiding emissions.

Concentrations of greenhouse gasses (GHGs) in the Earth's atmosphere have also become a global concern because of their potential to result in long-term climate changes. Evidence indicates that human activity has contributed to an increase in the Earth's surface temperature over the past century. Concentrations of GHGs continue to rise, with US emissions up 13% between 1990 and 2002 (US Environmental Protection Agency, 2004b). About 82% of US GHG emissions are from combusting fossil fuels used to generate electricity and operate vehicles. Other emissions are from methane, primarily from landfill wastes, livestock production (enteric fermentation), natural gas systems, coal mining, production of industrial chemicals and other sources.

US agriculture's contribution to air pollutants as a consumer of energy is small, as agricultural crop, livestock and poultry production accounts for only about 1% of direct energy use (US Department of Agriculture, 2004a). However, production agriculture accounts for about 7% of annual net GHG emissions. Increases in agricultural productivity and conservation can reduce energy intensity of agricultural production and contribute to reduced pollution. Greater conservation through adoption of alternative management practices and crop choices can reduce GHG emissions and increase carbon storage in biomass and soils (carbon sequestration).

There is a significant opportunity to reduce pollutants and GHG emissions by replacing fossil energy with renewable energy and bioproducts derived from farms. Ethanol and biodiesel are prime examples. Ethanol, which is 35% oxygen, improves combustion and reduces carbon monoxide emissions and exhaust VOC emissions. Although ethanol can increase VOC emissions when blended with conventional petrol, this effect can be reduced, as it is in reformulated petrol markets where smog is a problem, by reducing the volatility of the petrol used for blending. Ethanol is also harmless and biodegradable. It is increasingly favoured over methyl tertiary-butyl ether (MTBE) as a source of oxygen in fuel because it does not contaminate groundwater as MTBE does. Like ethanol, biodiesel has many desirable environmental properties. It is non-toxic, biodegradable, and biodiesel exhaust emits less toxic air emissions, carbon monoxide and particulate matter than petroleum diesel (Graboski and McCormick, 1998). Biodiesel also contains no sulphur. However, when blended with diesel, it can increase nitrogen oxide emissions (NO_x). This problem can be corrected with fuel additives.

GHG emissions can be reduced using ethanol and biodiesel compared with petrol and diesel. Biofuels have the advantage that the plants grown each year to produce the fuel sequester carbon, which offsets the carbon released during fuel combustion (National Renewable Energy Laboratory, 1998; Wang *et al.*, 1999; and Levelton Engineering Ltd. *et al.*, 1999). In contrast, the combustion of fossil fuels releases carbon that took millions of years to be removed from the atmosphere and there is no recycling effect. The GHG emissions reduction potential has improved over time as ethanol's positive net energy balance has increased. A 2003 study found that a gallon of ethanol contained 34% more energy than was required to produce it, up from 24% estimated in 1995 (Shapouri *et al.*, 1995; 2003). Updated information for 2004 provided by Shapouri for this chapter raise the estimate to 67%. Increased production of dedicated biomass crops, a potentially large source for the production of bioenergy and bioproducts that replace fossil energy based products, would also reduce GHG emissions. Another potentially large source of renewable energy is livestock waste that can be turned into electricity through anaerobic digestion, which reduces methane emissions from manure.

Agriculture's Response to Energy Market Developments

US agriculture has responded to incentives to conserve energy and produce energy. These incentives include higher energy prices relative to farm prices and government programmes stemming from national security, environmental and farm and rural economic concerns. Technological changes that have made energy conservation and renewable energy production more cost effective have also spurred the response in agriculture.

Energy Efficiency on the Farm

Energy consumption on farms trended up from the late 1960s to the late 1970s, as the strong farm economy of that period appeared to overshadow the oil crisis of the early 1970s. However, the downturn in the farm economy in the 1980s, combined with another round of oil price increases, initiated a decline in energy consumption in agriculture (and CO_2 emissions) especially between the late 1970s and the late 1980s, as energy use fell 30% from 2.5 quadrillion British thermal units (Btu) in 1978 to 1.7 quadrillion Btu in 1989 (Chapter 3, p. 70; US Department of Agriculture, 2004a). Since then, annual energy use in agriculture has been more stable, varying from 1.7 trillion Btu to 1.9 trillion Btu.

Notable changes accounting for the decline in energy consumption on US farms included the shift from petrol to more efficient diesel fuel, use of larger equipment, conservation tillage, more energy efficient irrigation systems and a decline in fertilizer use. Between 1970 and 2002, petrol use in agriculture declined from 0.54 to 0.15 quadrillion Btu, while diesel fuel consumption rose from 0.30 quadrillion Btu to 0.47 quadrillion Btu. The shift to diesel engines reflected the energy efficiency of diesel fuel as well as the increasingly larger, higher horsepower, multiple-function equipment purchased by farmers as farm size has grown. Conservation tillage, which uses much less energy than conventional tillage practices, rose sharply during the late 1980s and continued through the 1990s. For example, the portion of maize acres farmed under no-till residue management rose from 7.4% in 1989 to 17.9% in 2000, while soybeans under no-till rose from 7.7% to 32.8% during the same period (Heimlich,

2003). Fertilizer consumption by volume has been stable over the past decade although the energy content of fertilizer has declined as the energy efficiency of fertilizer production has increased. Thus farm energy consumption in the form of fertilizer has dropped by nearly 25% since 1995.

In addition to changes in the consumption of energy intensive inputs in production, the increases in farm productivity since the 1970s have caused a sharp increase in energy efficiency in agriculture. In 1978, the peak year of farm energy use since 1970, the direct and indirect consumption of energy in US agriculture was 12,550 Btu/US$ of real production (Chapter 3, p. 111). By 2002, energy use had declined to 7,600 Btu/US$ of real farm production – nearly a 40% decline in energy intensity of production. While production agriculture remains highly dependent on energy inputs, this improvement in energy efficiency reduces the vulnerability of US farms to energy supply shocks.

Renewable Energy Production

US production of ethanol from maize is a remarkable story of growth, with production rising from about 60 million gallons (gal) in the mid-1970s to 3.4 billion gal in 2004. National energy security concerns, cleaner petrol standards and government tax incentives have been the primary stimuli for this growth (Lee, 1993). The initial impetus for ethanol commercialization came during the 1970s in response to petrol shortages. About the same time, the US Environmental Protection Agency (US EPA) was looking for a replacement for lead additives to petrol used to boost the octane level. Because of its high octane and oxygen content, ethanol soon established a role as an octane enhancer (Lee and Conway, 1992). In 1990, ethanol production received a major boost with the passage of the *Clean Air Act Amendments of 1990* (CAA). Provisions of the CAA established the Oxygenated Fuels Program and the Reformulated Gasoline (RFG) Program to control carbon monoxide and ozone problems. Both programme fuels require oxygen, and blending ethanol has become a popular method for petrol producers to meet the new oxygen requirements mandated by the CAA.

Another factor helping ethanol growth has been technological improvements in production. Extracting more ethanol from a bushel (bu) of maize, improving the value of by-products and increasing the efficiency of production have reduced the costs of producing ethanol. While data are limited to draw conclusions about long-term trends, USDA cost of production surveys indicate that ethanol dry-mill plants saw a 15% decline in operating costs between 1987 and 1998 while wet-mill plants experienced little change (Shapouri *et al.*, 2002). Expansion of ethanol production has occurred with dry-mill plants due to their lower operating and capital costs.

More recently, ethanol increased its market share in the oxygenated fuels market when MTBE, the only other oxygenate used in the USA, was banned in California, effective January 1, 2004. California took this action when it was discovered that MTBE has a propensity to contaminate ground and surface water (Blue Ribbon Panel on Oxygenates in Gasoline, 1999). This created an additional market for ethanol estimated at 700 million gal to 800 million gal. Following California's lead, New York and Connecticut, also large RFG users, banned the use of MTBE at the beginning of 2004. Ethanol capacity has expanded rapidly to meet this new demand. Twenty-eight new plants have been constructed since 2001, increasing capacity by almost 1.7 billion gal.

The production of ethanol, now accounting for about 12% of US maize use, helps bolster farm prices and farm income (US Department of Agriculture, 2004b). If annual renewable fuel use were to grow to a minimum of 5 billion gal, as proposed by energy legislation under consideration by the US Congress calling for a Renewable Fuels Standard (RFS), the price of maize would increase by US$0.32/bu, according to an USDA analysis conducted in 2000 (US Department of Agriculture, 2000). The USDA study estimated that a 5 billion gal ethanol market would increase annual net farm income by almost US$3 billion. In addition, increasing ethanol production lowers the US trade deficit and creates new jobs, many in rural areas where employment growth is relatively slow (US Department of Agriculture, 2002). Higher maize prices from increased ethanol demand can also lower farm programme payments.

Biodiesel, a diesel fuel substitute and additive made from oil crops and animal fats, is just emerging as a commercial fuel in the USA (Duffield *et al.*, 1998). Most biodiesel is made from soybean oil. Soybean farmers have funded much of the research and development for biodiesel over the past decade. The second most common feedstock for biodiesel is recycled cooking oil that is supplied primarily by the rendering industry. Most biodiesel produced today is blended with petroleum diesel at a 20% blend (B20) and sometimes 1% to 2% biodiesel is added to diesel fuel to increase lubricity. Until recently, biodiesel production was used primarily for demonstration and research projects amounting to less than 200,000 gal/year. With implementation of USDA's Bioenergy Program, biodiesel production climbed to an estimated 6.5 million gal in fiscal year (FY) 2001 and exceeded 20 million gal in FY 2003. The Bioenergy Program makes cash payments to manufacturers who produce biodiesel from qualified feedstocks, including most oil crops, animal fats and recycled fats, oils and greases (Farm Service Agency, 2004). Biodiesel is used by trucks, buses, trains, electric generators, boats and other equipment operated by diesel engines. It can also be used to fuel furnaces and commercial burners. Government agencies are important purchasers of biodiesel as they earn credits toward alternative fuel vehicle requirements established by the *Energy Policy Act of 1992* and amended by the *Reauthorization Act of 1998* (US Department of Energy, 2004b).

A USDA analysis estimating the effects of biodiesel production on soybean oil demand showed that an increase in biodiesel demand sufficient enough to increase soybean oil use by an average of 1.5 billion pounds/year, over a 10-year period, would increase the average soybean oil price by 22% (US Department of Agriculture, 2001). The average farm price for soybeans would increase by 3%. The study also showed a 0.7% increase in annual net farm income over the 10-year period. The US trade deficit declined a total of US$1.55 billion and 13,000 jobs were created across the US economy, mostly in the agricultural sector.

Biodiesel may also create opportunities for the rendering industry that is positioned to become a leader in biodiesel production due to its access and control of large volumes of low-value feedstocks (Talley, 2004). Producing biodiesel with animal wastes, fats and rendered products, such as yellow and brown grease, can reduce feedstock cost significantly compared to soybean oil and other oil crops. For example, yellow grease is typically priced US$0.10/pound to US$0.15/pound lower than soybean oil, resulting in about US$1.00/gal savings. Recycled fats and oils are currently used in animal feeds and fatty acids are produced for use in plastics, lubricants, paints, soaps, cosmetics and many other commercial materials. However, using some animal products in food and feed markets is now prohibited because of health concerns over bovine spongiform encephalopathy, commonly known as mad cow disease. Food processors and renderers are currently searching for new uses for these products. Biodiesel has the potential of providing a significant market outlet. In 2004, USDA initiated a new programme that guarantees loans for manufacturers who would produce energy using high risk animal materials that are prohibited from being used for food or ruminant feed (Office of the Federal Register, 2004).

Another biofuel that is showing great promise, but has not yet been developed on a commercial scale, is cellulosic ethanol made from biomass such as agricultural forestry residues, industrial waste, municipal solid waste, trees and grasses. In the past decade, there has been considerable research conducted on advancing the technology that breaks cellulose and hemicellulose down into sugars that can be made into ethanol (US Department of Energy, 2004a). Hopefully, this research will lead to an economical process of converting large quantities of low-value plant materials and waste into fuel ethanol. If this technology were to become commercially viable, the size of the ethanol industry would grow significantly. The first feedstocks to enter this market would be crop residues and other low cost biomass materials, such as wood processing residues. As the industry expands, dedicated bioenergy crops, such as switchgrass (*Panicum virgatum*), hybrid poplar (*Populus* x) and willow (*Salix* sp.) could become economical feedstocks.

Biomass can also be used to generate electric power using gasification systems or mixing biomass with coal in coal-fired electrical generation facilities. Currently, biomass is the largest domestic source of renewable energy, surpassing hydroelectric power, supplying over 3% of the US total energy consumption – mostly through industrial heat and steam production by the pulp and paper industry and electrical generation with forest industry residues and municipal solid waste (US Department of Energy, 2004a). There is also interest in producing biopower from high-yielding trees and grasses to create a higher use value for marginal farmland.

Producing energy from biomass has many environmental advantages over petroleum-based energy. Cellulosic ethanol has the same air quality benefits as maize ethanol with the added benefits of being able to utilize crops that require less energy intensive cultivation practices than maize. Biomass power generation emits less pollution, such as sulphur and nitrogen oxide emissions, and growing energy crops reduces greenhouse gases (GHGs) since producing the plants uses less energy and the plants absorb carbon dioxide during their growth cycle. Growing biomass energy crops could provide new uses for large quantities of agricultural land, spurring development of new processing, distribution and service industries in rural areas. Developing a strong biomass industry in the USA would also increase income for farmers and foresters and stimulate rural employment.

There are also opportunities for developing biomass energy and other energy alternatives for on-farm use. Using anaerobic digestion, wind and solar energy to generate electricity and produce heating for crop and livestock operations can help conserve fossil energy, reduce input costs and increase farm revenue. Using alternative energy sources can also reduce air pollution and provide other environmental benefits. In fact, the recent interest in anaerobic digesters has often been driven more by environmental concerns than the digesters' energy potential. Anaerobic digesters are growing in popularity to help dairy farmers and other livestock producers meet new state and federal regulations for controlling animal waste. Anaerobic digesters can help control water pollution, GHGs, and odour from animal waste, as well as provide a source of electrical and thermal energy generation. Anaerobic decomposition converts manure into biogas and a digested solid. The biogas can be used in an engine generator or burned in a hot water heater. The material drawn from the digester is free of faecal coliform bacteria, a major source of water pollution. The end product is more environmentally friendly and may be more marketable than untreated manure, increasing the sales potential of manure products. In addition, the methane gas that is used to produce heat and electricity is prevented from entering the atmosphere, thereby reducing GHGs. US EPA has identified livestock operations as a principal source of methane gas, which is a significant contributor to potential global change.

Agriculture's Contribution to Energy Security

Fuel or Food

Agriculture interests and supporters of renewable fuels in general often claim that developing alternative fuels from domestic sources can reduce US dependence on energy imports. Others argue that biomass energy production diverts agricultural production away from food crops at a time when food supplies are needed to meet the needs of the rapidly growing world population (Pimentel, 1991). This debate over food versus fuel is long standing. The *Energy Security Act of 1980* ordered USDA and DOE to prepare a plan that would increase ethanol production to a level that would reach 10% of total petrol supply by the end of 1990. At the time it was estimated this would require 11 billion gal of alcohol fuel production. This large figure led USDA to evaluate the impact of such an expansion on the food supply and conclude that the food, fibre and fuel needs of the US could be met, but that would require commodity prices to rise 75% to 85% between 1980 and 1990 (Clayton and Fox, 1981).

Today, the strong growth trend in agricultural productivity, chronic periods of excess production and low prices, and the continuing consolidation in US agricultural production all suggest the US farm production sector is not stressing its capability to produce a sufficient and affordable supply of food and serve foreign customers. Long-run projections of US and global food supply and demand and other factors indicate that the situation determining the number of food-insecure people in the world is improving (Shapouri and Rosen, 2004; Bruinsma, 2003). Declining real food prices are expected to continue as a long-term trend. Some regions have arable land that could be brought into production, such as in Africa and Latin America. New technologies and management practices have the potential to increase food production in many regions. Reduced trade barriers from continuing trade liberalization can provide an additional source of food availability for countries facing food consumption-production gaps. While meeting the goal from the World Food Summit of halving the number of hungry people in the world by 2015 will be a challenge, more energy production from agricultural products does not appear likely to be a significant barrier. Barriers to achieving the food security goal are more likely to come from such factors as political upheaval, civil war, regional wars and disease outbreaks such as HIV/AIDS. Chronic problems such as weak civil and legal systems, banking systems and educational systems are additional constraining factors. In many countries facing food shortages, the problem is not that there is insufficient food in the world, but poor access to it by the countries due to governance and financial problems.

Similar to the study conducted by Clayton and Fox in 1981, more recent studies have evaluated the effects on the farm economy of increasing energy production in the future. For example, the USDA analysis of a Renewable Fuel Standard mentioned earlier projected price increases of major crops by 2011 in a range of 5% for maize and sorghum, 17% for soybean oil, but only 1% for soybeans (US Department of Agriculture, 2002). Another approach examined the effects of producing as much as 16.7 billion gal of ethanol annually using biomass from switchgrass, hybrid poplars and willows (De La Torre Ugarte *et al.*, 2003). Land for biomass production was drawn from the acres enrolled in the Conservation Reserve Program (with appropriate management practices to produce biomass while still achieving conservation goals), land in major crops and land that was idle or in pasture. Price increases for major crops ranged from 4% to 14%. The price effects estimated by these studies are within the historical range annual price variability.

Another concern with biofuels is that their supplies are limited. Maize ethanol, used primarily as a petrol additive, extended the petrol supply by about 3.4 billion gal or about 2% in 2004. About 12% of the US maize crop was utilized for ethanol in 2003. This amount could increase to about 15% if a renewable fuel standard were passed by the US Congress, mandating use of at least 5 billion gal of ethanol by 2012. This new demand is expected to be satisfied by annual increases in maize yield and commodity substitution. However, the supply of maize is relatively small compared to petrol demand, so it can only provide a partial solution to US energy problems. Likewise, biodiesel can extend diesel fuel supply, but it too is limited, when comparing the supply of animal fats and oil crops to US diesel fuel demand (Duffield *et al.*, 1998). A much larger quantity of energy feedstocks is needed for biofuel production to reach a higher scale.

The desire to extend biofuel production to other feedstocks has created a great deal of interest in cellulosic ethanol. A considerable amount of research has been done on cellulose conversion technologies, but production has not yet moved beyond the pilot scale. This technology, which can use a wide variety of low-costs feedstocks, holds the promise of potentially increasing annual ethanol supply by as much as 10 to 20 times the amount now being made from maize (De La Torre Ugarte *et al.*, 2003; Brasher, 2004). If biomass were to become more cost effective, there is expected to be some shifting of agricultural land among current uses and somewhat higher crop prices. The rapidly expanding demand could also be met by non-farm biomass, such as forest and green urban wastes and new technologies that could increase biomass production per acre.

Clearly, there is not a single solution to US energy problems. Along with biomass, other forms of alternative energy, such as wind, solar and geothermal power, are beginning to emerge as new sources of domestic energy. Largely due to technological advancements, the production of alternative energy sources has increased significantly over the past two decades. Government incentives and high energy prices will continue to encourage investment in new technologies that could greatly increase our potential to produce a wide variety of new domestic energy sources. Together these fuels can help diversify and expand the domestic energy supply.

Can the US Produce Enough Renewable Fuel to Make a Difference?

Is the potential supply of domestic renewable fuels large enough to replace a significant amount of oil imports and increase energy independence? In order to answer this question, first it is important to know how much renewable fuel is enough to make a difference. From an energy security standpoint, it may not be necessary to replace 100% of US oil imports. It can be argued that there is little to be concerned about with imports from countries like Canada and Mexico who have been very reliable trading partners over the years. These two countries are currently the top sources of US crude oil imports, exporting about 30% of daily US imports in March 2004 (Energy Information Administration, 2004b). Oil imports from the Middle East, including Saudi Arabia, Iraq and Kuwait, provided about 22% of US oil imports in March 2004. The USA imported crude oil from 10 other countries in March 2004. Nigeria and Venezuela, countries with unsteady political systems, were by far the largest exporters among this remaining group, providing about 27% of US daily oil imports. It seems reasonable that as the USA increases its energy independence, imports from the most unreliable sources would be reduced first. Thus, a 30% to 50% reduction in oil imports could make a meaningful difference.

Renewable fuels can replace a significant amount of imports and decrease US dependence on foreign oil; however, increases in energy efficiency and other technological advancements will also have to play an important role. The National Research Council reported that the Corporate Average Fuel Economy (CAFE) standards established by the *Energy Policy and Conservation Act of 1975* have reduced petrol consumption by roughly 2.8 million bbl/day from where it would be in the absence of the CAFE standards (US Department of Energy, 2002). The car industry and government researchers continue to pursue advanced combustion technologies to achieve even greater improvements in fuel efficiency. Currently, carmakers are on the verge of wide-scale commercialization of hybrid electric vehicles that are capable of achieving major fuel economy gains. Hybrid electric vehicles and other fuel saving technologies have the potential of doubling or tripling the efficiency of current vehicles. Major increases in fuel economy would help the USA use its renewable and other domestic fuel supplies more efficiently, greatly easing the task of replacing oil imports.

Increasing crop yields and advances in fuel conversion technologies must continue to increase the capacity of renewable fuel production. Eidman, in Chapter 2, points out that new technology is currently being developed for ethanol production that can utilize the maize germ to make additional ethanol from maize though lignocellulosic conversion. Improvements in cellulosic biomass conversion technology could lead to the availability of a wide range of other low cost feedstocks that could be used to increase US renewable capacity far beyond current levels. Despite recent advancements, scientists are still years away from developing an economic production process for cellulosic ethanol, according to the US DOE (Brasher, 2004). Others are more optimistic, arguing that with the creation of federal tax incentives and a well-focused government programme, the market penetration of cellulosic ethanol could come in a few years, rather than decades (Lave *et al.*, 2001; Energy Future Coalition, 2003).

Role of Policy

US Farm Policy

Farm policies have only recently been directed at energy. Farm price and income support programmes have been in operation since the 1930s in generally similar forms. Thus far, they have not been linked to energy consumption or production on farms. Similarly, farm conservation programmes historically have been directed at natural resource and environmental conservation and stewardship. At various times, alleviating farm financial stress due to low prices and high input costs, such as high energy prices, has served as a general motivation for political support for farm price and income support programmes and loan programmes for farmers.

Energy production became an explicit policy goal in farm programmes in the late 1990s with a provision in the USDA's FY 2000 appropriations act authorizing the establishment of pilot projects for harvesting biomass on lands set aside from crop production under the Conservation Reserve Program (CRP). Under the CRP at that time, up to 36.4 million acres annually could be enrolled under long-term contracts and put into conserving uses. The new pilot programme authorized up to six biomass projects with a total area not to exceed 250,000 acres. USDA subsequently approved six projects to harvest biomass from CRP land. Producers harvesting biomass were required to forego 25% of their annual CRP rental payment during the year the biomass was harvested. The approved projects were intended to test the economic and technical feasibility of using switchgrass, poplars, willows and other biomass to generate electric power. The pilots were approved in Iowa, Minnesota, New York, Pennsylvania, Illinois and Oklahoma.

In 2000, USDA initiated the Commodity Credit Corporation (CCC) Bioenergy Program. The CCC is the mechanism used to provide mandatory funding for farm price and income support programmes. The new programme was authorized under the *CCC Charter Act*. USDA implemented the programme to stimulate demand and alleviate crop surpluses, which were contributing to low crop prices and farm income and to encourage new production of biofuels. Ethanol and biodiesel producers who expanded production above year-earlier levels were eligible to receive the cash equivalent of one unit of feedstock for each three and one half units purchased for use in biofuel production by large plants, and one unit of feedstock for each two and one half units used by small plants. Feedstocks included a range of field crops, biomass and oilseeds. USDA made US$150 million available for FY 2001 and FY 2002. Most of the funds went to ethanol plants using maize as the feedstock.

Major agricultural disaster and crop insurance legislation, the *Agricultural Risk Protection Act of 2000* (ARPA), was signed into law in June 2000. Title III of ARPA, the *Biomass Research and Development Act of 2000*, directed the agriculture and energy secretaries to cooperate and coordinate polices to promote research and development leading to the production of biobased products. In particular, the Title III established a biomass research and development initiative that authorized financial assistance for public and private sector entities to carry out research on biobased products. The objectives of the initiative include enhancing the productivity and sustainability of biomass production and decreasing its cost. The statute authorized funding for the initiative under general US DOE authorities and US$49 million/year for USDA for FYs 2000-2005. While US DOE initially undertook research activities under the statute, USDA did not receive funding for the initiative until enactment of the *Farm Security and Rural Investment Act of 2002* (2002 farm bill).

Farm bills, enacted every 5 to 7 years, authorize most of the programmes that USDA operates, including farm price and income support programmes, trade programmes and rural development programmes. The 2002 farm bill, enacted in May 2002, contained the first energy title in farm bill history. The energy title, Title IX, created a range of programmes through 2007 to promote bioenergy and bioproduct production and consumption. Key provisions that had been funded and had been, or were in the process of being, implemented in mid-2004 include Section 9002, which mandates the Federal Biobased Product Procurement Preference Program (FB4P). Modelled on the existing programme for purchase of recycled materials, the FB4P requires all federal agencies to prefer biobased products in their procurements. The requirement for federal agencies begins upon designation of items by USDA. A voluntary labelling programme is also included.

Another programme, the Biodiesel Fuel Education Program created by Section 9004, awards competitive grants to educate governmental and private entities with vehicle fleets and the public about the benefits of biodiesel fuel use. In 2003, the initial grants were awarded to the National Biodiesel Board and the University of Idaho. Section 9006 created the Renewable Energy Systems and Energy Efficiency Improvements Program, a loan, loan guarantee and grant programme to assist eligible farmers, ranchers and rural small businesses in purchasing renewable energy systems and making energy efficiency improvements. In FY 2003, US$23 million in grants were awarded, including 35 wind power projects, 30 anaerobic digester projects, six solar energy projects and 16 projects totalling US$3.9 million for direct combustion such as fuel pellet systems and combined ethanol plant and anaerobic digesters.

The energy title of the 2002 farm bill also amended the *Biomass Research and Development Act of 2000* by extending its termination date to September 30, 2006, and by providing funding to USDA for the research initiative. Mandatory funding for USDA was US$5 million in FY 2002 and US$14 million annually for FYs 2003 through 2007. USDA and US DOE have issued joint solicitations for projects under the programme, including a total of US$21 million in FY 2003. A wide range of projects have been funded, from addressing biomass production issues to improvements in biorefinery production processes.

Section 9010 of the energy title of the 2002 farm bill codified the CCC Bioenergy Program and broadened it to include animal by-products and fat, oils and greases (including recycled fats, oils and greases). The programme was provided US$150 million/year in mandatory funding for FYs 2003 through 2006, although the US Congress limited funding to US$115.5 million in FY 2003.

The 2002 farm bill was also notable for greatly expanding natural resource conservation and environmental programmes. The CRP was continued and a new programme, the Conservation Security Program (CSP) was authorized. The CSP has turned out to be a controversial programme. It was conceived as a way to reward producers who have been good stewards in the past and those who can improve their conservation performance in the future. It has substantial funding for future years, although the US Congress capped funding in FY 2004 at US$41 million. The programme provides financial and technical assistance to producers for conservation and improvement of soil, water, air, energy, plant and animal life on cropland, grassland, prairie land, improved pasture and range land, as well as forested land that is an incidental part of an agriculture operation.

There has been some speculation that the CSP, or a successor programme, could serve as a basis for a new type of basic financial support programme for producers in the future. Thus, it is especially interesting that energy conservation measures are statutorily specified as one of the approved conservation practices for CSP participants. Energy enhancement activities in the CSP include cost sharing for a farm energy audit, cost sharing to encourage recycling of on-farm lubricants, payments to encourage more precise application of inorganic fertilizer, payments to encourage a reduction of tillage operations, payments to encourage the use of biobased fuels and payments to encourage energy efficiency by reducing energy consumption.

US Energy Policy

Federal energy polices have also played a major role in encouraging renewable energy production (Fig. 1.5). Historically, renewable energy policies were first adopted to establish domestic fuel reserves during emergencies, such as wartime, when imported and regional fuel supplies could be interrupted. As US dependence on foreign oil increased, energy policies promoting bioenergy began to be viewed as a way to improve energy security. For example, the primary goals of the *National Energy Act of 1978* and the *Energy Policy Act of 1992* were to increase the domestic production of alternative fuels to help reduce US dependence on foreign oil. More recently, President George W. Bush's National Energy Policy Group advocated the use of federal programmes to promote alternative fuels, such as ethanol and biodiesel, to help reduce US reliance on oil-based fuels.

There are many types of US government policy tools, including regulatory mandates, various types of tax credits and guaranteed loan programmes that have been used to encourage bioenergy development. One of the earliest energy policies aimed at increasing the domestic energy supply and addressing energy security concerns was the *National Energy Act of 1978* (NEA), signed into law by President Jimmy Carter. The NEA established the *Public Utility Regulatory Policies Act of 1978* (PURPA), a regulatory mandate that encouraged facilities to generate electricity from renewable energy sources (Gielecki *et al.*, 2001). A major goal of PURPA was to foster the development of biopower by requiring utilities to buy electricity generated from small power plants using renewable energy sources (Energy Information Administration, 1996).

Production tax credits have also been used to encourage electricity generated by qualified energy resources, including biomass and some animal wastes (Gielecki *et al.*, 2001). The *Energy Policy Act of 1992* (EPACT) established a 10-year US$0.015/kilowatt-hour (kWh) production tax credit for biomass plants, wind energy and other renewable energy production. This programme was restricted to production brought on-line before June 30, 1999. EPACT also instituted the Renewable Energy Production Incentive, which provides financial incentive payments for electricity produced and sold by new qualifying renewable energy generation facilities. Eligible technologies include landfill gas, wind and biomass. Funding is subject to annual US Congressional appropriations.

Much of the success of maize ethanol can be attributed to government incentive programmes starting back in the 1970s (Fig. 1.5). The motor fuel excise tax exemption was originally passed by the *Energy Tax Act of 1978*, giving ethanol blends of at least 10% by volume a US$0.40/gal exemption on the federal motor fuels tax. In 1980, the *Energy Security Act* was enacted that offered insured loans to small ethanol plants, producing less than 1 million gal/year. Under this act, the secretary of agriculture and secretary of energy were ordered to prepare a plan that would increase ethanol production to at least 10% of total petrol supply by the end of 1990. Also in 1980, the *Crude Oil Windfall Tax Act* extended the ethanol motor fuel excise tax exemption and provided blenders the option of receiving the same tax benefit by using an income tax credit instead of the fuel tax exemption. Since 1980, various tax laws have been adopted changing the level of the tax credit that currently stands at US$0.51/gal through 2010. The EPACT extended the fuel tax exemption and the blender's income tax credit to two additional blend rates containing less than 10% ethanol, effective January 1, 1993. The two additional blend rates were for petrol with at least 7.7% ethanol and for petrol with 5.7% ethanol. These additional blends were added to encourage blending of ethanol to make oxygenated petrol, requiring 7.7% ethanol and reformulated petrol, requiring 5.7% ethanol.

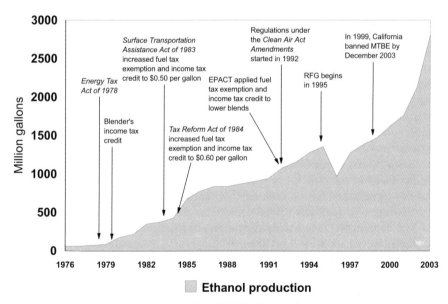

Source: National Corn Growers Association (1992); Energy Information Administration (2004c).

Figure 1.5. Effect of government policies on US ethanol production, 1976-2003.

When the US Congress adopted the motor fuel tax credit for ethanol, they also enacted a special duty on fuel ethanol imports to offset the value of the federal tax exemption so that foreign ethanol producers could not benefit from the exemption. Duty-free treatment for eligible articles, including ethanol, was granted to 22 Caribbean Basin countries and territories in January 1984 under the Caribbean Basin Initiative (CBI). The *Steel Trade Liberalization Program Implementation Act of 1989* limited the amount of ethanol CBI countries can import into the USA unless the ethanol is made from indigenous products. Ethanol not made from indigenous products can still be exported by CBI countries duty-free; however, the total amount exported to the USA cannot exceed 60 million gal/year or 7% (whichever is greater) of US annual ethanol consumption. If the base quantity is exceeded during the year, an additional 35 million gal may enter the USA duty-free, provided at least 30% of the ethanol is derived from indigenous products. After these additional 35 million gal are exceeded, any additional imports will be duty-free, only if 50% of the ethanol is derived from indigenous products.

US ethanol imports have increased under the CBI, but not enough to have a significant effect on the US market. However, Cargill and other US companies have recently shown interest in extending their ethanol operations to CBI countries. Building an ethanol plant in a CBI country would allow them to take advantage of low-cost feedstocks from Brazil and import ethanol into the USA under the CBI. If US companies follow through on their plans to channel investment funds to CBI countries for ethanol production, imports could grow well beyond historical levels. However, some members of the US Congress have expressed concern with this 'loophole' and have already drafted legislation that would prevent Brazilian ethanol imports from entering the country duty-free under the CBI.

The EPACT established a number of alternative-fuelled vehicle (AFV) requirements that have encouraged biofuel use. The act requires government and state motor fleets to purchase alternative-fuelled vehicles (75% of new heavy-duty vehicle purchases). Alternative fuel providers, including natural gas producers, must also comply. Ninety per cent of their new purchases of heavy duty vehicles must be AFVs. US DOE has the authority to implement private and local government programmes if necessary. The *Energy Conservation Reauthorization Act of 1998* amended EPACT to include biodiesel fuel use credits. The related rule, effective January 2001, gives fleet operators one alternative-fuelled vehicle credit for using 450 gal of biodiesel.

The use of alternative-fuelled vehicles is also increasing in the private sector, primarily due to the *Alternative Motor Fuels Act* that was passed in 1988 to encourage car manufacturers to produce cars that are fuelled by alternative fuels, including ethanol. The law provides credits to car makers towards meeting their CAFE standards. Car makers can lower their average fuel economy requirements by receiving credits for producing alternative-fuelled vehicles and dual-fuel vehicles that meet government requirements. Several car manufacturers offer various models that run on both ethanol fuel and petrol. About 3.5 million of these vehicles, called flexible fuel vehicles (FFVs), were on the road by mid-2004 (National Ethanol Vehicle Coalition, 2004). This programme, however, has been criticized because most FFV owners usually use petrol instead of ethanol.

Environmental polices can also have a major impact on the development of renewable fuels. As described earlier, ethanol was first used as a fuel additive in the late 1970s when US EPA began phasing out lead in petrol. Ethanol replaced lead as an octane enhancer. The Oxygenated Fuels and RFG programs created by the *Clean Air Act Amendments of 1990* require that certain urban areas in 'non-attainment' add oxygen to their petrol: 2.7% by weight for oxygenated fuel and 2.0% by weight for RFG.

US EPA diesel fuel regulations could increase the demand for biodiesel as a lubricity additive. US EPA's low-sulphur highway diesel fuel regulations begin July 2006 and the non-road diesel fuel regulations begin June 2010. Lowering the sulphur in diesel fuel also lowers the fuel's lubricity. As a result the demand for diesel fuel lubricity additives is expected to increase significantly. Research suggests that biodiesel is an excellent fuel lubricity agent. Only a small amount of biodiesel (1% to 2%) is needed to restore the lubricity level of ultra-low sulphur diesel fuel. The lubricity additive market could provide a much larger market than the niche markets that currently exist for biodiesel.

The *Clean Air Act Amendments of 1990* also has provisions for controlling stationary sources of air pollution, such as the Acid Rain Program that set tighter restrictions on sulphur dioxide and nitrogen oxides. Under this programme, utilities may apply for bonus emission allowances as a reward for undertaking energy efficiency or renewable energy measures. Qualified renewable energy sources include wind, solar, geothermal and biomass energy (US Environmental Protection Agency, 2004a); however, these energy sources are not widely used in the programme. Although US EPA does not regulate GHG emissions, global climate change is viewed as a serious potential problem and US EPA prepares an annual inventory of GHG emissions and sinks. This inventory has a number of purposes, the foremost of which is fulfilling US commitments to report under the *United Nations Framework Convention on Climate Change* (UNFCCC). US EPA's efforts are also part of a broader federal effort to conduct research aimed at understanding the relative contribution of different emission sources and sinks to overall emissions. In support of their programmes, US policy officials often take credit for reducing GHG emissions through the expanded use of ethanol and other renewable fuels. If the USA decides to intensify its efforts to reduce GHG emissions, some argue that the best approach is through government programmes that include financial incentives leading to the widespread use of biomass energy (Lave *et al.*, 2001; Wirth *et al.*, 2003).

US State Energy Programmes

There are also many US state programmes designed to encourage the growth of renewable energy use. The recent boost in ethanol demand is the result of California and other states banning MTBE. State bans of MTBE have allowed ethanol to become the dominate fuel in the oxygenate market. These state actions are important factors behind the ethanol industry's major expansion from 2001 to 2004 when capacity increased almost 1.7 billion gal.

Many states have been active in promoting bioenergy use through tax credits and production incentives. A detailed description of all state renewable energy policies would be too extensive to cover in this chapter. The best source for this information is the Database of State Incentives for Renewable Energy (DSIRE), developed by the Interstate Renewable Energy Council. DSIRE is a comprehensive source of information on state, local, utility and selected federal incentives (North Carolina Solar Center, 2004). A summary of state programmes indicates that 20 states provide tax credits and/or producer payments for ethanol and biodiesel. Eleven states have grant programmes, personal income tax credits and corporate income tax credits for constructing biofuel facilities, conducting research on renewable fuel technologies, developing renewable energy systems and investing in alternative energy development. Twenty-three states have a 'Renewable Energy Portfolio Standard' or other type of renewable electricity programme that provides incentives for biomass power, anaerobic digestion and other renewable fuels used to generate power. Only 12 states have no incentives for bioenergy.

New Energy Legislation

With recent record oil and natural gas prices in mid-2004 and increasing energy supply uncertainty, there is much interest in passing new energy legislation. In early 2001, President Bush's National Energy Policy Development Group laid out a proposal for a long-term, comprehensive strategy to lessen the impact of energy price volatility and supply uncertainty (National Energy Policy Development Group, 2001). The strategy takes a portfolio approach, recommending that in the short run, a variety of domestic energy supplies, including domestic oil, gas, coal, hydropower, nuclear power and biomass, be developed to diversify future energy supply. In the long run, the strategy sees the possibility of alternative energy technologies, such as hydrogen and fusion, providing much of US energy needs. The development of hydrogen has implications for renewable fuels, because ethanol is an excellent energy source for extracting hydrogen (Gray, 2004).

The US Congress responded to the energy situation and President Bush's energy strategy by developing energy legislation. The House and Senate struggled with alternative versions and did not pass legislation in 2004. The House and Senate versions reflected President Bush's general approach by attempting to develop a comprehensive energy policy aimed at increasing and diversifying domestic energy production. Environmentalists and others criticized the bills for focusing too much on energy production while ignoring conservation. There was much debate over whether to change environmental laws and other regulations to make domestic reserves of oil and gas more assessable. The issue of allowing oil drilling in Alaska's Arctic National Wildlife Refuge was particularly contentious. Many argued that federal tax incentives, loan programmes and other financial incentives were necessary to encourage more exploration and production of domestic energy reserves. There were several proposals to provide government support for investing in America's ageing energy infrastructure, such as providing funds to repair the existing pipeline system and to construct new pipelines where needed. The existing pipeline system for domestic oil and gas is getting old, and insufficient capacity has increased the potential risk of energy supply and price volatility. The US electricity infrastructure was also viewed as a system in much need of repair, as the interstate transmission system has long been neglected, with the most recent reminder coming from the 2003 blackout that shut off power to millions of customers in the Northeast.

Energy legislation proposed by the Senate included key provisions to help diversify domestic energy production through the development of renewable fuels. The bill included several tax incentives to stimulate growth in the renewable fuels industry and the RFS, described earlier in this chapter, that provided for a national ban on MTBE and mandated the use of at least 5 billion gal of renewable fuels by 2012. Ethanol and biodiesel would each receive a 1 gal credit toward meeting the RFS for each gallon used, while cellulosic ethanol would receive a 1.5 gal credit for each gallon used. Advocates argued that the RFS would result in a range of farm, rural, environmental and energy security benefits. Opponents of the RFS feared that forcing ethanol into the fuel market would strain the distribution system, thus raising fuel costs, and that the environmental benefits claimed by ethanol advocates were overstated.

The Senate bill also contained an energy tax incentive for diesel fuel blended with biodiesel made from oil crops and animal fats equivalent to US$1.00/gal of biodiesel. Biodiesel made from recycled fats and oils would receive US$0.50/gal. The biodiesel tax credit would be deducted from the general tax fund, unlike the ethanol tax credit that now comes out of the Highway Trust Fund (HTF). There was also an effort to change the tax code to have the ethanol tax credit transferred from the HTF to the general tax fund. This provision, referred to as the Volumetric Ethanol Tax Credit (VEETC), was intended to placate critics complaining that funding the ethanol tax credit through the HTF takes money away from state highway projects. In addition, proposed tax legislation extended the ethanol tax credit by three years to 2010. There were several proposals establishing tax credits to increase alternative-fuelled vehicle sales and stimulate renewable fuel demand. New mandates would be imposed to require federal fleets to increase their use of alternative fuels.

Proposed energy legislation also included provisions to encourage renewable fuel use in the power industry. A mandate was proposed to require the US government to purchase electricity from renewable fuel sources: 3% in FY 2003, increasing to 7.5% in 2010. In addition, a Renewable Portfolio Standard for electric utilities would increase their use of renewable energy from 1% in FY 2005 to 10% in FY 2020. The renewable electricity production tax credit, which expired in 2003, would be extended to 2007. This production tax credit that provides a US$0.018/kWh tax credit for renewable energy has been crucial to the recent development of the wind industry. Wind energy advocates provided testimony to the US Congress illustrating the need for the tax credit and warned that the financing of new wind power installations would stagnate without this support.

Although a US national energy bill was not passed in 2004, bioenergy advocates were able to include a few of the energy bill's tax incentives in the *American Jobs Creation Act of 2004* (H.R. 4520) signed into law by President Bush on October 20, 2004. The so-called JOBS bill, effective on January 1, 2005, extended the excise tax credit for ethanol to 2010. It was scheduled to be reduced from the current US$0.52/gal to US$0.51/gal in January 2005 and then expire in 2007. Under the JOBS bill, a US$0.51/gal excise tax credit for ethanol will be extended from 2007 to December 31, 2010. The Small Ethanol Producer Tax Credit of US$0.10/gal was also extended to the end of 2010. This tax credit was originally passed in 1990 to benefit small producers, such as farm cooperatives, with a capacity of 30 million gal or less. In addition, the VEETC provision was passed, which shifted the amount of the excise tax that was funded by the HTF to the general fund. The JOBS bill also adopted the US Senate's excise tax credit for biodiesel that provided a credit of US$1.00/gal of biodiesel made from oil crops and animal fats and a US$0.50/gal credit for biodiesel made from recycled fats and oils. The renewable electricity production tax credit for wind and biomass power was extended to the end of 2005.

Conference Discussion

This chapter concludes with a brief summary of major issues discussed at the conference. About 200 participants from academia, government, industry, consulting firms, associations, media and the US Congress attended the one and a half-day conference. The conference programme was organized to facilitate discussion with much of the time dedicated to general sessions that included the presentations of three commissioned papers, an industry panel, a luncheon speaker and a wrap-up session.

Topics related to agriculture as an energy producer dominated the discussion. Much of the early dialogue centred on Eidman's estimates of the potential supply of renewable fuels. Eidman concluded that the US could at least produce enough ethanol and biodiesel to replace MTBE, but some participants expressed doubt of agriculture's ability to replace a significant amount of oil imports, without the development of cellulosic production. There were others that discounted the importance of the current energy supply situation, arguing that the real price of petrol was relatively low compared with past price spikes and that supply disruptions are only temporary. However, there appeared to be general agreement that renewable fuels should be available to serve as an insurance policy in times of emergency. As Miranowski points out in Chapter 3, energy disruptions at critical points or periods can have devastating effects on specific sectors, including crop and livestock production, and greater availability of domestic renewable fuels can lower the risks of imported fuel disruptions.

There was much debate over the future of cellulosic ethanol. Industry representatives argued that Eidman's cost estimates (see Chapter 2) were too high and that the technology was well beyond the laboratory stage. Industry experts believe that with more capital investment in cellulosic technology, commercial development would only be a few years away. Boyden Gray, the luncheon speaker, provided a strong argument for supporting government research that would rapidly move cellulosic conversion technology towards economic efficiency. Government support for the ethanol industry should be contrasted with the trillions of dollars in government subsidies over the years that have gone to the petroleum industry, argued Mr. Gray in urging for greater support for the ethanol industry. Intensifying the US commitment to this emerging industry would accelerate the technological timetable for developing biological conversion techniques that can make cost-effective use of cellulosic plant material (Wirth *et al.*, 2003).

Roger Conway, the discussion leader for the wrap-up session, advised the group that they should not just focus on energy supply, but they need to look also at the effect renewable fuels could have on price stability. Research suggests that even small influxes of renewable fuels can decrease price volatility, i.e., diversifying our fuel supply can reduce risk, similar to adding diverse stocks to an investment portfolio (Tareen *et al.*, 2000). Adding ethanol, biodiesel and other non-petroleum fuels to the market could help reduce price shocks. Moreover, the effect that renewable fuels have on price stability may become more prominent over time, as the USA and other countries throughout the world further develop their renewable fuel resources. Brazil, the largest renewable fuel producer in the world, produced 3.6 billion gal of sugar-derived ethanol in 2003. Brazilian ethanol production has become so successful that its exports are expanding. Worldwide interest in biodiesel has also been expanding rapidly with Europe leading the way, producing over 1.4 million metric tons in 2003, up 35% from 2003. Canada, Australia, India, China, Uruguay, Argentina, Mexico, South Africa, Taiwan, Philippines, the United Kingdom and many other countries around the world are also interested in developing their renewable fuel resources (E.O. Lichts, 2004). Within a few years, these countries could be adding new non-petroleum-based fuels to the world market, having a calming effect on world oil prices.

Conway also pointed out that the higher cost of renewable fuels production does not reflect the environmental benefits of replacing petroleum fuels. Although the environmental benefits of renewable fuels have been well documented, little work has been done to estimate the monetary value of these benefits. More research is needed to quantify the non-monetary value of renewable fuels to help policymakers determine the appropriate level of public support for renewable fuel programmes.

Most of the discussion related to agriculture as an energy consumer took place during the industry panel session and centred around the effects of higher energy costs on farm profitability and the ability of US industries to compete in the global market. There were concerns that higher energy costs would make some livestock producers and other agribusinesses more vulnerable to foreign competition. For example, the cost of controlling temperatures in poultry and pig facilities has been increasing with rising energy prices. If this trend continues, US livestock producers may find it more profitable to move their operations to countries with lower energy prices. Likewise, other agribusiness firms with high energy costs, such as nurseries, greenhouses, food processing and fertilizer plants, may be attracted to countries with low-cost energy supplies. The existing competitive problem of labour cost disadvantages for some US industries that are also energy intensive could be exacerbated if they are unable to adjust to rising energy costs.

Panel discussant, Steve Wilson from CF Industries, provided a bleak outlook for the US fertilizer industry that in recent years has been adversely affected by higher natural gas prices. Higher production costs have caused the US fertilizer industry to constrict and domestic fertilizer production capacity has stagnated in recent years. US nitrogen production capacity has declined 20% since 2000. At the same time, higher US fertilizer prices have attracted more imports from around the world, from countries such as Trinidad and Tobago, Venezuela, Russia and Ukraine (US General Accounting Office, 2003). Natural gas is relatively cheap in these countries, so their production costs for nitrogen fertilizer are less compared with the USA. Imports of nitrogen increased 20% from July 2003 to March 2004. This influx of imports is causing much concern within the US fertilizer industry, which is rapidly losing its market share to foreign producers. They are particularly concerned with controlled pricing of natural gas in Russia, currently the world's largest nitrogen exporter.

With current US nitrogen trade policy and continuing high natural gas prices, it is likely that the US fertilizer industry will continue to shrink. A concern was expressed that as US fertilizer imports rise, US fertilizer supply may become increasingly uncertain as many of the supplying countries have a history of political unrest. Imports may increasingly come from Russia and the Middle East, where much of the proven reserves of natural gas are concentrated. The solution to this problem, expressed by US fertilizer industry representatives present, is passing federal legislation that streamlines regulations to allow more access to natural gas reserves in Alaska, the Rocky Mountains and other areas of the USA.

The importance of research was a common theme throughout the conference. There seemed to be a general consensus that the federal government should spend more on research to develop renewable fuels and increase energy efficiency on the farm. More research should be funded to develop technologies that would reduce the costs of producing renewable fuels. For example, the cost of producing ethanol could be substantially reduced through the development of conversion technologies, such as cellulosic conversion, that can utilize low-cost feedstocks. Advances in biotechnology could increase feedstock yields and possibly develop new crop varieties specifically for renewable energy production. Bob Dinneen from the Renewable Fuels Association identified co-product utilization as a critical research area. He explained that, as the production of ethanol increases, the industry might find it increasingly difficult to market ethanol co-products, such as distiller's dried grains, maize gluten meal and maize gluten feed. Research is needed to develop new uses for these co-products and create more value-added opportunities for ethanol. There was interest in funding more research to investigate the economics of the 'biorefinery concept'. Modelled after a petroleum refinery, a biorefinery would process biomass feedstocks into a wide range of bioenergy and biobased products. Research could help determine if this large-scale multiple product approach could help bioenergy and biobased products become more cost competitive with petroleum products.

As an oil industry representative, Kevin Lindemar saw a critical need to examine the logistics of integrating renewable fuels into the current US petroleum distribution system. Doering, also focusing on energy systems integration, points out that 'The systems themselves co-determine what fuels will be used. Any change often requires that the systems be modified or changed.' Research identifying logistic issues with the current energy system and developing cost-effective methods to address these issues could improve the marketability of bioenergy.

Several participants mentioned the need for more research aimed at developing technologies that would help increase energy efficiency on US farms and other agriculture operations. More effort should be made to gather data to help researchers evaluate farm energy use and invest in more research to develop new technologies that use less energy for irrigation, crop drying, tillage practices and fertilizer and pesticide use. Likewise, more data and research are needed to develop energy-saving strategies in livestock facilities, dairies, nurseries and greenhouses. Research that provides energy-efficient technologies would also help producers lower their production costs and increase their global competitiveness.

The conference provided an opportunity to foster better relationships and exchange information between the USDA, other federal and state agencies, the private sector, environmental and other non-profit organizations and academia. Each group has its own perspectives and information bases. Each group is driven by its own incentive system. These groups must work together to describe the best picture of the current relationships between energy and agriculture and to identify the key problems that need to be solved to help agriculture optimize its energy production and energy use. We are at a critical time in the history of energy and agriculture, and this conference helped mark the beginning of a renewed effort for the agricultural community to get involved in providing solutions to US energy problems. In the coming years, many fundamental federal budget and legislative choices will have to be made and they will hasten or retard US agriculture's progress toward greater energy production and efficiency. This book on energy and agriculture provides a unique source of information that can be used to help make the right choices.

References

Blue Ribbon Panel on Oxygenates in Gasoline. (1999) *Achieving Clean Air and Clean Water: The Report of the Blue Ribbon Panel on Oxygenates in Gasoline.* Commissioned by the US Environmental Protection Agency, Washington, DC.

Brasher, P. (2004) Biomass Ethanol is Still Years Away. *Des Moines Register* May 7. [Accessed 2004.] Available from http://www.dmregister.com/apps/pbcs.dll/article?AID=/20040507/BUSINESS01/405070366/1030

Bruinsma, J., Ed. (2003) *World Agriculture: Towards 2015/2030, An FAO Perspective.* Earthscan Publications Ltd., London.

Cambridge Energy Research Associates (2004). The Worst is Yet to Come: Diverging Fundamentals Challenge the North American Gas Market. *North American Natural Gas Watch Spring 2004.* Cambridge Energy Research Associates, Cambridge, Massachusetts.

Clayton, K. and Fox, A. (1981) Agriculture's Production Potential. *Agricultural-Food Policy Review, Perspectives for the 1980s.* AFPR-4. US Department of Agriculture, Economics and Statistics Service, Washington, DC, pp. 70-80.

Commodity Research Bureau. (2004) *Historical Data – Energy.* Commodity Research Bureau, Chicago, Illinois. [Accessed 2004.] Available from http://www.crbtrader.com/Marketdata/energy.asp

De La Torre Ugarte, D., Shapouri, H. Walsh, M. and Slinsky, S. (2003) *The Economic Impacts of Bioenergy Crop Production on U.S. Agriculture.* AER Report No. 816. US Department of Agriculture, Office of the Chief Economist, Office of Energy Policy and New Uses, Washington, DC.

Duffield, J., Shapouri, H., Graboski, M., McCormick, R. and Wilson, R. (1998) *Biodiesel Development: New Markets for Conventional and Genetically Modified Agricultural Products.* ERS Report No. 770. US Department of Agriculture, Economic Research Service, Washington, DC.

Economic Research Service. (2004a) *Data: Feed Grains Data Delivery System.* US Department of Agriculture, Economic Research Service, Washington, DC. [Accessed 2004.] Available from http://www.ers.usda.gov/db/feedgrains/default.asp?ERSTab=3

Economic Research Service. (2004b) U.S. and State Farm Income Data. *Data: Farm Income.* US Department of Agriculture, Economic Research Service, Washington, DC. [Accessed 2004.] Available from http://www.ers.usda.gov/Data/FarmIncome/Finfidmu.htm

Energy Future Coalition. (2003) *Challenge and Opportunity: Charting a New Energy Future.* Energy Future Coalition, Washington, DC.

Energy Information Administration. (1991) *Historical Monthly Energy Review 1973-1988.* DOE/EIA-0035(73-84). US Department of Energy, Energy Information Administration, Washington, DC.

Energy Information Administration. (1996) *Changing Structure of the Electric Power Industry: An Update.* DOE/EIA-0562. US Department of Energy, Energy Information Administration, Washington, DC.

Energy Information Administration. (2004a) *Annual Energy Outlook with Projections to 2025.* Report # DOE/EIA-0383. US Department of Energy, Energy Information Administration, Washington, DC.

Energy Information Administration. (2004b) *Crude Oil and Total Petroleum Imports: Top 15 Countries.* US Department of Energy, Energy Information Administration. [Accessed 2004] Available from http://www.eia.doe.gov/pub/oil_gas/petroleum/data_publications/company_level_imports/current/import.html

Energy Information Administration. (2004c) Historical Renewable Energy Consumption by Sector and Energy Source. *Renewable Energy Trends 2003.* US Department of Energy, Energy Information Administration, Washington, DC. [Accessed 2004] Available from http://www.eia.doe.gov/cneaf/solar.renewables/page/trends/tableb1.html

Energy Information Administration. (2004d) *Natural Gas Navigator.* US Department of Energy, Energy Information Administration. [Accessed 2004.] Available from http://tonto.eia.doe.gov/dnav/ng/hist/n3020us3m.htm

Energy Information Administration. (2004e) *Petroleum Prices.* US Department of Energy, Energy Information Administration. [Accessed 2004.] Available from http://www.eia.doe.gov/oil_gas/petroleum/info_glance/prices.html

Energy Information Administration. (2004f) *Petroleum Supply Monthly.* Energy Information Administration (EIA), 2004. DOE/EIA-0562. US Department of Energy, Energy Information Administration, Washington, DC.

Energy Information Administration. (2004g) *Short-Term Energy Outlook, July 2004.* US Department of Energy, Energy Information Administration, Washington, DC. [Accessed 2004] Available from http://www.eia.doe.gov/emeu/steo/pub/contents.html

Farm Service Agency. (2004) *Bioenergy.* US Department of Agriculture, Farm Service Agency. [Accessed 2004.] Available from http://www.fsa.usda.gov/daco/bio_daco.htm

Foreign Agricultural Service. (2004) *Export/Import Statistics for Bulk, Intermediate, and Consumer Oriented (BICO) Foods and Beverages.* US Department of Agriculture, Foreign Agricultural Service, Washington, DC. [Accessed 2004.] Available from http://www.fas.usda.gov/scriptsw/bico/bico_frm.asp

Gielecki, M., Mayes, F. and Prete, L. (2001) *Incentives, Mandates, and Government Programs for Promoting Renewable Energy.* US Department of Energy, Energy Information Administration. [Accessed 2004.] Available from http://www.cia.doe.gov/cneaf/solar.renewables/rea_issues/incent.html

Graboski, M. and McCormick, R. (1998) Combustion of Fat and Vegetable Oil Derived Fuels in Diesel Engines. *Progressive Energy Production Science* 24: 125-164.

Gray, B. (2004) Conference Presentation. Agriculture as a Producer and Consumer of Energy. June 24-25, 2004, Arlington, Virginia.

Gullickson, G. (2001) Nipping at Nitrogen Prices. *Farm Industry News.* (February 15) PRIMEDIA Business Magazines & Media Inc. [Accessed 2004.] Available from http://farmindustrynews.com/issue_20010215.

Heimlich, R. (2003) *Agricultural Resources and Environmental Indicators*. Agriculture Handbook No. (AH722). US Department of Agriculture, Economic Research Service, Washington, DC.

Lave, L., Griffin, W. and MacLean, H. (2001) The Ethanol Answer to Carbon Emissions. *Issues in Science and Technology Online*. National Academies and University of Texas at Dallas. (Winter) [Accessed 2004.] Available from http://www.issues.org/

Lee, H. (1993) Ethanol's Evolving Role in the U.S. Automobile Fuel Market. *Industrial Uses of Agricultural Materials: Situation and Outlook Report*. US Department of Agriculture, Economic Research Service, Washington, DC.

Lee, H. and Conway, R. (1992) New Crops, New Uses, New Markets. *1992 Yearbook of Agriculture*. US Department of Agriculture, Washington, DC.

Levelton Engineering Ltd., (S&T)[2] Consultants Inc. and J.E. & Associates. (1999) *Assessment of Net Emissions of Greenhouse Gases From Ethanol-Gasoline Blends in Southern Ontario*. Prepared for Agriculture and Agri-Food Canada. Levelton Engineering Ltd., Richmond, British Columbia and (S&T)[2] Consultants Inc., Vancouver, British Columbia.

Lichts, E.O. (2004) *World Ethanol and Biofuels Report*. Various Issues. Agra Informa Ltd., Tunbridge Wells, Kent, UK.

Morris, D. and Ahmed, I. (1992) *The Carbohydrate Economy: Making Chemicals and Industrial Materials from Plant Matter*. Institute for Local Self-Reliance, Washington, DC.

National Agricultural Statistics Service. (2004a) *Agricultural Prices April 2004*. US Department of Agriculture, National Agricultural Statistics Service, Agricultural Statistics Board, Washington, DC.

National Agricultural Statistics Service. (2004b) *Crop Production Acreage Supplement June 2004*. CrPr 2-5. US Department of Agriculture, National Agricultural Statistics Service, Washington DC. [Accessed 2004.] Available from http://usda.mannlib.cornell.edu/reports/nassr/field/pcp-bba/acrg0604.pdf

National Corn Growers Assocation. (1992) *The World of Corn, a Comprehensive Look*. National Corn Growers Assocation and National Corn Development Foundation, Washington, DC.

National Energy Policy Development Group. (2001) *National Energy Policy*. Superintendent of Documents, US Government Printing Office, Washington, DC.

National Ethanol Vehicle Coalition. (2004) *For All the Right Reasons*. National Ethanol Vehicle Coalition, Jefferson City, Missouri. [Accessed 2004.] Available from http://www.E85Fuel.com

National Renewable Energy Laboratory. (1998) *Life Cycle Inventory of Biodiesel and Petroleum Diesel for Use in an Urban Bus*. NREL/SR-580-24089. US Department of Energy, National Renewable Energy Laboratory, Golden, Colorado, and US Department of Agriculture, Office of Energy, Washington DC.

North Carolina Solar Center. (2004) *Database of State Incentives for Renewable Energy*. A project of the North Carolina Solar Center and the Interstate Renewable Energy Council. [Accessed 2004.] Available from http://www.dsireusa.org

Office of the Federal Register. (2004) Notice of Funds Availability Inviting Applications for the Specific Risk Materials and Certain Cattle Renewable Energy Guaranteed Loan Pilot Program. In: *Federal Register*. Vol. 69, No. 96 (Tuesday, May 18), pp. 28111-28119.

Ostich, J. (2003) History of Ethanol as a Motor Fuel. US Department of Energy, Energy Information Administration, Washington, DC. [Accessed 2004.] Available from http://www.eia.doe.gov/oog/ethanol/page/fact_sheets.html

Paarlberg, D. and Paarlberg, P. (2000) *The Agricultural Revolution of the 20th Century*. Iowa State University Press, Ames, Iowa.

Pimentel, D. (1991) Ethanol Fuels: Energy Security, Economics, and the Environment. *Journal of Agricultural and Environmental Ethics* 4 (February): 1-13.

Shapouri, H., Duffield, J. and Graboski, M. (1995) *Estimating The Energy Balance of Corn Ethanol*. Agricultural Economic Report No. 721. US Department of Agriculture, Economic Research Service, Washington, DC.

Shapouri, H., Gallagher, P. and Grabowski, M. (2002) *USDA's 1998 Ethanol Cost-of-Production Survey*. Agricultural Economic Report No. 808. US Department of Agriculture, Office of the Chief Economist, Office of Energy Policy and New Uses, Washington, DC.

Shapouri, H., Duffield, J. and Wang, M. (2003) The Energy Balance of Corn Ethanol Revisited. *American Society of Agricultural Engineers* 46: 959-968.

Shapouri, S. and Rosen, S. (2004) *Food Security Assessment.* Agriculture and Trade Report GFA-15. US Department of Agriculture, Economic Research Service, Washington, DC.

Talley, D. (2004) Biodiesel: a Compelling Business for the Rendering Industry. *Render, The National Magazine for Rendering.* February. Render Magazine, Camino, California.

Tareen, I., Wetzstein, M. and Duffield, J. (2000) Biodiesel as a Substitute for Petroleum Diesel in a Stochastic Environment. *Agricultural & Applied Economics* 32(2) (August): 373-381.

US Department of Agriculture. (1997) *Agricultural Resources and Environmental Indicators, 1996-97.* US Department of Agriculture, Economic Research Service, Washington, DC.

US Department of Agriculture. (2000) *Analysis of Ethanol Production Under a Renewable Fuels Requirement.* Staff Analysis. US Department of Agriculture, Office of the Chief Economist, Washington, DC.

US Department of Agriculture. (2001) *Economic Analysis of Increasing Soybean Oil Demand Through the Development of New Products.* Staff Analysis. US Department of Agriculture, Office of the Chief Economist, Washington, DC.

US Department of Agriculture. (2002) *Effects on the Farm Economy of a Renewable Fuels Standard for Motor Vehicle Fuel.* Staff Analysis. US Department of Agriculture, Office of the Chief Economist, Washington, DC.

US Department of Agriculture. (2004a) *U.S. Agriculture and Forestry Greenhouse Gas Inventory: 1990-2001.* Technical Bulletin No. 1907. US Department of Agriculture, Office of the Chief Economist, Global Change Program Office, Washington, DC.

US Department of Agriculture. (2004b) *USDA Agricultural Baseline Projection Tables.* US Department of Agriculture, World Agricultural Outlook Board. [Accessed 2004.] Available from http://usda.mannlib.cornell.edu/data-sets/baseline/2004/index.html

US Department of Energy. (2002) *Fact #208: CAFE Standards Reduce Petroleum Use.* US Department of Energy, Energy Efficiency and Renewable Energy, Freedom Car and Vehicle Technologies Program. [Accessed 2004.] Available from http://www.eere.energy.gov/vehiclesandfuels/facts/favorites/fcvt_fotw208.shtml

US Department of Energy. (2004a) *Biomass Program.* US Department of Energy, Energy Efficiency and Renewable Energy. [Accessed 2004.] Available from http://www.eere.energy.gov/biomass/biomass_today.html

US Department of Energy. (2004b) EPAct Fleet Information & Regulations. *Freedom Car & Vehicles Technology Program.* US Department of Energy, Energy Efficiency and Renewable Energy. [Accessed 2004.] Available from http://www.eere.energy.gov/vehiclesandfuels/epact/

US Environmental Protection Agency. (2004a) Clean Air Markets – Programs and Regulations. *Conservation and Renewable Energy Incentives.* [Accessed 2004.] Available from http://www.epa.gov/airmarkets/arp/crer/index.html

US Environmental Protection Agency. (2004b) *Inventory of U.S. Greenhouse Gas Emissions and Sinks: 1990-2002.* US Environmental Protection Agency, Washington, DC.

US General Accounting Office. (2003) *Natural Gas: Domestic Nitrogen Fertilizer Production Depends on Natural Gas Availability and Prices.* GAO-03-1148. US General Accounting Office, Washington, DC.

Wang, M., Saricks, C. and Santini, D. (1999) *Effects of Fuel Ethanol Use on Fuel-Cycle Energy and Greenhouse Gas Emissions.* ANL-38. US Department of Energy, Argonne National Laboratory, Center for Transportation Research, Argonne, Illinois.

Wirth, T., Gray, B. and Podesta, J. (2003) The Future of Energy Policy. *Foreign Affairs* 82(4) July/August: 133-155.

Yergin, D. (1991) *The Prize.* Simon & Schuster, New York, New York.

Chapter 2

Agriculture as a Producer of Energy

Vernon R. Eidman

Introduction

The US agricultural sector currently plays a relatively small, but increasingly important role in energy production. A century ago agriculture played a much larger role in producing the nation's energy supply. In the early 1900s the sector provided much of the fuel, in the form of grain and hay, for a major part of the nation's transportation needs and much of the wood used for heating and cooking. Changes in technology and the concurrent development of inexpensive petroleum and natural gas reduced the demand for the traditional sources of energy agriculture provided. Now that world demand for petroleum and natural gas is increasing relative to world supplies, the higher prices for petrol, diesel fuel and natural gas are making renewable sources of energy more economically attractive. Increasing concern with the environmental impacts of fossil fuel consumption and concerns about energy scarcity are also increasing interest in renewable energy, suggesting that agriculture's role as an energy producer will expand.

The purpose of this chapter is to document the role agriculture is playing as a supplier of energy for the US economy and to discuss the potential contributions agriculture can make to the nation's energy supply in the near future. More specifically, this chapter 1) documents the amount of energy used by the US economy and agriculture's role in supplying it; 2) discusses the current state of technology in energy production from agricultural biomass; 3) summarizes some recent literature on the potential contribution agriculture can make to the US energy supply; and 4) provides a brief overview of the developing ethanol and biodiesel industries in other countries.

Renewables as a Part of Current Supply

The USA consumed 97.1 quadrillion British thermal units (Btu) of energy during 2001 (Table 2.1). Net imports made up 27% of the total, with the largest imports being crude oil and natural gas. Renewables made up 5.8% of consumption in 2001, down somewhat from the customary 6% or more of recent years, because of reduced availability of water for hydroelectric power generation during 2001.

The sources of renewable energy consumed during 2001 are shown in Table 2.2. Of the total renewables, almost one-half was provided by hydroelectric, geothermal and solar. The remaining 2.93 quadrillion Btu (quad), about 3% of the total energy consumed in the USA, was obtained from biomass and wind. The biomass category is composed of wood, waste

Table 2.1. US energy flows (quadrillion Btu), 2001.

	Domestic production[a]	Imports[c]	Exports[d]	Domestic consumption
Coal	23.44	0.95	1.47	22.02[e]
Petroleum and petroleum products[b]	14.93	24.88	2.06	38.23[e]
Natural gas	19.84	4.12	0.4	23.22[e]
Nuclear electric power	8.03	N/A[f]	N/A	8.03
Renewable energy	5.43	0.16	N/A	5.59
Total	71.67	30.11	3.93	97.09

[a] Source: Energy Information Administration (2002), Table 2.
[b] Includes crude oil, petroleum products, and natural gas liquids.
[c] Derived from data reported in Energy Information Administration (2002), Table 1.3.
[d] Derived from data reported in Energy Information Administration (2002), Table 1.4.
[e] Production plus imports minus exports does not equal consumption reported. The difference is due to stock changes, losses, gains, miscellaneous blending components and unaccounted for supply. The aggregate adjustment is a loss of 0.75 quadrillion Btu.
[f] Not applicable.

Table 2.2. US renewable energy flows (quadrillion Btu), 2001.

	Domestic production	Imports	Domestic consumption
Hydroelectric power	2.13	0.16	2.29
Geothermal	0.31	N/A[a]	0.31
Solar	0.06	N/A	0.06
Wood, waste, and alcohol	2.87	N/A	2.87
Wood	2.17		
Waste[b]	0.55		
Ethanol and biodiesel	0.15		
Wind	0.06	N/A	0.06
Total renewables	5.43	0.16	5.59

Source: Energy Information Administration (2002), Table 2.
[a] Not available or applicable.
[b] Municipal solid waste, landfill gas, sludge waste, tyres, agricultural by-products and other biomass.

(municipal solid waste, landfill gas, sludge waste, tyres, agricultural by-products and other biomass), ethanol and biodiesel. Of the 3%, wood (2.24%) and waste (0.57%) provide the dominant share. The liquid fuels, ethanol and biodiesel, provided about 0.15% of US energy consumption in 2001. Agriculture's share of this total includes ethanol, biodiesel and part of the wind, waste and wood. These data suggest agriculture's current share is between 0.3% and 0.5% of the energy consumed in the USA.

While these data indicate that the percentage of consumption currently supplied by agriculture is relatively small, the quantities of ethanol, biodiesel and electricity from wind produced have been growing rapidly (Table 2.3). The amount of biodiesel produced in 2003 was 58% higher than the previous year. The increases for electricity produced from wind and for ethanol are 50% and 27%, respectively. Ethanol production doubled from 1998 to 2003. Double-digit percentage increases are expected again for 2004.

In spite of the growth, renewables make up a small part of the total petrol, diesel and electricity consumption in the USA (Table 2.4). Ethanol comprised about 2.1% of petrol consumption in 2003, while biodiesel accounted for less than 0.1% of diesel consumption. Electricity generated by wind power had increased to 0.3% of the US total for 2002. Given

Table 2.3. Production of ethanol, biodiesel, and electricity from wind and wood, 1980-2003.

Year	Ethanol[a] (million gallons)	Biodiesel[b] (million gallons)	Wind[c] (million megawatt-hours)	Wood energy[d] (trillion Btu)
1980	175	N/A[e]	N/A	N/A
1985	610	N/A	N/A	N/A
1990	900	N/A	N/A	N/A
1994	1,350	N/A	3.5	N/A
1995	1,400	N/A	3.2	N/A
1996	1,100	N/A	3.4	N/A
1997	1,300	N/A	3.3	N/A
1998	1,400	N/A	3.0	2,175
1999	1,470	0.5	4.5	2,224
2000	1,630	2	5.6	2,257
2001	1,770	5	6.7	2,017
2002	2,130	15	10.5	2,032
2003	2,810	20	N/A	N/A

[a] Source: Renewable Fuels Association (2004), p. 4.
[b] Source: National Biodiesel Board (2004).
[c] Source: Energy Information Administration (2003), Table 4.
[d] Source: Energy Information Administration (2003), Table 7.
[e] Not available.

Table 2.4. Estimated consumption of traditional energy sources in the USA, 1995-2003.

Year	Petrol with oxygenates[a] (million gallons)	Diesel[a] (million gallons)	Natural gas[b] (billion cubic feet)	Electricity[c] (million megawatts)
1995	115,943	28,555	N/A[d]	3,353
1996	117,783	30,101	N/A	3,444
1997	119,336	31,949	22,737	3,492
1998	122,849	33,665	22,246	3,620
1999	125,111	35,797	22,405	3,695
2000	125,720	36,990	23,333	3,802
2001	127,768	37,085	22,239	3,737
2002	131,299	38,305	23,018	3,858
2003	132,961	39,930	21,940	3,848

[a] Source: Energy Information Administration (2004c).
[b] Source: Energy Information Administration (2004e).
[c] Source: Energy Information Administration (2004b).
[d] Not available.

this base, this chapter addresses the question, how much energy can agriculture produce and how can society maximize the benefits from the resulting production of renewable fuels?

The prices of fossil fuels have been relatively low over much of the past decade, limiting the opportunities to invest profitably in the production of renewable fuels (Table 2.5). However, nominal annual average wholesale prices of petrol, diesel fuel and natural gas have been higher during the last 4 years (2000 through 2003) and electricity prices moved to higher nominal levels beginning in 2001. In real terms, petrol and diesel have averaged about 35% higher during the last 4 years of the period. Natural gas has averaged 26% higher during 2000-2003, but electricity prices have been relatively constant in real terms over the past 8 years.

Table 2.5. US wholesale petrol and diesel fuel, commercial natural gas and electricity prices, 1994-2003.

	Petrol with oxygenates[a] (US cents/ gallon)	Diesel[b] (US cents/ gallon)	Natural gas[c] (US$/thousand cubic feet)	Electricity[d] (US cents/ kilowatt-hour)	
Nominal energy prices					
1994	50.2	53.8	5.43	6.91	
1995	53.4	54.6	5.05	6.89	
1996	62.2	66.7	5.40	6.85	
1997	61.5	61.6	5.80	6.85	
1998	44.9	45.4	5.48	6.74	
1999	54.2	55.2	5.33	6.64	
2000	87.3	90.4	6.59	6.81	
2001	78.6	79.1	8.43	7.32	
2002	73.5	73.5	6.64	7.21	
2003	88.5	89.2	8.26	7.40	
Real energy prices	Petrol with oxygenates (US cents/ gallon)	Diesel (US cents/ gallon)	Natural gas (US$/thousand cubic feet)	Electricity (US cents/ kilowatt-hour)	GDP implicit price deflator
1994	52.2	56.0	5.65	7.19	96.09
1995	54.4	55.6	5.15	7.02	98.12
1996	62.2	66.7	5.40	6.85	100.00
1997	60.4	60.5	5.70	6.73	101.75
1998	43.6	44.1	5.32	6.54	102.98
1999	52.0	52.9	5.11	6.36	104.33
2000	82.0	84.9	6.19	6.40	106.44
2001	77.2	72.6	7.74	6.72	108.93
2002	66.3	66.3	5.99	6.50	110.86
2003	78.5	79.2	7.33	6.57	112.67

[a] Source: Energy Information Administration (2004g), Table 6.
[b] Source: Energy Information Administration (2004g), Table 16.
[c] Source: Energy Information Administration (2004f).

These prices reflect the changes in the world markets for petroleum and natural gas. Increasing world demand for petroleum and natural gas is pushing prices higher at the same time as the USA is importing a larger proportion of its energy supplies. The increases in energy prices, environmental concerns and energy security have increased interest in alternative fuels that can be produced from domestic resources. These events and the changes in technology for producing alternative fuels make this an appropriate time to assess the potential for agriculture as a producer of energy.

This chapter provides an overview of the current state of the technology and the feasibility of producing renewable fuels from agricultural resources. Some technologies, such as the production of ethanol from starch, have been used over a longer period of time and significant gains have been made in efficiency. While further gains are expected for the future, the rate of improvement is likely to be slower than it has been in the past. Other technologies, including the generation of electricity from wind and the production of biodiesel, are at earlier stages of their commercialization, suggesting larger gains in efficiency over time are likely. Still others, such as producing ethanol from lignocellulosic biomass, are in their infancy and have not been commercialized. This creates more uncertainty about the production efficiencies and cost levels that will be associated with commercial production. The next several sections of this chapter evaluate the relative costs of producing liquid fuels, solid fuels and electricity from agricultural resources and summarize estimates of the potential supply of agricultural biomass for energy production.

Ethanol

The US fuel ethanol industry began developing in the late 1970s as a result of restrictions in world petroleum supplies and a rapid increase in oil prices. An oil embargo in 1973 and some political instability in petroleum exporting countries created much concern about the security of national energy supplies. Fuel ethanol became attractive as a means to increase national supplies of liquid fuels. About the same time, the US Environmental Protection Agency (US EPA) was seeking a replacement for lead additives used in petrol to increase octane level. Because of its high octane, ethanol began to fill the role of an octane enhancer (Shapouri *et al.*, 2002a).

As the industry developed, more information on the environmental benefits of fuel ethanol became available. Passage of the *Clean Air Act Amendments of 1990* (CAA) gave the industry a major boost. The provisions of the CAA established the Oxygenated Fuels Program and the Reformulated Gasoline Program (RFG) in an effort to reduce carbon monoxide levels (CO). Both programmes require certain oxygen levels in petrol: 2.7% by weight for oxygenated fuel and 2.0% by weight for reformulated petrol. Ethanol and methyl tertiary-butyl ether (MTBE) are the two primary oxygenates used in the USA. The use of MTBE is being reduced, and in some states banned, because of its propensity to contaminate ground water. Thus, there is a great deal of interest in the environmental benefits of ethanol and the nation's ability to produce an adequate supply at acceptable prices. Williams *et al.* (2003) summarize a number of studies of the estimated emissions from ethanol-fuel blends relative to conventional fuel. They note the results vary depending on the test vehicles and conditions, with the benefits in many studies being greater for 'high emitters' than for the 'fleet average.' Thus, the studies indicate the general direction of the change in emissions resulting from ethanol-fuel blends, but the percentage changes in emissions are specific to a particular test situation. A few of these specific results are given below.

Ethanol-petrol blends have several advantages compared to regular petrol.

- Ethanol contains 35% oxygen. Petrol with 10% ethanol by volume reduces CO about 13% to 15% relative to conventional petrol. Many studies also show that ethanol-fuel blends reduce benzene emissions (Williams *et al.*, 2003).
- The greenhouse gasses (GHGs) are considered to include methane, nitrous oxide (N_2O) and carbon dioxide (CO_2). A study by Wang *et al.* (1997) estimated the reduction of GHGs from using ethanol-fuel blends. They found that petrol with 10% ethanol reduces GHGs about 2.4% to 2.9% in cars and light trucks. An E85 blend (85% ethanol and 15% petrol) reduces GHGs by 30.9% to 36.4%.
- Ethanol-petrol blends reduce volatile organic compounds emissions, reducing ozone-forming potential of the fuel (Williams *et al.*, 2003).
- Ethanol reduces toxic emissions (Williams *et al.*, 2003).

Table 2.6. Co-product yields in ethanol production.

Co-product	Units	Amount	
Dry-milling maize[a]			
Anhydrous ethanol	Gallons/bushel	2.6-2.8	
DDGS	Pounds/bushel	18.0	
CO_2	Pounds/bushel	18.0	
Wet-milling maize[b]			
Anhydrous ethanol	Gallons/bushel	2.6	
Maize gluten meal	Pounds/bushel	3.3	
Maize gluten feed	Pounds/bushel	12.7	
Maize oil	Pounds/bushel	1.6	
CO_2	Pounds/bushel	17.3	
Cellulosic – maize stover		Base case	Future case
Anhydrous ethanol	Gallons/ton	67.8[c]	89.7[d]
Excess electricity credit	kWh/gallon[d]	2.28	2.28

[a] Source: Tiffany and Eidman (2003).
[b] Source: Butzen and Hobbs (2002), p. 4.
[c] Source: BBI International (2001), p.39.
[d] Source: Aden *et al.* (2002), Appendix D.

Ethanol-petrol blends also have some disadvantages.

- Ethanol-fuel blends have higher evaporative emissions of some air contaminants compared to petrol. Williams *et al.* (2003) note that acetaldehyde emissions are higher with ethanol-petrol blends.
- The energy content of ethanol is lower than conventional petrol which results in lower fuel efficiency.

Ethanol from Grain

Commercial ethanol production facilities include both wet- and dry-milling operations. Wet-milling operations that process maize separate the kernel into its component parts, including the starch, fibre, gluten and germ. The fibre, gluten and germ are used to produce maize oil, maize gluten feed and maize gluten meal. The starch is converted to sugars and fermented to ethanol. Some wet-mills capture and sell the CO_2 produced during fermentation. The ethanol is recovered by distillation and dehydration to produce anhydrous alcohol. The quantity of ethanol and the co-products produced per bushel of maize for wet-mill and dry-mill operations are given in Table 2.6.

Dry-mill operations grind the maize and mix it with water to form a mash. The mash is cooked and enzymes are applied to convert the starch to sugar. Yeast is added to ferment the sugars, producing a mixture containing ethanol, water and solids. The ethanol-water mixture is distilled and dehydrated to produce anhydrous alcohol. The residual mash, called stillage, is dried to produce distillers dried grains and solubles (DDGS), which is sold as animal feed. Some dry-mill operations also capture and market the CO_2 produced. The quantity of ethanol and the by-products produced per bushel of maize are given in Table 2.6. Some firms also use dry-milling operations to produce ethanol from waste starch and sugar from food manufacturing operations.

The anhydrous ethanol produced by both wet- and dry-milling operations is denatured by adding a poison (2% to 5%) to make it unfit for human consumption. Natural petrol or raffinate, derived from natural gas, is often used because it is low octane and less expensive

Table 2.7. Ethanol from starch and sugar.

	1998 USDA Survey[a]			Tiffany and Eidman[b]
Feedstock	Grain	Grain	Non-grain	Maize
Process	Wet-mill	Dry-mill	Dry-mill	Dry-mill
Yield (gallons/bushel)	2.6800	2.6400	N/A	2.8900
Feedstock cost (US$/gallon)	0.9065	0.8151	0.2000	0.7600
By-product credits (US$/gallon)	0.4270	0.2806	0.0009	0.2674
Net feedstock cost (US$/gallon)	0.4795	0.5345	0.1910	0.4926
Operating cost (US$/gallon)	0.4597	0.4171	0.4264	0.4408
Net feedstock and operating cost (US$/gallon)	0.9392	0.9516	0.6174	0.9334
Capital cost (US$/gallon)	c	c	c	0.1958[d]
Breakeven cost (US$/gallon)	N/A[e]	N/A	N/A	1.1292

[a] Source: Shapouri et al. (2002b).
[b] Source: Tiffany and Eidman (2003).
[c] Reported levels of investment per gallon of annual capacity ranged from US$1.07 to US$2.39. No capital costs per gallon were reported.
[d] The investment cost was US$1.25/gal. Annual depreciation of US$0.0833/gal plus interest on debt and return to equity capital of US$0.1125/gal.
[e] Not available.

than petrol. The cost comparisons for commercial plants discussed in this chapter are for the production of denatured ethanol, which is typically 95% ethanol and 5% denaturant.

The total cost of owning and operating an ethanol plant can be organized into three major categories: net feedstock costs, operating costs and capital-related charges. Table 2.7 summarizes two sources of data on these costs. The first is a US Department of Agriculture (USDA) survey of 28 ethanol plants to estimate their 1998 costs of production (Shapouri et al., 2002b). The survey provided detailed data on operating expenses, feedstock costs and credit for the by-products produced. The plants participating in the survey produced over 1.1 billion gallons (gal) of ethanol in 1998. Wet-mill operations produced 66% of the total, while dry-mill operations processing grain produced 32.8% and dry-mill operations processing waste sugar and starch produced the remaining 1.2%.

The second source of data is an economic engineering study of the costs of a well-managed, state-of-the-art dry-mill operation processing maize in the US Upper Midwest (Tiffany and Eidman, 2003). The study estimated 2003 costs and profitability of producing ethanol in the type of plant the industry is building today. This study is particularly useful as a comparison with other technologies that are being proposed to produce liquid fuels. It is also possible to alter assumptions, including the price of feedstock and fuels, to analyse the impact of these and other conditions on the costs of producing ethanol.

The average yield in gallons/bushel, shown in the first row of Table 2.7, is an important factor in the profitability of an ethanol plant. The wet-mill operations surveyed averaged 2.68 gal of denatured ethanol, while the dry-mill operations surveyed averaged 2.64. Feed stock costs and by-product credit reflect local grain and feed markets and the marketing agreements of the individual ethanol plants. The wet-mill operations produce a set of higher value by-products and the by-product credit per gallon was higher for wet-mill than for dry-mill firms, as expected. The by-product credit for dry-mill operations reflects the market for DDGS, which has been burdened with relatively large supplies. In 1998 only six of the 28 firms surveyed sold CO_2 and the revenue obtained by those six firms was combined with the other by-products.

Operating expenses include electricity, fuels, waste management, water, enzymes, yeast, chemicals, denaturant, repairs and maintenance, labour, management, administration, taxes, licence fees, insurance and miscellaneous costs. Electricity and fuel are the largest components. Shapouri *et al.* (2002b) report costs of energy for plant operations of US$0.112/gal for wet-mills and US$0.131/gal for dry-mills surveyed. The other major category of variable costs includes enzymes, yeast and chemicals. These averaged US$0.1146/gal for wet-mill plants surveyed and US$0.0897/gal for dry-mill operations. Total operating expenses per gallon averaged US$0.4597 for wet-mill operations and US$0.4171 for dry-mill operations. The sum of operating expenses and net feedstock costs per gallon total US$0.9392 for wet-mill and US$0.9516 for dry-mill plants. Dry-mill operations processing waste sugar and starch reported similar operating expenses, but lower feedstock costs. Hence the sum of operating and net feedstock costs is much lower for operations processing waste starch and sugar than for those processing grain.

The surveyed plants reported an investment of US$1.07 to US$2.39/gal of capacity. While no data were obtained on capital charges for the surveyed firms, capital charges are expected to add US$0.15 to US$0.20/gal for firms with the lower investment per gallon and US$0.35 to US$0.50/gal for firms with investment at the higher end of the range. Thus, including capital costs is expected to make the breakeven costs per gallon somewhat higher than the total shown in Table 2.7.

The second study estimates costs for a plant producing 48 million gal of denatured ethanol per year. The plant is assumed to achieve 2.75 gal of anhydrous ethanol per bushel of maize and with the 5% denaturant, this is 2.89 gal of denatured ethanol per bushel. The by-products, DDGS and CO_2, are marketed at US$80/ton and US$6/ton, respectively. These by-product credits per gallon are US$0.2487 for DDGS and US$0.0187 for sale of CO_2.

Tiffany and Eidman (2003) listed energy costs of US$0.2142/gal, somewhat higher than the surveyed firms. This difference may reflect the use of higher priced natural gas (US$4.50/million Btu) and electricity (US$0.05/kWh). Enzymes, yeasts and chemicals total US$0.095/gal. The authors estimated the capital for investment and start up would be US$1.25/gal of annual capacity. Depreciation and interest (with 60% debt capital at 7.0% and a 12% return on equity capital) adds US$0.1958 to the cost of producing a gallon of denatured ethanol. The sum of feedstock and operating costs total US$0.9334/gal. Adding the capital costs results in a breakeven cost of US$1.1292/gal.

These data suggest the breakeven cost per gallon from well managed, state of the art plants is about US$1.13/gal of denatured ethanol. The cost per gallon is sensitive to many factors. For example, an increase of US$0.03 in the cost of maize per bushel (US$2.20 to US$2.23) raises the breakeven cost of ethanol approximately US$0.01/gal. More specifically, increasing the maize price to US$3.00/bu would increase the breakeven price by US$0.28, to US$1.41/gal. Similarly, either a US$0.29 increase in the price of natural gas (from US$4.50 to US$4.79)/million Btu, or a decrease of 0.05 in the yield of anhydrous ethanol per bushel (from 2.75 to 2.7) will increase the breakeven cost US$0.01/gal. As these sensitivities suggest, the profitability of plants producing ethanol from grain fluctuates widely over time, primarily depending on the cost of maize and the market price of ethanol.

The average US rack (spot) price of ethanol and the average US wholesale petrol prices have followed similar paths over the past decade (Fig. 2.1). For the early part of the decade ethanol-fuel blends were essentially a niche fuel market. However, the occurrence of MTBE in water supplies caused US EPA to form the Blue Ribbon Panel on Oxygenates in Gasoline. The panel issued its final report in September 1999, recommending the use of MTBE be reduced and with some members supporting its complete phase-out (Blue Ribbon Panel on Oxygenates in Gasoline, 1999). About the same time, wholesale petrol prices began moving up reflecting tightening world petroleum markets. As wholesale petrol prices have increased over the past 3 years, ethanol prices have moved with them. The demand for ethanol as an oxygenate can be expected to command a premium price when ethanol supplies are short, but its market price is driven largely by its value as a fuel extender, which depends primarily on the wholesale price of petrol.

In addition to the effect of wholesale petrol prices on the price of ethanol, the price of maize and the willingness of society to use MTBE as an oxygenate have been major factors in the development of the ethanol fuel industry. The average US price of maize and average US rack price of ethanol for the past decade are shown in Fig. 2.2. The high maize price from mid-1995 through mid-1996 reduced ethanol output from 1.4 billion gal in 1995 to 1.1 billion gal in 1996 and ethanol production plateaued near 1995 levels through mid-1998. As a fuel extender, the market would not support a higher ethanol price. During the next 5 years, 1998-2002, lower maize prices combined with somewhat higher ethanol prices and expectations of an expanding market for ethanol as an oxygenate spurred investment in production facilities. Additional production capacity began coming on line beginning in late 1999. Since that time ethanol production has grown rapidly, to sales of 2.8 billion gal in 2003 (Table 2.3). Opportunities for further growth in producing ethanol from grain are explored in a later section of this chapter.

New Technology

New technology is being developed for dry-mill ethanol production that fractionates the maize kernel into its pericarp or hull, the germ and the starch (St. Anthony, 2004; National Renewable Energy Laboratory, 2004). The maize germ can be used to produce maize oil and the hull can be processed into maize gluten meal and possibly other products, including additional ethanol via lignocellulosic conversion processes. Removing the hull and germ leaves a relatively pure starch that can be processed to produce ethanol. The developers claim the process will result in a higher value for the by-products (or more ethanol per bushel

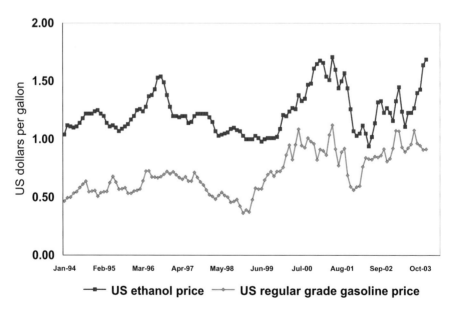

Source: Hart Energy Publishing (2004); Energy Information Administration (2004c).

Fig. 2.1. US ethanol and regular grade petrol prices, 1994-2003.

if the fibre is processed to produce ethanol) and more efficient processing of the starch because less energy is required to process the reduced volume. The technology is being tested on a commercial scale during 2004. Commercial adoption depends on the economic feasibility of applying the technology to existing as well as new plants, but could begin as early as 2005.

Ethanol from Lignocellulosic Biomass

The wet- and dry-mill operations referred to above produce ethanol from sources of soluble starch or sugar. New technologies are being developed to produce ethanol from lignocellulosic biomass, the leafy or woody part of plants, on a commercial scale. Development of these processes will permit the production of ethanol from maize stover, wheat straw, switchgrass (*Panicum virgatum*), woody biomass, waste wood and many other cellulosic waste products.

The primary components of lignocellulosic biomass are cellulose, hemicellulose and lignin. The processes being developed produce fermentable sugars from the cellulose and hemicellulose and then ferment the sugars to produce ethanol. The remaining component, lignin, has nearly the same energy content as coal, but does not have the sulphur content of coal. Lignin is a clean-burning source of energy that can be recovered from the conversion process and burned to supply high-pressure steam for process heat and production of electricity.

Several processes to produce ethanol from lignocellulosic biomass are under development. DiPardo (2000) describes and contrasts several alternative processes and indicates that enzymatic hydrolysis of cellulose has the greatest long-run potential for commercialization. The National Renewable Energy Laboratory (NREL) is pursuing a research programme to make this process commercially feasible. NREL has published two studies of the process design and economics of producing ethanol from lignocellulosic biomass

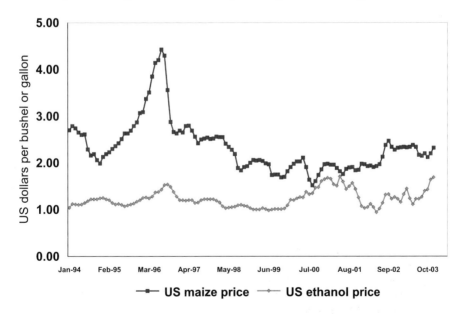

Source: US Department of Agriculture (2004); Hart Energy Publishing (2004).

Fig. 2.2. US average maize prices per bushel and rack ethanol prices per gallon, 1994-2003.

based on the co-current dilute acid prehydrolysis and enzymatic hydrolysis process they are developing. One study presents an analysis for a plant that uses hardwood (yellow poplar) chips as the feedstock (Wooley et al., 1999). The second assumes maize stover is the feedstock (Aden et al., 2002). The process design and economic analysis in both studies includes feedstock handling and storage, wastewater treatment, lignin combustion and product storage.

The process design begins at the point where the feedstock is delivered to the feeding area for storage, cleaning and size reduction. From there the biomass is pre-treated with dilute sulphuric acid at a high temperature for a short time to liberate the hemicellulose sugars and other compounds. The acid and certain compounds that would be toxic to the fermenting organism are removed at the end of this stage, before moving the hydrolysate to saccharification (or enzymatic hydrolysis) and co-fermentation. A purchased cellulase enzyme preparation is added to the hydrolysate and the tanks are maintained at a temperature to promote conversion of the cellulose to glucose. Several days are required for the conversion to glucose and for fermentation of the glucose and other sugars to ethanol. The resulting beer is distilled to separate the ethanol from the water and solids. The remaining water is removed using a vapour-phase molecular sieve to produce anhydrous ethanol. The waste products from the process are combusted in a fluidized bed combustor to produce high-pressure steam for electricity production and process heat. The majority of the high-pressure steam demand is required by the pre-treatment and distillation areas. The excess steam is converted to electricity for use in the plant and for sale to the electrical power grid. See Aden et al. (2002) for a detailed description of the process.

Evaluation of the appropriate technology and research to increase the conversion rate has been ongoing throughout the 1990s and that work continues today. Much of the development work related to the co-current dilute acid prehydrolysis and enzymatic hydrolysis process is focused on improving the activity of the enzyme, to increase the yield of ethanol per ton of feedstock and to reduce the amount of time required for the conversion to be completed. Increasing the yield of ethanol per ton reduces the per-gallon cost of ethanol. If the more active enzyme also reduces the amount of time required to complete saccharification, less equipment is required to process a given flow of biomass through the facility. With an enzyme that results in a higher and more rapid conversion rate, the cost per gallon will be reduced because of lower capital cost and because of more production. However, there is some uncertainty concerning when the improved enzyme will be available for commercial use. There is also a lack of experience constructing these plants suggesting that the investment and operating cost estimates are more uncertain than for plants producing ethanol from grain. The authors assume the estimates are for the nth plant, so that all of the equipment and material flows have been designed and produced previously for other plants, reducing investment costs for the plant analysed. Finally, the cost of acquiring the large amount of feedstock needed is also uncertain. Given these uncertainties, a range of conversion rates, enzyme costs and feedstock costs is considered in estimating the cost of ethanol produced.

The estimated production cost of ethanol from lignocellulosic biomass is based on the study by Aden et al. (2002). The Aden study incorporates several advances in design and is based on more recent price data than the Wooley et al. (1999) study. The plant is designed to process 2205 tons (2000 metric tons) of dry organic matter per day and to operate 350 days per year. The size of the plant was selected considering the economies of larger plant size and the diseconomies of transporting biomass to the plant from a larger area. Much of the literature assumes producing about 68 gal of anhydrous ethanol per ton of biomass is a reasonable base case for the early commercial plants (Wooley et al., 1999; BBI International, 2001). Using Wooley et al. (1999)'s and Aden et al. (2002)'s terminology, these would be the nth, n+1, n+2, etc. plants, where n is eight or ten. (Thus the first several plants built are

Table 2.8. Estimated production cost per gallon of denatured ethanol produced from maize stover.[a]

	Conversion rate (gallons/ton)	Cost of feedstock (US$/ton)	Enzyme cost per gallon	Plant output, million gallons per year Anhydrous	Plant output, million gallons per year Denatured	Cost/gallon denatured
Base case	67.8[b]	30	0.10	52.3	55.1	1.34
			0.40			1.62
		50[c]	0.10			1.65
			0.40			1.93
Future case	89.7	30	0.10	69.3	72.9	1.04
			0.40			1.32
		50[c]	0.10			1.25
			0.40			1.54

[a] Derived from capital costs and operating costs presented in Aden *et al.* (2002), Appendix D.
[b] Conversion rate based on BBI International (2001), p. 39.
[c] US$50 cost of feedstock based on US Department of Energy (2003), p. 13.

expected to have lower yields.) Assuming 350 days of operation per year, a plant producing 68 gal of anhydrous ethanol per ton would produce 52.3 million gal of anhydrous ethanol per year. Assuming 5% denaturant is added, this is equivalent to 55.1 million gal of denatured ethanol per year. The future case analysed by Aden assumed 89.7 gal of anhydrous ethanol per ton, resulting in an annual production level of 72.9 million gal of denatured ethanol. The plant is also assumed to produce an excess of 2.28 kWh of electricity/gal of ethanol produced and that the excess electricity is sold to the grid for US$0.041/kWh.

The conceptual design of the plant and costs presented are based on a start-up date after 2010. The economic analysis assumes 30 months are required to move from planning the project through completion of the initial performance testing. The total project investment is estimated to be US$197.4 million with an additional US$9.9 million in operating capital required to begin operating the plant. The annual operating cost is estimated to be US$73.8 million, including the purchase of the feedstock at US$30/ton and the cellulase enzyme at US$0.101/gal of anhydrous ethanol. The detailed costs are given by Aden *et al.* (2002), Appendix D.

The cost of producing denatured ethanol is summarized in Table 2.8, for alternative conversion rates per ton, alternative feedstock costs and alternative enzyme cost levels. The base case assumes a conversion rate of 67.8 gal/ton, a US$30/ton cost of feedstock and a cost of the cellulase enzyme of US$0.101/gal. The resulting cost of denatured ethanol is US$1.34/gal. Currently the cost of cellulase is reported to be higher than US$0.101/gal, in the US$0.30 to US$0.50 range (US Department of Energy, 2004a). Increasing the cellulase cost to US$0.40/gal raises the cost of denatured ethanol to US$1.62/gal. If the cost of the feedstock increases to US$50/ton, the breakeven costs are increased to US$1.65/gal with a cellulase cost of US$0.101, or US$1.93 with a cellulase cost of US$0.40.

With the development of improved enzymes, the breakeven cost is expected to decrease. The assumptions used by Aden *et al.* (2002) are a conversion rate of 89.7 gal/ton, a feedstock cost of US$30/ton and an enzyme cost of US$0.101/gal. Modelling these assumptions results in a breakeven cost of US$1.04/gal of denatured ethanol. Raising the cost of the enzyme and/or the cost of the feedstock increases the breakeven cost per gallon of denatured ethanol as shown in Table 2.8.

The results are based on maize stover as the feedstock. Similar analyses can be completed for other sources of lignocellulosic biomass. The conversion rates will differ based on the amount of lignin in the feedstock used, but until some commercial plants are built and operated it will be difficult to estimate the investment cost and conversion rate more accurately. In the example the capital investment is US$3.58/gal of denatured ethanol produced with the lower conversion rate and US$2.71/gal of annual output with the higher conversion rate. For example, the combination of the higher conversion rate, an enzyme cost of US$0.40/gal and a feedstock cost of US$50/ton would result in a breakeven cost of US$1.54/gal of denatured ethanol.

How rapidly will commercial plants be built to produce ethanol from lignocellulosic biomass? A Canadian firm announced in April 2004 that they had begun producing ethanol from wheat stover in a one million gallon per year demonstration plant. Few details were provided on conversion rates and costs (Iogen Corporation, 2004). Industry sources indicate some other companies, perhaps with access to niche sources of biomass at favourable costs, may begin construction of pilot or small commercial plants in the next few years. Information available in early 2004 suggests these first several small plants are likely to have conversion rates below 67.8 gal of anhydrous ethanol/ton, which can be expected to raise the cost per gallon to well above US$2.00 with a US$50/ton feedstock cost.

The initial plants may use an older processing technology, however, until more active enzymes become available (J. Sheehan and M. Ruth, National Renewable Energy Laboratory, Golden, Colorado, 2004, personal communication). The process being considered is two-stage dilute acid hydrolysis. With this process, the two stages accommodate the differences between hemicellulose and cellulose. The first stage maximizes yield from the more readily hydrolysed hemicellulose. The second stage hydrolyses the more resistant cellulose. The hydrolysates are recovered from each stage and fermented to ethanol (US Department of Energy, 2004b). The two-stage dilute acid hydrolysis is currently able to produce higher ethanol yields than the dilute acid prehydrolysis and enzymatic hydrolysis process and the processing facilities can be retrofitted as the newer technology is proven to be more economic. If it takes 2.5 years to complete construction and testing of a new plant as the Aden study assumes, it may be possible to have several small plants in production by 2010. As the new technology (the dilute acid prehydrolysis and enzymatic hydrolysis process) is proven, larger commercial plants are expected to be constructed. Given the expected construction time of 2.5 years, it seems unlikely that we will have large-scale plants using the new technology in production in less than 10 years.

The production of a single product, ethanol, is seen as an intermediate run technology on the way to biorefining. A biorefinery will produce a range of products including ethanol and other higher value products (Van Dyne *et al.*, 1999; J. Sheehan and M. Ruth, National Renewable Energy Laboratory, Golden, Colorado, 2004, personal communication). If the by-products of biorefining are more valuable than the electricity produced by the ethanol plant, the industry may move directly to biorefining as the most economically feasible approach to process lignocellulosic biomass.

New Technology

Much of the research for improving lignocellulosic biomass is focused on developing cellulase enzyme that will provide higher conversion rates in a shorter period of time and that can be produced at a lower cost per gallon of ethanol produced (National Renewable Energy Laboratory, 2004). Recently Novozymes and NREL announced improvements that will bring the cost of the enzyme below US$0.50/gal of ethanol produced to approximately the US$0.40/gal used in Table 2.8 (BBI International, 2001). Other efforts are being devoted to developing a biorefinery for cellulosic biomass that is capable of producing liquid fuels, power and chemicals, much like an oil refinery (National Renewable Energy Laboratory, 2004).

Biodiesel

Interest in using vegetable oils as a substitute for diesel fuel has fluctuated over the years, peaking when limited petroleum supplies resulted in high prices. Many US farmers and their commodity groups have been strong advocates of the development of renewable fuels, including biodiesel, as a means of creating new markets for agricultural commodities. More recently, environmental concerns and energy security have increased interest in biodiesel as an alternate fuel that can be produced from domestic resources. Legislation, including the *Clean Air Act Amendments (CAA) of 1990* and the *Energy Policy Act of 1992* and some state legislation (such as legislation enacted in Minnesota in 2002 requiring the state's diesel for over-the-road use to include 2% biodiesel) have opened markets for biodiesel as a substitute for petroleum diesel fuel.

Biodiesel is an oxygenated fuel made from vegetable oils or animal fats. The fats and oils are modified through a process called transesterification. This process reacts the liquid fat or oil with a solution of an alcohol (typically methanol) together with a catalyst such as potassium hydroxide (KOH) or sodium hydroxide (NaOH) to produce biodiesel and glycerine. The by-product, glycerine, is used in many products including toothpaste, paint, rubber, cosmetics, explosives and pharmaceuticals. Biodiesel can be used as an alternative to petroleum diesel in its pure form (B100) or as a blend with petroleum diesel at various ratios, such as B20 (20% biodiesel).

Potential biodiesel feedstocks produced in the USA include soybean, maize, oilseed rape (canola), peanut, cottonseed, yellow grease and animal fats, including beef tallow and pork lard (Duffield *et al.*, 1998). Waste yellow grease is soybean oil (or substitute), which has been used for cooking. Kinast (2003) notes that yellow grease, tallow and lard tend to be less uniform than the processed vegetable oils and require more processing to provide a uniform biodiesel product.

Biodiesel has several advantages compared to petroleum-derived diesel fuel.

- Biodiesel has greater lubricating qualities, reducing engine wear and increasing engine life. Kinast (2003) reported dramatic increases in lubricity with even small amounts of biodiesel blended with petroleum diesel. Lubrication improved 10% with just 0.25% biodiesel and 30% with 0.5% biodiesel. Kinast (2003) reported tests of blends composed of biodiesel from various feedstocks and found that blends made from all feedstocks tested exceeded 50% improvement in lubricity at 3% biodiesel.
- Biodiesel contains no sulphur and has 11% to 12% oxygen (compared to 0% for diesel) and essentially no aromatic compounds (Kinast 2003). As a result, biodiesel and biodiesel blends have lower emissions of CO, sulphur dioxide (SO_2), particulate emissions and less odour. Lindjhem and Pollack (2000) analysed heavy-duty vehicle emissions due to the use of a biodiesel fuel over a standard diesel fuel. They report a 13.1% reduction in CO with B20 and a 42.7% reduction with B100. For SO_2 emissions, they found that B20 reduces emissions by 20% and B100 reduces them by 100%. Thus biodiesel and biodiesel-blends offer a method to meet the lower sulphur emissions enacted for on-highway and off-highway diesel engines in the USA. Lindjhem and Pollack (2000) found particulate emissions are reduced 8.9% with B20 and 55.3% with B100. Biodiesel usage emits lower volatile organic compounds (VOC). VOC were reduced 17.9% with B20 and 62.3% with B100. Diesel emits substantial aromatic compounds while biodiesel

is essentially free of aromatic compounds. The lower aromatic compound levels of biodiesel and biodiesel blends also reduce odour.
- Biodiesel has a substantially higher flash point, making it safer to transport and store than diesel.
- Biodiesel has a higher cetane number, a measure of ignition delay, than petroleum diesel. Fuels with a higher cetane number have a shorter time between injection and ignition, resulting in smoother engine operation (Kinast, 2003).

Biodiesel also has some disadvantages.

- Biodiesel produces higher nitrogen oxide (NO_x) emissions than petroleum diesel. Lindjhem and Pollack (2000) report NO_x emissions increase 2.4% with B20 and 13.2% with B100.
- Biodiesel and biodiesel blends have higher freezing points and less desirable cold flow properties than diesel. Kinast (2003) reports that these problems are greater for biodiesel produced from animal sources than from vegetable sources (soybean and oilseed rape). He suggests using additives to improve cold flow problems in areas with low ambient temperatures.
- Fuel oxidation occurs more rapidly with biodiesel, reducing the length of time the fuel can be stored without deterioration.

The feedstock used for biodiesel production depends largely on the available supply and its price. Duffield *et al.* (1998) list the 1993-1995 US annual average production of the potential feedstocks for biodiesel. They found that soybean oil made up 52% of the total US potential feedstock supply while yellow grease was 9%. Maize oil and cottonseed oil made up 7% and 4%, respectively. Canola and rapeseed comprised about 1%. Among animal sources, inedible and edible tallow made up 13% and 5%, respectively. Large proportions of the inedible animal tallow and yellow grease are exported with the remainder used in animal feed and industrial products, suggesting these oils may be candidates for biodiesel production. As noted earlier, yellow grease and tallow tend to be less uniform than the processed vegetable oils and require more processing to produce a uniform biodiesel product. In addition, soybean oil prices have compared favourably with the price of other oils in recent years (Table 2.9). The large amount of soybean oil produced, its uniform quality and competitive price have contributed to an interest in the production of biodiesel from soybean oil. There has also been some interest in producing biodiesel from rapeseed in the USA (Withers and Noordan, 1996) and in Europe where government programme subsidies provide incentives for its production and use. However, supplies would have to be expanded to support a viable biodiesel industry.

The literature on the investment and operating costs of biodiesel plants has been evolving rapidly over the past 5 years. The reader is referred to Bender (1999) for a review of 12 studies of the economic feasibility of biodiesel production involving several feedstocks and operational scales. Tiffany (2001) summarizes data on investment and operating costs from the National Renewable Energy Laboratory. The data from these reviews suggest the investment cost per gallon of annual capacity ranges from US$3.00/gal for small plants to US$1.00/gal for plants in the 5 to 10 million gal range and to US$0.50/gal of annual capacity for plants producing 30 million gal/year (Tiffany, 2001). The amount of feedstock required depends on the type and quality of oil used. For soybean oil, 7.3 to 7.4 pounds (lb) of degummed soybean oil is required to produce one gallon of biodiesel.

Table 2.9. US vegetable oil and fats prices[a] (US cents per pound), 1994-2004.

Marketing year[b]	Soy-bean oil	Cotton-seed oil	Sunflower oil	Peanut oil	Maize oil	Lard	Edible tallow	Rape oil
1994/95	27.51	29.23	28.10	28.90	26.47	N/A	N/A	28.55
1995/96	24.70	26.53	25.42	40.30	25.24	21.70	21.56	29.05
1996/97	22.51	25.58	22.58	43.70	24.05	23.02	23.01	25.68
1997/98	25.83	28.85	27.00	49.00	28.94	19.46	20.69	28.83
1998/99	19.80	27.32	20.15	39.74	25.30	14.66	15.14	22.48
1999/00	15.59	21.56	16.68	35.39	17.81	13.64	13.21	17.11
2000/01	14.15	15.98	15.88	34.81	13.54	14.61	13.43	17.56
2001/02	16.46	17.98	23.25	32.52	19.14	13.55	13.87	23.45
2002/03	22.04	37.75	33.11	46.70	28.17	18.13	17.80	29.75
2003/04[d] (Oct-Jan)	28.76	32.80	32.07	63.59	28.14	26.40	27.07	N/A

[a] Source: Ash et al. (2003).
[b] The year beginning in October.
[c] Not available.
[d] Source: Ash and Dohlman (2004), Table 9.

A detailed analysis of capital investment, operating costs and capital costs was published in a feasibility study of producing biodiesel from soybean oil in North Dakota (Van Wechel et al., 2003). This study lists investment costs of about US$5.7 million for a plant producing 5 million gal annually. The authors estimated operating costs of US$0.4917 and capital costs of US$0.1842/gal when the plant is operated at capacity. The value of the glycerol co-product was left out of the analysis because of the lack of buyers in the area. With the plant operating at capacity, the estimated cost per gallon ranges from US$2.16 with degummed soybean oil costing US$0.20/lb to US$2.90/gal with soybean oil costing US$0.30/lb.

A recent study describes a computer model developed to estimate the capital and operating costs of a moderately-sized industrial biodiesel production facility (Haas et al., 2004). The model assumes current production practices, equipment and supply costs. The model is based on continuous-process vegetable oil transesterification and ester and glycerol recovery. The authors state that purifying the glycerol to US Pharmacopoeia standards for food, pharmaceutical and personal care products is too expensive for a plant of this size. They model a more cost-effective alternative of partially purifying the glycerol, removing methanol, fatty acids and most of the water and selling the product (80% glycerol by mass) to industrial glycerol refiners. The article presents results for a plant producing 10 million gal of biodiesel annually, using crude, degummed soybean oil as feedstock. They estimate a total investment cost of US$11.5 million, or US$1.15/gal of annual capacity. The operating costs are estimated to be US$0.2550 and the capital costs, assuming a 10 year life and 15% rate of return on capital, are US$0.2292/gal. Sale of the co-product at US$0.15/lb provides a credit of US$0.128/gal. With the plant operating at capacity, the estimated cost/gal ranges from US$1.83 with degummed soybean oil costing US$0.20/lb to US$2.58 with degummed soybean oil costing US$0.30/lb.

The average annual US price of low sulphur #2 petroleum diesel at the refinery over the past decade has fluctuated from US$0.454 to US$0.904/gal (Table 2.5). The average annual price has been higher during the past 4 years, averaging US$0.831/gal. However, with neat biodiesel (100% biodiesel) at US$2.00/gal, it would be an expensive substitute for diesel fuel. While biodiesel blends will also be higher than 100% petroleum diesel, the lubricity and other benefits may justify the increased cost. With petroleum diesel at US$0.80/gal and neat biodiesel at US$2.00/gal, each 1% biodiesel in the blend increases the cost of the blended fuel US$.012/gal. For example, the cost of a B2 blend (2% biodiesel) would be US$0.824/gal and the cost of a 5% blend would be US$0.86/gal.

Electricity from Wind Power

While a small proportion of the US total, the amount of electricity generated from wind power has increased rapidly, reaching 10.5 million megawatts in 2002 (Table 2.3). Not all of this is generated on farms and ranches, but many landowners have invested in one or more turbines and others have leased the rights to place turbines on their land.

Generating electricity with wind turbines has several advantages relative to other methods of generating electricity.

- There is potential for large-scale electricity production without emissions.
- Generating capacity can be built in small modular units within a relatively short period of time (2 years).
- There are no fuel costs and maintenance costs for wind turbines have declined.
- Generating electricity with wind power protects against some types of unforeseen cost increases, such as more emission regulations and fossil fuel price increases.

There are also some disadvantages.

- Capital costs have declined as turbines have increased in size. However, the capital costs remain relatively high and connecting to the existing transmission grid results in high transmission costs for electricity generated with wind power.
- Production is intermittent. The Electric Power Research Institute (1997) reports capacity factors ranging from 26% to 36% in 1997.[1] The intermittent production makes the task of balancing the grid more complex.
- Many of the areas with the best wind resources on agricultural land are in the middle of the continent and a considerable distance from the major consumption areas.
- Wind turbines create visual obstruction and noise pollution.

A comparison of the cost of generating electricity using wind and other sources of energy was reported by the Energy Information Administration (EIA) (2004a). The study estimates the cost per kilowatt-hour (kWh) of electricity from new plants to come on line in 2010 and 2025. The comparison for 2010 shows that wind is very competitive with coal, gas and nuclear (Table 2.10). Wind has higher capital, operation and maintenance and transmission costs than either coal- or gas-fired plants in both time periods. Wind has no fuel costs, keeping the total cost/kWh close to the total for natural gas and below both coal and nuclear. Nuclear is the high-cost alternative with the highest capital costs. It also has about the same operating and maintenance costs as wind. The report predicts that much of the replacement capacity in the near term will be natural gas-fired and new wind production is very competitive with new gas-fired plants. The report indicates that the price of natural gas is expected to increase more rapidly than coal after 2010 and coal plants will become the low-cost alternative for

---

[1]Capacity factor is the ratio of the electrical energy produced by a generating unit for the period of time considered to the electrical energy that could be produced at continuous full power during the same period.

Table 2.10. US levelized[a] electricity costs for new plants, 2010 and 2025.

	Capital	O&M[b]	Fuel	Transmission	Total
		(2002 mills[c] per kilowatt-hour)			
2010					
Coal	33.77	4.58	11.69	3.38	53.42
Gas combined cycle	12.46	1.36	32.95	2.89	49.66
Wind	35.93	7.69	0	6.92	50.54
Nuclear	46.15	7.51	4.87	2.79	61.32
2025					
Coal	33.62	4.58	11.74	3.26	53.20
Gas combined cycle	12.33	1.36	37.91	2.78	54.38
Wind	44.34	7.47	0	6.52	58.33
Nuclear	48.42	7.51	4.67	2.69	63.29

[a] Costs were amortized (levelized) over 20 years.
[b] Operating and maintenance costs.
[c] 1/1000 of a US dollar.
Source: Energy Information Administration (2004a), Figure 72, p. 4.

Table 2.11. Wind power: US installed capacity, 1981-2003.

	Megawatts	Annual average capacity factor
1981	10	N/A
1985	1,039	N/A
1990	1,525	N/A
1994	1,656	0.241
1995	1,697	0.214
1996	1,698	0.229
1997	1,706	0.221
1998	1,848	0.185
1999	2,511	0.205
2000	2,578	0.248
2001	4,275	0.179
2002	4,685	0.256
2003	6,374	N/A

Source: American Wind Energy Association (2003).
N/A = not available.

2025. Note, however, that the EIA estimates indicate wind will continue to be competitive with coal- and gas-fired generating capacity in 2025. The EIA projections assume no new nuclear units will become operable between 2002 and 2025, because natural gas and coal plants are projected to be more economical.

The growth in wind-generating capacity over the past 20 years has been very rapid, as shown in Table 2.11. The American Wind Association estimates wind can meet 6% of the nation's electrical needs by the year 2020. After the large increase in capacity during 2003, an annual growth rate of 18% is needed to meet this goal (American Wind Association, 2003). Comparing the production levels in Table 2.3 with the capacity in Table 2.11 indicates the national capacity factor has been in the range of 18% to 25% over the 1994-2002 period. Better operation and reduced maintenance are expected to increase the capacity factor over time. With the rapid rate of expansion in capacity, it is difficult to verify from these data if that is happening.

Seven states accounted for almost all of the electricity generated by wind turbines in 2001 (Table 2.12). California produced over half of the nation's electricity generated from wind that year, followed by Texas with one-sixth. Minnesota and Iowa were third and fourth in terms of electricity production from wind turbines.

Table 2.12. Electricity generated by wind by state (million megawatt-hours), 2001.

	2001
California	3.500
Iowa	0.488
Minnesota	0.897
Oregon	0.089
Texas	1.188
Wisconsin	0.072
Wyoming	0.365
All other states	0.139
US Total	6.737

Source: Energy Information Administration (2003), Table C6.

Much of the development of wind resources on agricultural land is being completed by wind energy developers. In this situation the developer negotiates a lease with the land owner. The specific terms depend on the desirability of the site for a wind turbine and other factors. Current agreements typically pay the landowner 2% to 4% of the gross turbine revenue (Energy and Environmental Research Center, 2001). For a 750 kilowatt (kW) turbine with a capacity factor of 0.25 and selling power for US$0.045/kWh, this would provide a payment to the land owner of US$1478.00 (at the 2% rate) to US$2956.00 (at the 4% rate) annually.

With this type of return, many landowners are interested in leasing the wind rights on their land. Historically, the rate of new capacity development has been tied closely to the availability of a federal incentive, the Federal Production Tax Credit (FPTC). The FPTC provides a per kWh tax credit for wind-generated electricity. Available during the first 10 years of operation, the rate was 1.8 cents/kW before it expired December 31, 2003 (Windustry, 2003). The rate of new capacity development slowed during the early part of 2004 until a 2-year extension (through December 31, 2005) of the tax credit was passed by Congress and signed by President Bush in October, 2004. Another federal subsidy for the development of wind energy is the renewable energy production incentive that provides financial incentives for electricity produced from qualifying renewable sources and sold by state and local governments and not-for-profit electric cooperatives. In addition, wind-powered units are eligible for rapid depreciation (double-declining balance on a 5-year schedule).

Electricity from Methane

A very small number of anaerobic digester systems has been installed on farms. Some of the early adopters installed systems in the 1970s and few of those continue in operation today. Ernst *et al.* (1999) report an exception, an Iowa pig farm that installed a digester in 1972 and was continuing to use it 27 years later. The operator installed the digester primarily to deal with an odour problem that neighbours found objectionable. No attempt was made to operate the system in a manner to maximize the returns from the methane produced, but the authors estimate the value of the methane would have offset only about 1/4 of the investment and operating costs of the system. They also note the digester did accomplish the operator's objective of solving the odour problem that existed in 1972 and allowed the farmer to remain in the swine business.

Lazarus and Rudstrom (2003) report more favourable results for an anaerobic digester installed on an 800-cow Minnesota dairy farm in 1999. An odour problem also prompted the initial interest in adding an anaerobic digester to the farm business. The farm had been paying a high price for electricity, 7.25 cents/kWh, and was able to negotiate a contract to sell the excess electricity to the electric cooperative at the same price. The digester has produced enough methane gas to substitute for much of the liquefied petroleum (LP) gas formerly used for heat on the farm and to generate electricity. The savings on the purchase of the LP gas, the savings in electricity purchases and the cash income from electricity sales has made the investment quite profitable.

Both studies note that there are significant economies of size in anaerobic digesters and that operation of the digesters requires a great deal of managerial expertize. Thus many operations of an adequate size may be hesitant to invest in an anaerobic digester because of the time and effort required to operate it efficiently.

Ernst *et al.* (1999) also analysed the application of an anaerobic digester to treat wastewater at a meat processing plant. The anaerobic digester reduced the surcharges the plant was paying the city for wastewater treatment and the methane was used to replace natural gas in the plant. The authors report that the digesters reduced the odour problems and were profitable using the usual criteria.

These studies suggest that anaerobic digesters may provide a way for large livestock operations and agricultural processors to deal with a major social problem and to generate another source of income for the business. More work is needed before we can estimate the contribution of this source of energy to savings in natural gas and LP gas consumption and to the supply of electricity.

The Energy Information Administration (2004a) projects a significant increase in generation of electricity from municipal waste and landfill gas. They project an increase to about 0.5% of US electricity generation capacity by 2025.

Wood Used as Fuel

Wood is the largest component of biomass used as fuel (Table 2.2). During 2002, 17% of the total was consumed by the residential sector for home heating and domestic purposes (Table 2.13). Much of this is in the form of traditional firewood, although an increasing proportion is composed of sawdust and other waste products that are sold in the form of manufactured fireplace logs, briquettes and pellets. Two per cent of wood used as fuel was consumed by the commercial sector and 7% was used in electrical power generation. The largest user, the industrial sector, consumed the remaining 74%. The largest user of wood as fuel in the industrial sector is the forest products industry.

The forest products industry includes both primary wood processors and secondary mills (Energy Information Administration, 1998). The primary wood processors include the production of sawed lumber (saw milling), pulp and paper manufacturers and the production of primary engineered wood products (plywood and panels). The secondary products include flooring, siding, moulding and other finish-milling products. An extended group of secondary wood processors includes manufacturers of cartons, pallets, transmission poles, boats, mobile homes, furniture, musical instruments and other durable goods and value-added nondurable goods made from wood.

The forest products industry satisfies a large portion of its energy needs for processing by using waste wood and the by-products of production. Large amounts of energy are required for the drying process, the operation of the kilns and steam and electricity production to power mill processes. Residues from harvesting and excess wood are commonly converted

Table 2.13. US utilization of wood as fuel by energy usage sector (trillion Btu), 1998-2002.

	1998	1999	2000	2001	2002
Residential	387	414	433	407	350
Commercial	48	52	53	41	41
Electric power	137	136	134	126	135
Industrial	1,603	1,620	1,636	1,443	1,506
Total	2,175	2,224	2,257	2,017	2,032

Source: Energy Information Administration (2003), Table 7.

to chips for fuel. Sawdust and other waste not used for fuel on site are commonly sold for fuel in the form of pressed fireplace logs, briquettes and pellets.

A survey of additional sources of wood residue is beyond the scope of this chapter. Collecting waste wood from forested land economically is clearly limited by the costs of collection and transportation. Other potential sources of wood include the collection of urban tree and landscape residue, construction and demolition debris, wood pallets and containers. Some data on an inventory of the quantities in each category for earlier years is available from the Energy Information Administration (1995).

The role of short-rotation forestry on agricultural land can be considered in an appropriate manner using the supply function approach suggested by De La Torre Ugarte et al. (2003) discussed in the following section. Some short rotation forestry on forested land is already included in the totals in Table 2.13, but the amounts are not broken down.

Resource Base and Potential Growth of Bioenergy

Alternative approaches have been used to assess the amount of biomass that can be produced in an area or country. Walsh (2004) groups the alternatives into three types: 1) an inventory approach that estimates the quantity of biomass that can be produced without reference to the cost or impacts on other parts of the sector; 2) a quantity of biomass and associated cost approach that estimates either an average or range of quantities with an associated average or range of production costs; and 3) a supply curve approach that estimates the biomass quantities as a function of the price paid. Walsh (2004) notes that the estimated amount of biomass available for bioenergy typically decreases as one shifts from an inventory to the supply curve approach. Simple supply curve approaches estimate the quantity that can be produced for each of several price levels. More sophisticated approaches integrate supply curves within a dynamic framework that allocates resources to different commodities based on changes in profitability of the alternative uses of land.

Expansion of demand for grains to produce ethanol, increasing use of vegetable oils and fats to produce biodiesel and/or expanding markets for cellulosic biomass will create an incentive for producers to shift land and other resources away from existing uses. The preferred approach to estimate the amount of biomass forthcoming as markets expand is with a national supply and demand model that includes the major agricultural commodities. This is the more sophisticated approach Walsh (2004) refers to. The next best approach is to estimate supply functions for each source of biomass.

Several studies of these types have been completed in recent years, mostly in response to proposed renewable energy legislation (US Department of Agriculture, 2000, 2001; Food and Agricultural Policy Research Institute, 2001; Raneses et al., 1999; Gallagher et al., 2000). Each of these studies analysed the amount of certain types of energy that could be produced from agricultural biomass, holding the production of other types constant. It is important to

Table 2.14. Annual volume of ethanol and biodiesel modelled for 2003-2012 (million gallons).

Calendar year	Projected biodiesel	Projected grain ethanol
2002	13	1,854
2003	22	1,930
2004	33	2,162
2005	44	2,346
2006	55	2,588
2007	67	2,858
2008	79	3,123
2009	91	3,456
2010	103	3,810
2011	114	4,160
2012	124	4,430

Source: US Department of Agriculture (2002), Table 1.

analyse carefully what these studies suggest with respect to land-use changes, as well as the estimates of the quantities of biomass that can be produced. The most recent study for each type of biomass is reviewed here.

Ethanol and Biodiesel from Grain

A recent US Department of Agriculture (2002) study estimated the effect of increasing the amount of ethanol and biodiesel produced from current levels to 4.43 billion gal of ethanol and 124 million gal of biodiesel in 2012.[2] The study was conducted to analyse the effect of a renewable fuels standard for motor vehicle fuel in the proposed *Energy Policy Act of 2002* on the farm economy. The assumed level of ethanol and biodiesel production by year is given in Table 2.14. The results provide some insight into the effect of various levels of ethanol and biodiesel production on the agricultural sector. The analysis, completed using the USDA/Economic Research Service's Food and Agricultural Policy Simulator (FAPSIM), estimates changes in major crop and livestock prices, adjustments in land use and changes in net farm income and trade over the 10-year period. The analysis assumes that 93% of the ethanol is produced from maize and 7% is from grain sorghum throughout the 10-year period, the same proportion that was used in the base period. The model assumes that all biodiesel is manufactured from soybean oil.

The increased usage of maize and sorghum increases the price of the commodities 30% and 23%, respectively, over the 10-year period (Table 2.15). The higher commodity prices result in producers increasing the area planted to maize and sorghum by 3.4 and 0.2 million acres, respectively.

The projected increase in the demand for soybean oil needed to produce biodiesel leads to an increase in the domestic price of soybean oil. The soybean oil price is projected to increase by 85% over the decade. The higher prices reduce other domestic uses of soybean oil and exports. The higher soybean oil price leads to an increase in the farm price of soybeans of about 45% over the decade. The change in feedgrain and protein prices results in relatively minor changes in livestock production and profitability over the decade.

---

[2] The analysis also assumes some ethanol would be produced from cellulosic biomass, increasing from 15 million gal in 2003 to 241 million gal in 2012. Given the uncertainty of the development of commercial cellulosic production capacity, these totals are not considered in this summary.

Table 2.15. Projected market prices and income with the increased production of ethanol and biodiesel, 2001-2011.

Marketing year	Maize (US$/bushel)	Sorghum (US$/bushel)	Soybeans (US$ bushel)	Soybean meal (US$/ton)	Soybean oil (US cents /pound)	Net farm income (billion US$)	Agricultural exports (billion US$)
2001	2.10	2.05	4.31	154.3	15.82	N/A	53.00
2002	2.10	1.95	4.37	153.8	16.83	40.6	54.50
2003	2.20	2.05	4.54	150.8	18.57	41.4	56.10
2004	2.25	2.05	4.80	151.8	20.30	41.3	59.00
2005	2.30	2.10	5.16	158.2	22.04	41.0	61.40
2006	2.31	2.11	5.51	164.2	23.57	44.1	64.00
2007	2.44	2.23	5.77	169.1	24.71	47.3	66.50
2008	2.51	2.30	5.97	168.8	26.14	50.7	69.60
2009	2.64	2.44	6.13	170.8	27.19	54.2	72.30
2010	2.66	2.45	6.23	168.5	28.43	56.3	75.00
2011	2.73	2.52	6.25	165.7	29.32	58.5	77.00

N/A = not available.
Source: US Department of Agriculture (2002), Table 3.

Under the 2002 farm bill, higher prices for maize, sorghum and soybeans result in lower government outlays associated with the marketing assistance loan programme and lower counter-cyclical payments. Thus the higher commodity prices are largely offset by lower farm programme payments during the first half of the period. During the second half of the period, net farm income increases more rapidly (Table 2.15). With the higher prices of feedgrains, soybeans and soybean oil, the value of US agricultural exports increases 45% over the decade.

While the price levels would be different starting with current stocks and market conditions, the model's results suggest that the USA could produce enough maize and sorghum to produce over 4 billion gal of ethanol from grain without major disruption of the feedgrain and livestock markets. However, producing as much as 124 million gal of biodiesel from soybean oil as the single feedstock results in major changes in domestic use and exports of soybean oil, suggesting some production would need to come from other feedstocks if larger quantities of biodiesel are to be produced.

Meeting the Clean Air Mandate

The petroleum industry began producing MTBE in the early 1980s to meet the octane demand that resulted from the phase-out of lead additives. In later years, MTBE was used to help meet the demand for premium petrols. The oxygenated petrol programme stimulated demand for MTBE and the additive was the industry's oxygenate of choice. MTBE could be produced and blended with petrol at the refinery and the blended fuel distributed as a ready-to-sell product. Using ethanol as the oxygenate requires purchasing ethanol on the market, transporting the quantity needed to distributors and blending the fuel at the distributor level. Offering a product of consistent quality with minimal problems must have appeared to be easier using MTBE as the oxygenate. With the prices at the time it was also less expensive to blend MTBE than ethanol. With current environmental concerns about using MTBE, how do the two oxygenates compete?

The data in Table 2.16 indicate that ethanol has a higher percentage of oxygen (34.73%) than MTBE (18.15%) and a smaller amount of ethanol is required per gallon of fuel to meet the oxygen requirement for either reformulated petrol or the higher oxygen requirement of

Table 2.16. Properties of oxygenates and amounts needed to meet *US Clean Air Act* requirements for reformulated petrol and oxygenated petrol.

	Ethanol	MTBE
Oxygen content, % by weight	34.73	18.15
Octane	115	110
Reformulated petrol 2.1% oxygen by weight	5.8% by volume	11.7% by volume
Oxygenated petrol 2.7% by weight	7.4% by volume	15% by volume
Cost of reformulated fuel per gallon with 2003 prices[a]	US$0.849	US$0.876

[a] Ethanol at US$1.35/gal, a blenders credit of US$0.52/gal, petrol at US$0.85/gal and MTBE at US$1.07/gal.
Source: Based on data from the Energy Information Administration (2004d).

oxygenated petrol. Reformulated petrol must have 2.1% oxygen by weight, which requires 0.058 gal of ethanol and 0.942 gal of petrol in each gallon of blended fuel. Using MTBE as the oxygenate, each gallon of blended fuel is composed of 0.117 gal of MTBE and 0.883 gal of petrol to meet the requirement. Oxygenated petrol must have 2.7% oxygen by weight. Meeting this level of oxygen requires either 0.074 gal of ethanol or 0.15 gal of MTBE in each gallon of blended fuel. The cost of substituting ethanol for MTBE depends on the relative price of the two oxygenates and other adjustments that must be made in the refining process.

Substituting MTBE for ethanol requires some additional modifications in the refining and distribution system. Ethanol results in higher emissions of smog-forming VOC than MTBE. Refiners typically remove other highly volatile components of petrol, such as butane and pentanes, to meet the emission standards. Ethanol has a higher octane percentage than MTBE, but because only half as much is used, a net loss in octane occurs that must be made up. The net effect of these adjustments is that about 11% of the volume must be made up and some additional refining costs are incurred. In addition to the effect on the cost of producing a gallon of reformulated petrol, it is important to recognize that making this substitution requires obtaining an additional 11% of petrol to meet the same total demand. Substituting ethanol for MTBE also requires some modification of the distribution system. MTBE is blended with petrol at the refinery, but ethanol is shipped separately from petrol (typically by rail or truck but not by pipeline) and is blended with petrol at the distribution terminal.

Presently, approximately one-third (45 billion gal/year in 2002 and 2003) of petrol consumption in the USA is reformulated fuel and about 4% (4.9 billion gal in 2002 and 2003) is oxygenates. The industry used 2.4 billion gal of MTBE during 2002, mostly in reformulated petrol. Replacing all of the MTBE with ethanol will require about 1.2 billion gal of ethanol in addition to its traditional demand. Production of ethanol in 2002 totalled 2.1 billion gal. The US Department of Agriculture (2002) suggests we can produce as much as 4.4 billion gal without major distortions to the commodity markets. Thus it appears that agriculture can supply enough oxygenate to replace MTBE in reformulated petrol.

How will shifting to ethanol influence the cost of petrol? The national average cost of the two oxygenates over the past decade is shown in Fig. 2.3. As the figure shows, MTBE had a lower price per gallon than ethanol from 1994 through early 2000. Since mid-2000 the price per gallon of the two oxygenates has fluctuated more widely and the gap in prices has narrowed. This suggests the cost of adding oxygen to petrol has increased and that the cost of using MTBE has increased more than the cost of using ethanol as the oxygenate. How do they compare today? For illustration, consider the average prices and the blender's credit for

2003. As shown in Table 2.16, the cost of reformulated petrol with ethanol is US$0.016/gal higher than reformulated petrol with MTBE. This analysis does not include the costs of adjustments to meet the octane level and to use the volatile liquids removed to compensate for the higher volatility of ethanol. However, it suggests that the increased cost will be relatively small, probably in the 2 to 3 cent/gal range.

Blending Biodiesel and Diesel Fuel

Blending biodiesel and diesel fuel is commonly suggested as a way to improve lubricity and lower emissions of CO, SO_2, particulate matter and odour. Blends of 2% to 5% or more are commonly recommended. What do the results of the USDA study suggest about our ability to supply this potential market?

The US consumed 39,930 million gal of diesel fuel in 2003 (Table 2.4). Biodiesel production would need to be much larger than the 124 million gal to meet the demand for even a 2% blend.

Bioenergy Crop Production

A second study analysed the potential to convert cropped, idle, pasture and Conservation Reserve Program (CRP) acres to bioenergy crop production (De La Torre Ugarte *et al.*, 2003). The three bioenergy crops considered were switchgrass, willow (*Salix* sp.) and hybrid poplar (*Populus* x). The analysis considered two management scenarios on CRP acres, high wildlife diversity and high biomass production. The wildlife management scenario assumes fewer fertilizer and chemical inputs and that only one-half of the switchgrass is harvested each year. The wildlife scenario assumes a farm gate price of US$30/dry ton (dt) for switchgrass, US$31.74/dt for willow and US$32.90/dt for hybrid poplar. A second scenario, referred to as

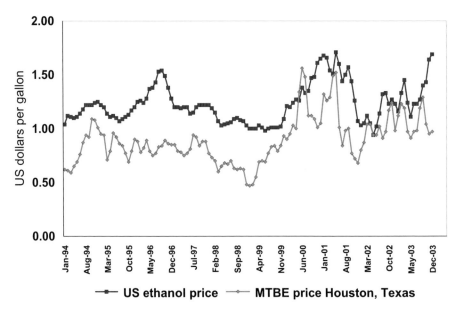

Source: Hart Energy Publishing (2004).

Figure 2.3. US ethanol prices and MTBE prices (Houston, Texas), 1994-2003.

Table 2.17. Projected plantings with the introduction of bioenergy crops, 2008.

Wildlife management scenario (units in million acres)

	All cropland	Major crops	CRP	Idle	Pasture
Switchgrass	12.32	10.44	1.10	0.23	0.55
Hybrid poplar	7.10	0	7.10	0	0
Willows	0	0	0	0	0
All bioenergy	19.42	10.44	8.20	0.23	0.55

Production management scenario (units in million acres)

	All cropland	Major crops	CRP	Idle	Pasture
Switchgrass	41.87	23.37	12.91	2.09	3.49
Hybrid poplar	0	0	0	0	0
Willows	0	0	0	0	0
All bioenergy	41.87	23.37	12.91	2.09	3.49

Source: De La Torre Ugarte *et al.* (2003), Table 4.

Table 2.18. Projected changes in cropland use with the introduction of bioenergy crops (million acres), 2008.

	1999 acres USDA baseline (Feb. 1999)	2008 acres Wildlife management scenario	2008 acres Production management scenario
Maize	82.0	80.6	78.3
Sorghum	10.6	10.6	10.2
Oats	4.7	4.6	4.6
Barley	7.0	7.0	7.0
Wheat	73.1	71.7	66.9
Soybeans	71.8	69.9	68.4
Cotton	12.8	12.3	12.1
Rice	3.2	3.1	3.1
Lucerne	27.1	26.1	24.9
Other hay	33.1	30.6	29.4
Bioenergy	0.0	19.4	41.9
Total CRP	29.8	21.6	16.9
Total planted	325.4	335.9	346.8
Idle and pasture	79.3	77.0	70.8

Source: De La Torre Ugarte *et al.* (2003), Tables 5 and 6.

the management scenario, assumes standard fertilizer and chemical inputs with all of the switchgrass harvested each year. It also assumes somewhat higher prices: US$40/dt for switchgrass; US$42.32/dt for willow; and US$43.87/dt for hybrid poplar. Both management scenarios assumed farmers would be required to forfeit 25% of their CRP rental rate to receive permission to harvest the CRP acreage for biomass production.

The analysis was completed using POLYSYS, an agricultural policy simulation model of the US agricultural sector. The study used a 1999 baseline and estimated the impact that would occur under each scenario for 2008. Under the wildlife management scenario 19.4 million acres are planted to bioenergy crops (Table 2.17). This is composed of 10.44 million acres coming from land devoted to traditional crops, 8.2 million acres of CRP land, 0.23 million acres of idle land and 0.55 million acres of pasture. With the wildlife management scenario 7.1 million acres are planted to hybrid poplar and the remaining 12.32 million acres are planted to switchgrass. The cropland that moves into bioenergy production comes primarily from reductions in the planted acreage of maize, soybeans, wheat, cotton, lucerne and other hay (Table 2.18).

Table 2.19. Projected price changes for major crops in 2008 with introduction of bioenergy crops.

			Bioenergy crop management scenario					
		Baseline	Value		Difference			
		price	Wildlife	Production	Wildlife		Production	
Crop	Unit	(US$)	(US$)	(US$)	(US$)		(US$)	
Maize	bushel	2.55	2.65	2.79	0.10	4%	0.24	9%
Sorghum	bushel	2.44	2.57	2.77	0.13	5%	0.33	14%
Oats	bushel	1.50	1.58	1.67	0.08	5%	0.17	11%
Barley	bushel	2.35	2.43	2.55	0.08	3%	0.20	9%
Wheat	bushel	4.25	4.40	4.74	0.15	4%	0.49	12%
Soybeans	bushel	6.10	6.42	6.71	0.32	5%	0.61	10%
Cotton	pound	0.68	0.74	0.77	0.06	9%	0.09	13%
Rice	100 pounds	10.37	11.23	11.37	0.86	8%	1.00	10%

Source: De La Torre Ugarte et al. (2003), Table 12.

Table 2.20. Projected changes in net farm income (billion US$) in 2008 with the introduction of bioenergy crops.

	USDA baseline (Feb. 1999)	Wildlife management scenario	Production management scenario
Crops and livestock	50.5	52.6	54.2
Bioenergy crops	0	0.7	2.3
Total net farm income	50.5	53.3	56.5

Source: De La Torre Ugarte et al. (2003), Table 13.

Table 2.21. Annual energy production equivalents for bioenergy crops, 2008.

	Wildlife management scenario		Production management scenario	
	Switchgrass	SRWC[a]	Switchgrass	SRWC[a]
Million acres	12.30	7.10	41.90	0
Million dry tons	60.50	35.50	188.10	0
Quadrillion Btu[b]	0.94	0.60	2.92	0

[a] SWRC – Short rotation woody crop (poplar and willow).
[b] Energy content of energy crops assumes 15.5 MBtu/dt for switchgrass and 17 MBtu/dt for hybrid poplar.
Source: De La Torre Ugarte et al. (2003), Table 15.

With the production management scenario, all of the 41.87 million acres of bioenergy crop production are planted to switchgrass. Of this total, 23.37 million acres are from land formerly in crop production, 12.91 million acres are from land in the CRP, 2.09 million are from idle acres and 3.49 million acres are from pasture. The cropland that shifts to bioenergy production comes primarily from larger reductions in the planted acreage of the same crops listed for the wildlife management scenario (Table 2.18). The price and income impacts are summarized in Tables 2.19 and 2.20. Under the wildlife management scenario, commodity prices increase 5% or less for all crops except cotton and rice which increase 9% and 8%, respectively. However, the price impacts for the production management scenario are in the 9% to 14% range for all commodities listed. The higher prices result in an increase in net farm income of US$2.8 billion for the wildlife management scenario and US$6.0 billion for the production management scenario.

The analysis indicates the wildlife scenario will produce 96.0 million dt of biomass, equivalent to 1.54 quad, or 1.6% of the energy consumed in 2001 (Table 2.21). The production management scenario is expected to produce 188.1 million dt, equivalent to 2.92 quad, or

3.0% of the energy consumed in 2001. This lignocellulosic biomass could be converted to ethanol, used to produce electricity, or converted to other products through biorefining.

Biomass from Crop Residues

Another important potential source of biomass from agriculture is crop residue. A recent study estimated supply functions for crop residue by region of the USA (Gallagher et al., 2003). The study used the 1997 USDA baseline for crop acres planted to each crop. The 1997 baseline acreages are very similar to the baselines that were used in assessing the impact of increasing ethanol, biodiesel and producing energy crops.

The analysis excludes the residue needed for conservation and erosion control. The calculations consider residue harvest on land only if erosion is below tolerance. The required cover for erosion control was maintained on the remaining acres in calculating harvestable residue.

The study assumes that the residue would be made available to a biomass processing industry for its opportunity cost. For residue fed to livestock, the opportunity cost is the cost of an alternative source of livestock feed, typically low-grade hay. For residue remaining in the field, the opportunity cost is the sum of the cost of replacing nutrients removed, the cost of machinery operations to bale the residue, the on-farm transportation cost and the cost of transporting the biomass from the farm to the plant. Notice that this approach pays the farmer and other firms for the costs they incur, but it does not increase the farmer's net income.

The estimated quantities of residue for five regions in the USA are given in Table 2.22 (See Fig. 2.4 for regions). The study estimated the net production of residue as the amount available after allowing for the conservation adjustments and erosion-based adjustments to yield for the area. Feed use indicates the amount of residue used by livestock in the region. The industry supply is an estimate of the biomass available to a biomass processing industry at a price near its harvest cost.

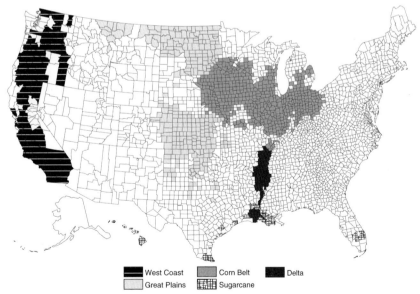

Source: Gallagher et al. (2003), p. 5.   Note: Boundary lines are counties and parishes.

Figure 2.4. US crop regions.

Table 2.22. Estimated biomass from crop residue (million tons).

Region	Types of crop residue	Net residue production[a]	Feed use[b]	Industry supply[c]	Price range to harvest up to 90% of industry supply and transport to processing plant
Corn Belt	Maize stover	103.6	4.7	98.9	US$16.50 to US$22.00 per ton for large plant
Great Plains	Maize and sorghum stover; wheat, barley and oat straw	40.5	5.0	35.5	US$14.00 to US$35.00 per medium plant
West Coast	Maize stover; wheat, barley and oat straw	3.7	1.3	2.4	US$12.00 to US$20.00
Delta	Rice straw	5.2	0.6	4.6	US$20.00 to US$25.00
Southeast	Sugarcane bagasse	3.6	0.0	3.6	Available at bid above US$34.65/ton
Total		156.6	11.6	145.0	

[a] Potential production net of conservation adjustments to yield and erosion based restrictions.
[b] Crop residues that will only be available for the biomass processing industry if price exceeds the cost of replacing these residues with other feeds.
[c] The unused residue that would be available to a biomass processing company.
Source: Gallagher *et al.* (2003), Table 4 and Fig. 5a-5d.

These estimates are presented as supply functions in the publication, increasing from the lowest price at which some biomass would be available, and show the relationship between the price and the estimated amount that would be supplied. These results, summarized in Table 2.22, indicate that about two-thirds of the total industry supply in the US would be available in the Corn Belt and an additional one-fourth could be supplied by producers in the Great Plains. The study indicates that some biomass would be made available and delivered to a large plant in the Corn Belt for US$16.50/ton and that raising the price to US$22/ton would cause farmers to make 90% of the region's industry supply available to biomass processing firms. Estimates for the Great Plains assume delivery to a mid-sized plant because of the lower density of harvestable crop residue in the region. The estimates indicate that about 90% of the total could be delivered for US$35/ton. The quantities available in the other three regions are much smaller, but the price range for delivery is similar to the Corn Belt and the Great Plains. The sugarcane bagasse in the Southeast is burned for energy in sugar refineries. Sugarcane provides about 35% more energy than a modern plant requires and some plants install generating equipment and sell electricity by burning the excess bagasse. Other users of bagasse must pay more than the breakeven price of using the bagasse for this purpose, which the authors estimate to be US$34.65/ton.

Aggregating over the five regions, the industry supply of crop residue for the USA is 145.0 million tons. The study indicates 90% of the total could be delivered to a processor at a cost of US$35 or less.

Some analysts have questioned whether farmers would be willing to supply biomass for the cost of harvesting and the cost of replacing the nutrients. They suggest farmers would require an increase in net returns to either harvest the residue themselves or permit others to harvest residue on their farm. Another recent study provides some additional information on the price at which farmers are willing to deliver crop residue. The report discusses the experience of a firm that collected maize stover for an industrial processor in Harlan, Iowa (Glassner *et al.*, 1998). The firm collected over 50,000 tons, paying US$31.60 to US$35.73/ton dry weight, depending on the amount of residue removed and the distance to the plant. They report that more stover was available at that price.

Summary of Bioenergy Supplies From Agriculture

The three studies cited used appropriate methodology to estimate the supplies of ethanol, biodiesel and cellulosic biomass that US agriculture could produce in the near future. Each study starts from about the same baseline, holds certain things constant and proceeds to analyse what happens if specified changes are made.

The US Department of Agriculture (2002) study assumes that an increase in the demand for ethanol and biodiesel will occur and estimates the impact on commodity prices, resource adjustments, farm income and agricultural exports. The results indicate that maize and soybean prices will increase, making production of these two commodities more profitable. The model shifts an additional 4.9 million acres into maize production and an additional 0.6 million acres into soybean production. Other changes are an 85% increase in the price of soybean oil, reducing other domestic uses and exports of soybean oil. This analysis does not assume there is competition for the use of land from bioenergy crop production.

The De La Torre Ugarte *et al.* (2003) study of bioenergy crop production assumes the 1999 USDA baseline, which includes the use of maize to produce 2.0 billion gal of ethanol for 2008 and no biodiesel production (P. Westcott, USDA/Economic Research Service, Washington, DC, June 2, 2004, personal communication). This is a much smaller commitment of maize to ethanol than in the USDA study. The De La Torre Ugarte study also assumes a set of prices that would be paid for switchgrass and woody biomass and estimates the number of acres that would shift into production of bioenergy crops under each of two scenarios, wildlife management and production management. With the wildlife management scenario, 10.44 million acres of cropland shift from major crops to switchgrass. The adjustment includes reductions in the acreage of maize and soybeans of 1.4 and 1.9 million acres, respectively. The reductions in maize and soybean acres with the management scenario are larger, 3.7 and 3.4 million acres, respectively. The estimated maize price is very similar for the two studies, but the soybean price is higher for the bioenergy crop production study than the USDA study. A primary difference between the two studies is the combination of renewable fuels produced. The first produces higher levels of ethanol and biodiesel, while the second produces more lignocellulosic biomass.

The Gallagher *et al.* (2003) study holds the baseline acreage constant and calculates the amount of crop residue that could be harvested. If the price paid to the farmer for the residue is the opportunity cost of the residue, then the collection will not change the relative profitability of the crops and acreage shifts would not occur. However, if higher prices are paid for crop residue, the profitability of crops with residue will increase relative to those

Table 2.23. Summary of lignocellulosic supplies from energy crops and crop residue assuming ethanol and biodiesel from grains continue to expand.

	Wildlife management scenario	Production management scenario
Total lignocellulosic biomass (millions of dry tons)	191.1	261.2
Ethanol @ 67.8 gallons anhydrous/ton and electricity[a]		
Gallons denatured (billion gallons)	13.6	18.6
Percentage of US 2003 petrol use[b]	6.4	8.7
Electricity (million megawatts)	29.5	40.4
Percentage of US 2003 electricity use	0.75	1.05
Ethanol @ 89.7 gallons anhydrous/ton and electricity[c]		
Gallons denatured (billion gallons)	18.0	24.7
Percentage of US 2003 petrol use[b]	8.4	11.6
Electricity (million megawatts)	39.1	53.4
Percentage of US 2003 electricity use	1.0	1.4
Electricity (gasifier combined cycle)[b]		
(million megawatts)	312.4	427.1
Percentage of US 2003 electricity use	8.1	11.1

[a] Yields of ethanol and electricity based on data in Table 2.6.
[b] Conversion made on a Btu basis.
[c] Electricity estimates assume gasifier combined cycle with 36% conversion efficiency and 90% operating rate.

without residue and some additional acreage can be expected to shift into residue-producing crops.

Given the difference in assumptions, it is inappropriate to simply sum the estimated quantities of biomass for energy from the three studies to obtain an estimate of agriculture's potential contribution to the nation's energy supply. A simple sum bases the estimate on more acres than are available. A more definitive answer to the question of how much ethanol, biodiesel and lignocellulosic biomass US agriculture is capable of producing could be provided by rerunning one of the national models including both the alternative to harvest and sell crop residue and the opportunity to produce bioenergy crops. With this approach, one can estimate the amounts and types of lignocellulosic biomass that would be made available for a series of ethanol and biodiesel production levels. Until that result is available, we will have to approximate the amount with less sophisticated methods.

With stronger world petroleum demand, it is reasonable to assume the demand for ethanol and biodiesel will remain high, supporting the projected production of ethanol and biodiesel, (and the resulting acreage of the two crops) from the USDA study as a starting point to estimate the amount of energy agriculture can produce. These totals are 4.43 billion gal of ethanol and 0.124 billion gal of biodiesel. Second, the amount of bioenergy crop production would be reduced by the acreage of additional maize and soybeans required for the higher production levels of ethanol and biodiesel, 6.1 million acres for the wildlife management scenario and 9.9 million acres for the production management scenario. This reduces the amount of lignocellulosic biomass produced by US agriculture for the two scenarios by 35.4 and 57.4 million tons, respectively. Adding the amount of crop residue estimated by Gallagher *et al.* (2003) (90% of the industry supply), suggests US agriculture could produce as much as 191.1 million tons of lignocellulosic biomass with the wildlife management scenario (Table 2.23). If policy is put in place to implement the production management scenario, the US could produce 261.2 million tons of lignocellulosic biomass in addition to the 4.43 billion

Table 2.24. World fuel ethanol production (million gallons), 2003.

Brazil	3,660
USA and Canada	3,130
EU	440
Other	336
World total	7,566

Source: European Biodiesel Board (2004).

gal of ethanol and 0.124 billion gal of biodiesel. These totals may overstate the amount of lignocellulosic biomass that could be produced, because the acreages of other crops, particularly wheat, lucerne hay and other hay were also reduced in estimating the production of bioenergy crops. More information is needed on the location of the acres shifted from major crops to bioenergy crops to make a more accurate adjustment.

The lignocellulosic biomass can be used to produce various combinations of ethanol and electricity using the technology discussed in this paper. If the entire annual harvest is used to produce ethanol with the base case technology (67.8 gal of anhydrous ethanol/ton and 2.28 kWh/gal), the 191.1 million tons harvested under the wildlife management scenario would produce 13.6 billion gal of ethanol and 29.5 million megawatts of electricity. These quantities are equal to 6.4% of the petrol consumption and 0.75% of the electricity consumption in 2003. Applying the same conversion rates to the 261.2 million tons under the production management scenario would produce an amount of ethanol and electricity equal to 8.7% and 1% of the 2003 use, respectively. Applying the higher conversion rates results in producing a larger proportion of the petrol and electricity usage as shown in Table 2.23.

Another way of using the biomass is to produce a larger amount of electricity (and no ethanol) using the gasifier combined cycle technology. The 191.1 million tons produced under the wildlife management scenario could be used to generate 312 million megawatts, an amount equal to 8.1% of 2003 electricity use in the USA. The 261.2 tons produced under the production management scenario would generate 427 million megawatts, an amount equal to 11% of 2003 electricity use.

Worldwide Ethanol and Biodiesel Production

Previous parts of this chapter have focused on the production of renewable fuels in the USA for domestic consumption. This section briefly notes the status of renewable liquid fuel production in other parts of the world. The discussion is limited to liquid fuels because they have physical characteristics that allow them to be easily transported and traded.

Brazil and the USA dominate fuel ethanol production, but other countries are expanding as well. Brazil continues to be the leader, producing 3.7 billion gal of fuel ethanol from sugar in 2003 (Table 2.24). The USA and Canada are the second highest production area in the world, but fuel ethanol production is also increasing in Europe and Asia. France and Spain have established fuel ethanol industries and Germany and the United Kingdom are expanding capacity (Fulton, 2004). China announced a fuel ethanol programme in 2000 to utilize its grain surplus and reduce the amount of oil imports. Thailand announced a programme the same year to produce ethanol from sugarcane and tapioca. India, producing ethanol from sugar, recently mandated that petrol sold in four Union Territories contain a minimum of 5% ethanol. Japan is also a producer and an important importer in the region.

Table 2.25. Biodiesel production by country (million gallons), 2003.

Germany	215.1
France	107.4
Italy	82.1
Austria	9.6
Spain	1.8
Denmark	12.3
United Kingdom	2.7
Sweden	0.3
World total	431.3

Source: Fulton (2004).

Most countries are producing ethanol primarily for the domestic market. Producing countries typically provide subsidies to help establish the industry. These subsidies are often in the form of reduced tax on motor fuel containing ethanol. Countries reducing tax on ethanol blends are reluctant to open the borders to imports because each gallon sold reduces tax revenues. As a result many countries, including the USA, have imposed measures to discourage importation of ethanol for fuel use. The USA imposes a tariff of US$0.54/gal on imported ethanol, except ethanol imported through the Caribbean Initiative. This initiative allows ethanol processed in the Caribbean to enter duty free, up to 7% of US ethanol consumption. The USA has been importing ethanol from Brazil, typically through the duty-free Caribbean Initiative. The ethanol is shipped to the Caribbean, reprocessed and then shipped to the US without paying the tariff. Some of these imports have occurred to meet seasonal demands and to fill in the expanding demand resulting from restrictions on MTBE use. Brazil also exports to other countries. As more countries develop markets for ethanol fuels, more trade can be expected to develop to balance the seasonal shortages and surpluses that develop.

Biodiesel developed earlier in Europe than in the USA largely because of changes made in the European Union (EU)'s Common Agricultural Policy (CAP). Other policy commitments in recent years are continuing to encourage development of the biodiesel industry.

A set-aside programme was established in 1992 that obligated farmers not to grow food or feed crops on a portion of their arable cropland (Foreign Agricultural Service, 2003). However, they were allowed to plant rapeseed, sunflowers or soybeans for industrial purposes, making the production of vegetable oils for use in biodiesel an option and the biodiesel industry grew rapidly. The agreement limited production on set-aside land to 1 million metric tons of soybean meal equivalent. The industry has passed this limit and additional expansion of the biodiesel industry is dependent on purchasing oilseeds from non-set-aside land, importing oilseeds, or importing vegetable oil. Some biodiesel producers without enough contract rapeseed purchase rapeseed oil. The reformed CAP adopted in 2003 provides a carbon credit payment per hectare for farmers growing non-feed crops. These payments are for non-set-aside land only.

The EU Commission also adopted the Promotion Directive in 2003. This directive provides a target for member states that 2% of petrol and diesel used for motor transport should be from renewable sources, with the percentage increasing to 5.75% by 2010 (Foreign Agricultural Service, 2003). The percentages are of sales, not production, suggesting countries can import biofuels to meet the target. These provisions may lead to more trade in biodiesel among the European countries. Germany, France and Italy had the highest production levels in 2003 (Table 2.25). The production in Europe totalled about 431 million gal in 2003.

Summary and Conclusions

One purpose of this chapter is to document agriculture's contribution to the US energy supply. Renewables provided 5.8% of the nation's energy supply in 2001. Hydroelectric generation, solar, wood and waste made up almost all of this total. Liquid fuels, ethanol and biodiesel, contributed 0.15% of the energy consumed in the USA and electricity generated with wind power made up 0.06% of the nation's energy supply that year. It is difficult to provide a precise estimate of agriculture's current contribution to the nation's energy supply, but it is probably about 0.3% of the total Btu consumed. The production of ethanol and biodiesel, as well as electricity from wind turbines, has grown rapidly in recent years, but they continue to contribute a relatively small percentage of the transportation fuels and electricity used annually. Ethanol production was 2.1% of the US petrol consumption in 2003, while biodiesel was 0.05% of diesel fuel consumed and electricity generated by wind power was 0.3% of US total electricity production.

The increase in real energy prices in recent years has provided an incentive to develop renewables. Real petrol and diesel prices have averaged 35% higher during the 2000-2003 period than they did during the previous 6 years. Natural gas prices have been 26% higher, but electricity prices have been essentially constant in real terms for the past 8 years. New technologies becoming available and higher energy prices indicate a potential for agriculture to play a much greater role in producing renewable fuels.

A second purpose of the chapter is to assess the current production costs of renewables made from agricultural biomass. Ethanol is currently produced from grain, primarily maize, and some from grain sorghum. The cost of producing ethanol in a state-of-the-art dry-mill plant is estimated to be US$1.13/gal, when the cost of the feedstock, maize, is US$2.20/bu and other input costs are at early 2003 levels. Changes in the market price of ethanol are driven primarily by the price of petrol. Higher petrol prices increase the value of ethanol as a fuel extender and drive up its price. The relationship between the price of ethanol and the price of the feedstock is a major determinant of profitability of an ethanol plant, but not a determinant of ethanol price. New technology is being introduced in dry-milling operations that fractionates the maize kernel into the hull, germ and the starch at the start of the process, which will enable dry-mills to produce higher-value by-products than DDGS. Adoption of this new technology may modestly reduce the cost of producing ethanol from grain over the next several years.

NREL is developing the co-current dilute prehydrolsis and enzymatic hydrolysis process to produce ethanol from lignocellulosic biomass. Research is ongoing to develop enzymes that will provide higher conversion rates and more rapid conversion. Some pilot or small commercial plants may be built in the next several years. These initial plants are expected to achieve relatively low conversion rates, suggesting the cost of the ethanol from these initial plants may exceed US$2/gal with feedstock prices of US$50/ton. It appears unlikely that construction of large-scale plants will begin until after 2010.

The feedstock used for biodiesel depends largely on the available supply, the uniformity of the feedstock and its price. Soybean oil makes up about half of the vegetable and animal oils produced in the USA. Degummed soybean oil is of uniform quality and it has been available at a competitive price with other vegetable and animal oils over the past decade. A recent study estimated the cost of producing biodiesel in a plant producing 10 million gal annually using soybean oil as the feedstock. The study estimated the investment in the plant would be US$1.15/gal of capacity and that the annual operating and capital costs would be US$0.4842/gal. The estimated cost/gal of biodiesel ranges from US$1.83 with soybean oil at 0.20/lb to US$2.58/gal with soybean oil at US$0.30/lb. The relatively high cost of biodiesel suggests it is more likely to be used as a relatively low percentage blend (2 to 5%) with diesel fuel to achieve lubricity and environmental benefits.

The US Department of Energy estimates that the cost of generating and distributing electricity with new generation facilities is about the same for wind turbines, natural gas and coal-fired generating plants (Energy Information Administration, 2004a). Much of the wind power development is being managed by wind energy developers, who lease the rights to install the turbines on farms and ranches. The rate of development has been highly dependent on a Federal Production Tax Credit, which expired on December 31, 2003. This slowed development of new generation capacity until a 2-year extension (until December 31, 2005) was signed by President Bush in October, 2004.

Recent studies of anaerobic digesters indicate they can be used by large livestock operations and processing plants to deal with a major social issue and possibly increase the profitability of the business. In many cases the benefits of a digester include reducing sewer charges, substituting the gas produced for natural gas and LP gas formerly used in production and generating electricity that substitutes for purchased electricity used by the business and produces additional electricity to sell. While these benefits have been demonstrated, more information is needed on the factors required for successful operation to assess their contribution to energy supplies.

Wood as a fuel is used largely by the residential sector (17%) and the industrial sector (74%). The major user within the industrial sector is the forest products industry that uses waste wood as a source of heat for drying and to produce electricity. More than one-half of the electricity used by the forest products industry is self-generated.

The third purpose of the chapter is to assess the potential contribution agriculture can make to the nation's energy supply. Recent studies of the amount of ethanol, biodiesel and lignocellulosic biomass suggest US agriculture may be able to produce as much as 4.43 billion gal of ethanol from grain, 0.124 billion gal of biodiesel from soybean oil and 261.2 million tons of lignocellulosic biomass from a combination of bioenergy crops and crop residues. The results suggest the nation can supply its needs for ethanol to replace MTBE in producing reformulated and oxygenated petrol, using ethanol produced from grain. However, more biodiesel would be needed if a mandated blend is implemented for a large proportion of the diesel supply. The lignocellulosic biomass is potentially useful to produce additional ethanol, to use as a feedstock for biorefining, for generation of electricity and combinations of these three uses. The development of technology, energy prices and energy policy will determine the outcome.

Ethanol and biodiesel production is developing in other parts of the world as well as in the USA. Brazil is the leading producer of ethanol, while the USA is second. Several European and Asian countries are developing a fuel ethanol industry. The EU began developing a biodiesel industry in 1993 and is continuing to encourage development with additional incentives. Trade among countries in ethanol and biodiesel has been limited. Brazil has exported some ethanol to the USA and other countries, primarily to meet seasonal demands and to deal with shortages during periods of rapid expansion in demand. Some expansion in trade can be expected to deal with seasonal fluctuations in demand as more countries develop an ethanol industry. Rapid expansion in ethanol trade is unlikely until countries remove the protective measures imposed to stimulate development of the industry.

References

Aden, A., Ruth, M., Idsen, K., Jechura, J., Neeves, K., Sheehan, J., Wallace, B., Montague, L., Slayton, A. and Lukas, J. (2002) *Lignocellulosic Biomass to Ethanol Process Design and Economics Utilizing Co-Current Dilute Acid Prehydrolysis and Enzymatic Hydrolysis for Corn Stover.* NREL/TP-510-32438. US Department of Energy, National Renewable Energy Laboratory, Golden, Colorado.

American Wind Energy Association. (2003) *Wind Power Outlook 2004.* American Wind Energy Association, Washington, DC. [Accessed 2004.] Available from http://www.awea.org/pubs/documents/Outlook2004.pdf

Ash, M. and Dohlman, E. (2004) *Oil Crops Outlook.* OCS-04b. US Department of Agriculture, Economic Research Service, Market and Trade Division, Washington, DC.

Ash, M., Dohlman, E. and Davis, W. (2003) *Oil Crops Situation and Outlook Yearbook.* OCS-2003. US Department of Agriculture, Economic Research Service, Market and Trade Division, Washington, DC.

BBI International. (2001) *Ethanol Plant Development Handbook.* 3rd edition. BBI International, Cotopaxi, Colorado.

Bender, M. (1999) Economic Feasibility Review for Community-Scale Farmer Cooperatives for Biodiesel. *Bioresource Technology* 70: 81-87.

Blue Ribbon Panel on Oxygenates in Gasoline. (1999) *Achieving Clean Air and Clean Water: The Report of the Blue Ribbon Panel on Oxygenates in Gasoline.* Commissioned by the US Environmental Protection Agency, Washington, DC.

Butzen, S. and Hobbs, T. (2002) *Crop Insights 12(10)* Pioneer Hi-Bred International, Inc., Des Moines, Iowa.

De La Torre Ugarte, D., Walsh, M., Shapouri, H. and Slinsky, S. (2003) *The Economic Impacts of Bioenergy Crop Production on U.S. Agriculture.* US Department of Agriculture, Office of the Chief Economist, Office of Energy Policy and New Uses, Washington, DC.

DiPardo, J. (2000) *Outlook for Biomass Ethanol Production and Demand.* US Department of Energy, Energy Information Administration, Washington DC. [Accessed 2004.] Available from http://www.eia.doe.gov/oiaf/analysispaper/biomass

Duffield, J., Shapouri, H., Graboski, M., McCormick, R. and Wilson, R. (1998) *Biodiesel Development: New Markets for Conventional and Genetically Modified Agricultural Products.* ERS Report No. 770. US Department of Agriculture, Economic Research Service, Washington, DC.

Electric Power Research Institute. (1997) *Renewable Energy Technology Characteristics.* Electrical Power Research Institute and US Department of Energy, Palo Alto, California.

Energy and Environmental Research Center. (2001) *Harvesting the Wind.* University of North Dakota, Grand Forks, North Dakota.

Energy Information Administration. (1995) *Renewable Energy Annual 1995.* US Department of Energy, Energy Information Administration, Washington, DC.

Energy Information Administration. (1998) *Renewable Energy 1998: Issues and Trends.* US Department of Energy, Energy Information Administration, Washington, DC.

Energy Information Administration. (2002) *Annual Energy Review 2001.* US Department of Energy, Energy Information Administration, Washington, DC.

Energy Information Administration. (2003) *Renewable Energy Annual 2002.* US Department of Energy, Energy Information Administration, Washington, DC.

Energy Information Administration. (2004a) *Annual Energy Outlook 2004 with Projections to 2025.* US Department of Energy, Energy Information Administration, Washington, DC. [Accessed 2004.] Available from http://www.eia.doe.gov/oiaf/aeo/electricity.html

Energy Information Administration. (2004b) *Electric Power Industry Summary Statistics for the US, 1991-2004.* US Department of Energy, Energy Information Administration, Washington, DC. [Accessed 2004.] Available from http://www.eia.doe.gov/cneaf/electricity/epa/epates.html

Energy Information Administration. (2004c) *Estimated Consumption of Vehicle Fuels in the United States, 1995-2004.* US Department of Energy, Energy Information Administration, Washington, DC. [Accessed 2004.] Available from http://www.eia.doe.gov/cneaf/alternate/page/datatables/afvtable10_03.xls

Energy Information Administration. (2004d) *MTBE, Oxygenates and Motor Gasoline.* US Department of Energy, Energy Information Administration, Washington, DC. [Accessed 2004.] Available from http://www.eia.doe.gov/emeu/steo/pub/special/mtbe.html

Energy Information Administration. (2004e) *Natural Gas Consumption in the US, 1999-20004.* US Department of Energy, Energy Information Administration, Washington, DC. [Accessed 2004.] Available from http//:www.eia.doe.gov/pub/0.1_gas/natural_gas/data_publication/natural_gas_monthly/historical

Energy Information Administration. (2004f) *Natural Gas Navigator.* US Department of Energy, Energy Information Administration, Washington, DC. [Accessed 2004.] Available from http://tonto.eia.doe.gov/dnav/ng/hist/n3020us3a.htm

Energy Information Administration. (2004g) *Petroleum Marketing Monthly.* US Department of Energy, Energy Information Administration, Washington, DC.

Ernst, M., Rodecker, J., Luvaga, E., Alexander, T., Kliebenstein, J. and Miranowski, J. (1999) *The Viability of Methane Production by Anaerobic Digestion on Iowa Swine Farms.* Iowa State University, Department of Economics, Iowa State University, Ames, Iowa.

European Biodiesel Board. (2004) *Statistics, The EU Biodiesel Industry, 2003 Production by Country.* Brussels, Belgium. [Accessed 2004.] Available from http://www.ebb-eu.org/stats.php

Food and Agricultural Policy Research Institute. (2001) *Impacts of Increased Ethanol and Biodiesel Demand.* FAPRI-UMC Report #13-01. University of Missouri-Columbia, Food and Agricultural Policy Research Institute, Columbia, Missouri.

Foreign Agricultural Service. (2003) *EU: Biodiesel Industry Expanding Use of Oilseeds.* US Department of Agriculture, Foreign Agricultural Service, September 20, 2003. [Accessed 2004.] Available from http://www.fas.usda.gov/pecad2/highlights/2003/09/biodiesel3/index.htm

Fulton, L. (2004) *Biofuels for Transport: An International Perspective.* International Energy Agency, Organization for Economic Cooperation and Development, Paris, France.

Gallagher, P., Otto, D. and Dikeman, M. (2000) Effects of an Oxygen Requirement for Fuel in Midwest Ethanol Markets and Local Economies. *Review of Agricultural Economics* 22(2): 292-311.

Gallagher, P., Dikeman, M., Fritz, J., Wailes, E., Gauther, W. and Shapouri, H. (2003) *Biomass from Crop Residues: Cost and Supply Estimates.* Agricultural Economic Report 819. US Department of Agriculture, Office of the Chief Economist, Washington, DC.

Glassner, D., Hettenhaus, J. and Schechinger, T. (1998) Corn Stover Collection Project. In *BioEnergy 98 – Expanding Bioenergy Partnerships: Proceedings, Volume 2.* Great Lakes Regional Biomass Energy Program, Chicago, Illinois.

Haas, M., McAloon, A., Yee, W. and Foglia, T. (2004) *A Process Model to Estimate Biodiesel Production Costs: Fats, Oils and Animal Coproducts.* US Department of Agriculture, Agricultural Research Service, Eastern Regional Research Center, Wyndmoor, Pennsylvania.

Hart Energy Publishing. (2004) *Renewable Fuel News.* (various) Hart Energy Publishing, Rockville, Maryland.

Iogen Corporation. (2004) *Cellulose Ethanol is Ready to Go.* Press Release, 21 April 2004, Iogen Corporation, Ottawa, Canada. [Accessed 2004.] Available from http://www.iogen.ca

Kinast, J. (2003) *Production of Biodiesels from Multiple Feedstocks and Properties of Biodiesel/Diesel Blends.* NREL/SR-51D-31460. US Department of Energy, National Renewable Energy Laboratory, Golden, Colorado.

Lazarus, W. and Rudstrom, M. (2003) *Financial Feasibility of Dairy Digestor Systems Under Alternative Policy Scenarios, Valuation of Benefits and Production Efficiencies: a Minnesota Case Study.* Presented paper, Anaerobic Digestor Technology Applications in Animal Agriculture: A National Summit, June 2-4, Raleigh, North Carolina.

Lindjhem, C. and Pollack, A. (2000) *Impact of Biodiesel Fuels on Air Quality – Task 1 Report: Incorporate Biodiesel Data into Vehicle Emission Databases for Modeling.* US Department of Energy, National Renewable Energy Laboratory, Golden, Colorado.

National Biodiesel Board. (2004) *FAQs.* Jefferson City, MO. [Accessed 2004.] Available from http://www.nationalbiodieselboard.org/resources/faqs/

National Renewable Energy Laboratory. (2004) *Biomass Research.* US Department of Energy, National Renewable Energy Laboratory, Golden, Colorado. [Accessed 2004.] Available from http://www.nrel.gov/biomass

Raneses, A., Glaser, J. and Duffield, J. (1999) Potential Biodiesel Markets and Their Economic Effects on the Agricultural Sector of the United States. *Industrial Crops and Products* 9: 151-162.

Renewable Fuels Association. (2004) *Synergy in Energy Production: Ethanol Industry Outlook 2004.* Renewable Fuels Association, Washington, DC.

Shapouri, H., Duffield, J. and Wang, M. (2002a) *The Energy Balance of Corn Ethanol: An Update.* Agricultural Economic Report 813. US Department of Agriculture, Office of the Chief Economist, Office of Energy Policy and New Uses, Washington, DC.

Shapouri, H., Gallagher, P. and Grabowski, M. (2002b) *USDA's 1998 Ethanol Cost-of-Production Survey.* US Department of Agriculture, Office of the Chief Economist, Office of Energy Policy and New Uses, Washington, DC.

St. Anthony, N. (2004) More Profit, Less Waste From Ethanol: Biorefining, Inc. Extracts More Value from Corn. *Star Tribune,* Minneapolis, Minnesota (March 16), p. D1.

Tiffany, D. (2001) *Biodiesel: A Policy Choice for Minnesota.* University of Minnesota, Department of Applied Economics, St. Paul, Minnesota.

Tiffany, D. and Eidman, V. (2003) *Factors Associated with Success of Fuel Ethanol Producers.* University of Minnesota, Department of Applied Economics, St. Paul, Minnesota.

US Department of Agriculture. (2000) *Analysis of Ethanol Production Under a Renewable Fuels Requirement.* US Department of Agriculture, Office of the Chief Economist, Office of Energy Policy and New Uses, Washington, DC.

US Department of Agriculture. (2001) *Economic Analysis of Increasing Soybean Oil Demand Through the Development of New Products.* US Department of Agriculture, Office of the Chief Economist, Office of Energy Policy and New Uses, Washington, DC.

US Department of Agriculture. (2002) *Effects on the Farm Economy of a Renewable Fuels Standard for Motor Vehicle Fuel.* US Department of Agriculture, Office of the Chief Economist, Office of Energy Policy and New Uses, Washington, DC.

US Department of Agriculture. (2004) *QuickStats: Agricultiral Statistics Data Base.* National Agricultural Statistics Service, US Department of Agriculture, Washington, DC. Available from http://www nass.usda.gov:81/ipedb/grains.htm

US Department of Energy. (2003) *Roadmap for Agriculture Biomass Feedstock Supply in the United States.* US Department of Energy, Washington, DC.

US Department of Energy. (2004a) *Cellulose Enzyme Research.* Office of Energy Efficiency and Renewable Energy, Washington, DC. Available from http://www.ott.doe.gov/biofuels/cellulose.html

US Department of Energy. (2004b) *Dilute Acid Hydrolysis.* Office of Energy Efficiency and Renewable Energy, Washington, DC. Available from http://www.ott.doe.gov/biofuels/dilute.html

Van Dyne, D., Blase, M. and Clements, L. (1999) A Strategy For Returning Agriculture and Rural America to Long-term Full Employment Using Biomass Refineries. *Perspectives on New Crops and New Uses.* ASHS Press, Alexandria, Virginia.

Van Wechel, T., Gustafson, C. and Leistritz, F. (2003) *Economic Feasibility of Biodiesel Production in North Dakota.* Presented paper, American Agricultural Economics Association Annual Meeting, July 27-30, Montreal, Canada.

Walsh, M. (2004) Biomass Resource Assessment. In *Encyclopedia of Energy.* Elsevier, Inc., London.

Wang, M., Saricks, C. and Wu, M. (1997) *Fuel-Cycle Fossil Energy Use and Greenhouse Gas Emissions of Fuel Ethanol Produced from U.S. Midwest Corn.* US Department of Energy, Argonne National Laboratory, Center for Transportation Research, Argonne, Illinois, and Illinois Department of Commerce and Community Affairs, Springfield, Illinois. [Accessed 2005.] Available from http://www.transportation.anl.gov/pdfs/TA/141.pdf

Williams, P., Cushing, C. and Sheehan, P. (2003) Data Available for Evaluating the Risks and Benefits of MTBE and Ethanol as Alternative Fuel Oxygenates. *Risk Analysis* 23: 1085-1115.

Windustry. (2003) *Wind Energy Policy: Federal Incentives and Policies.* Windustry, Minneapolis, Minnesota. [Accessed 2004.] Available from http://www.windustry.com/resources/legislation.htm

Withers, R. and Noordam, M. (1996) Biodiesel Canola: An Economic Feasibility Analysis. In *Liquid Fuels and Industrial Products from Renewable Resources, Proceedings of the Third Liquid Fuels Conference.* September 15-17, Nashville, Tennessee.

Wooley, R., Ruth, M., Sheehan, J., Ibsen, K., Majdeski, H. and Galvez, A. (1999) *Lignocellulosic Biomass to Ethanol Process Design and Economics Utilizing Co-Current Dilute Acid Prehydrolysis and Enzymatic Hydrolysis Current and Futuristic Scenarios.* US Department of Energy, National Renewable Energy Laboratory, Golden, Colorado.

Chapter 3

Energy Consumption in US Agriculture

John A. Miranowski

Introduction

The US agriculture sector responds to increased real (inflation adjusted) energy prices through a number of adjustments. Given sufficient real energy price incentives, farmers will change what and how much they produce, how they produce it, when they produce it and where they produce it. Higher energy prices will not only mean adjustments in production strategies for farmers, but also higher costs of production and decreased net returns as well, at least in the short run. Beyond the primary production sector, similar adjustments to higher energy prices will also face food processors. However, unless increases in energy prices are significant in real terms, they likely will not have a significant impact on the food supply or retail food prices.

Not only real energy prices drive energy consumption in agriculture, but the relative costs of other production inputs and output prices also play an important role. In addition, both US government commodity programmes and energy programmes can play an important role. Although to a lesser extent than during previous farm bills, commodity programmes influence the acres planted to particular crops, the mix of crops, the location of crop production and the level of inputs applied to these crops, especially through policy tools such as loan deficiency payments (LDPs) which can impact output prices. Programmes to promote energy conservation, efficiency and use of renewable energy forms in agriculture can also have significant impacts on the sector.

What are the goals for this chapter on energy consumption in the agricultural sector? First, establish a baseline for energy consumption in agricultural production, both in the aggregate and by output categories. Second, evaluate the responsiveness of agricultural producers to real energy price increases, other relative price changes and associated input substitution possibilities. Third, assess energy efficiency of agriculture production and why efficiency changes over time. Fourth, review energy use in food processing and compare energy efficiency in agriculture with food processing and other sectors of the US economy. Fifth, identify the effects of agriculture and rural programmes on energy use in agriculture. Finally, discuss roles of technology, government agricultural programmes and rural energy security in agriculture energy consumption.

Aggregate Farm Energy Consumption (Direct and Indirect)

Different types of energy are utilized for different production activities on US farms. Energy use on farms is typically categorized as direct and indirect energy. Direct energy, including diesel fuel, petrol, natural gas and liquid petroleum (LP) gas is used largely for planting, tillage, harvesting, drying and transportation. Electricity consumption, also classified as direct energy, is primarily used for irrigation, climate control in livestock facilities and dairy operations. Indirect energy is used off-farm in producing inputs that are ultimately consumed on the farm. Some indirect energy-intensive farm inputs include pesticides and commercial fertilizer nutrients, which can use large amounts of energy in their production processes. As illustrated in Fig. 3.1, when considering both direct and indirect energy consumption, farm production used approximately 1.7 quadrillion British thermal units (Btu), or 1.7% of total energy consumed in the USA in 2002 (Energy Information Administration, 2004d). US farmers consume about 1% of US motor petrol and 6% of diesel and other distillate fuel. On-farm operations account for about 2.3% of LP gas, 0.25% of natural gas and about 1% of electricity consumed in the USA annually. In addition to on-farm energy consumption, substantial amounts of energy are consumed off-farm in processing and handling food.

Changes in Aggregate Energy Use over Time

Agricultural energy consumption (Fig. 3.2) grew throughout the 1960s and 1970s, peaking at about 2.4 quadrillion Btu (quad) in 1978. In the late 1970s and 1980s, farmers responded to the higher prices caused by the oil crises in 1974, 1975 and 1979 by substituting relatively cheaper inputs for the relatively more costly energy inputs. Thus, farm production became more energy-efficient. Farm energy consumption declined throughout most of the 1980s, increased slightly in the early 1990s due to declining real energy prices and returned to late 1980s levels during the past 5 years. During recent decades, farmers have switched from less fuel-efficient petrol-powered to more fuel-efficient diesel-powered engines, adopted energy-conserving conservation tillage practices, sized machines more appropriately to tasks and adopted energy-saving methods for crop drying and irrigation (Uri and Day, 1991; Economic

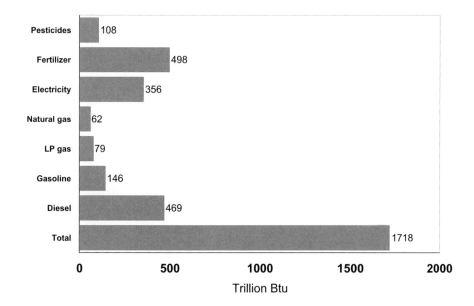

Source: see Appendix Tables 3.1 and 3.2.

Figure 3.1. Total energy consumed on US farms, 2002.

Research Service, 1997). As reported in Fig. 3.3, energy-saving measures have reduced on-farm direct energy consumption by 30% since 1978 and indirect on-farm energy consumption (i.e., fertilizer and pesticide use) by about 38% since 1980.

One of the most notable changes in on-farm energy consumption over the past 30 years has been the substitution of diesel fuel for petrol (Fig. 3.4). Total on-farm fuel (petrol plus diesel fuel) consumption peaked in 1978 at about one quad, decreased to a low of slightly over 500 trillion Btu in 1991 and was at about 600 trillion Btu in 2002. Petrol consumption decreased from just over 40% of total energy consumed on farms in 1965 to 9% in 2002, while diesel's share of total on-farm energy consumption increased from 13% to 28%. Producers switched to diesel-powered equipment as farms increased in size and diesel technology improved. As the scale of farm operations increased, farmers began to substitute larger-scale diesel-powered equipment with more horsepower that was also more energy-efficient. Larger-scaled equipment and vehicles powered by diesel engines are more energy-efficient than petrol engines (Uri and Day, 1991).

Farm energy consumption began decreasing after 1978 (Fig. 3.2) as farmers responded to higher energy prices by purchasing more fuel-efficient vehicles to replace the older, less fuel-efficient farm vehicles and further substituting fuel-efficient diesel-powered trucks for petrol-powered versions. More fuel-efficient vehicles were becoming available because manufacturers were responding to consumer demand and because they were forced to meet US Environmental Protection Agency (EPA) fuel economy fleet standards for automobiles. Diesel-powered trucks were not only more fuel-efficient, but diesel was also relatively less costly per Btu than petrol. These gains in fuel efficiency are reflected in Fig. 3.4.

The demand for refined petroleum products, natural gas and electricity in farm production activities is largely a derived demand explained by the price of energy, the price of the crop, acres planted and harvested and the weather. All prices are taken into account by the producer before making production decisions, which in turn creates the demand for both direct and

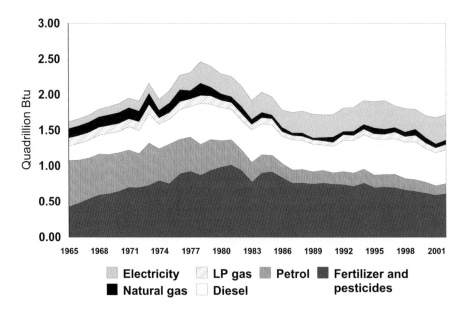

Source: see Appendix Tables 3.1 and 3.2.

Figure 3.2. Total energy consumed on US farms, 1965-2002.

indirect energy inputs. At the start of the growing season, the demand for energy is largely a function of the crops grown and the acres planted to each crop. Throughout the growing season, the demand for energy in agriculture can vary due to price expectations, pest problems and weather conditions, all impacting the potential value of the crop.

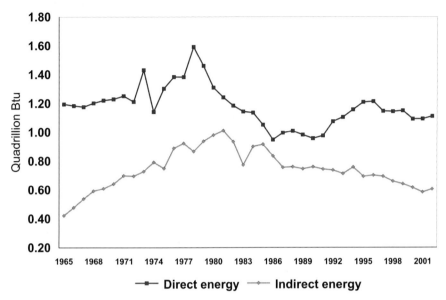

Source: see Appendix Tables 3.1 and 3.2. Direct energy includes diesel, petrol, LP gas, natural gas, and electricity. Indirect energy includes fertilizers and pesticides.

Figure 3.3. Direct and indirect energy consumed on US farms, 1965-2002.

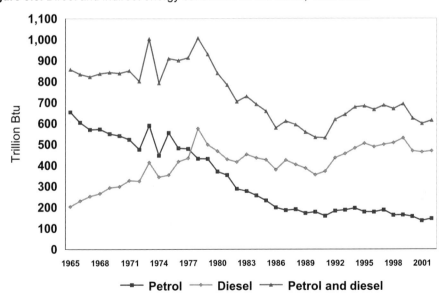

Source: see Appendix Tables 3.1.

Figure 3.4. Petrol and diesel consumed on US farms, 1965-2002.

Energy Used Indirectly by Farmers

Most inputs used in the agricultural production process use significant amounts of energy. Chemical manufacturers use natural gas, electricity, fuel oil and other fossil fuels to produce fertilizers and pesticides. Commercial fertilizers (nitrogen, phosphate and potash) are the most energy-intensive farm inputs, accounting for 29% of total energy consumed in farm production in 2002 (Fig. 3.1). As illustrated in Fig. 3.5, fertilizer consumption by farmers increased throughout the 1960s and 1970s and peaked at 23.7 million nutrient tons in 1981 (The Fertilizer Institute, 2004b). Since the mid-1980s, fertilizer use has remained relatively steady, ranging from 18 million tons in 1983 to 21 million tons in 2001 (Appendix Table 3.3). Use declined after the 1981 peak because application rates stabilized as the price of fertilizer increased relative to the prices of crops and fewer crop acres were planted to fertilizer-intensive crops under government programme set-asides and lower market prices.

The fertilizer industry experienced a major technological breakthrough in the 1980s, reducing indirect energy use through more energy efficient nitrogen and phosphorus fertilizers (Bhat *et al.*, 1994). From 1979 to 1987, the energy consumed in producing nitrogen fertilizer declined about 11% and the energy requirements for phosphorus fertilizer decreased 27%. Energy requirements for potash remained about the same. Improved energy-efficiency in nitrogen fertilizer production is critical since nitrogen has a significantly higher energy requirement than phosphorus and potash fertilizers. Even though farmers consumed about the same amount of fertilizer since the mid-1980s, the amount of energy embodied in this fertilizer has declined significantly over the past 20 years (The Fertilizer Institute, 2004a; Shapouri *et al.*, 2002).

About 108 trillion Btu of petroleum and natural gas were required to produce, process, package and distribute pesticides (including herbicides, insecticides, fungicides and other pesticides) consumed on-farm in 2002 (Donaldson *et al.*, 2002). Pesticides used on crops,

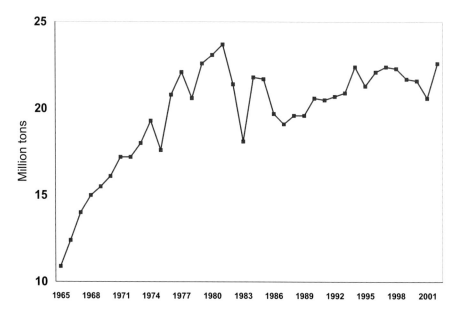

Source: The Fertilizer Institute (2004b).

Figure 3.5. Total commercial fertilizer consumed on US farms (nutrient consumption), 1965-2002.

especially herbicides, increased rapidly between 1960 and 1980 (Economic Research Service, 1994, 1997, 2003). However, as commodity prices fell and large amounts of land were removed from production by farm programmes, pesticide use on cropland decreased during the 1980s. Pesticide use decreased from about 1.05 billion pounds per year in 1980 to 846 million pounds per year in 1987 (Lin *et al.*, 1995). Pesticide use on farms grew in the early 1990s, but has been stable at around 955 million pounds since 1996 (Donaldson *et al.*, 2002). Like the fertilizer industry, pesticide manufacturing is becoming more energy-efficient over time (US Census Bureau, 2004).

Herbicides, used for weed control, are the largest pesticide class and accounted for almost 45% of active ingredients applied to crops (Economic Research Service, 2003). Herbicide use has remained relatively unchanged since 1990. Insecticides accounted for about 10% of the pesticides used on crops. Their use increased from 77 million pounds in 1991 to 93 million pounds in 1999. Fungicides, used for disease control on fruits and vegetables, remained stable during the 1990s and accounted for 5% of pesticide use.

The type of tillage and crop management systems employed can also have important impacts on demand for both direct and indirect energy consumption (Conservation Technology Information Center, 2004). Reductions in tillage intensity frequently reduce nutrient and sediment runoff, reduce diesel and petrol use and increase pesticide use (Economic Research Service, 2003), implying potential trade-offs between direct and indirect energy consumption (Miranowski, 1980; Zinser *et al.*, 1985).

Direct Energy Expenditures in Major Field Crop Production

Farmers depend on a variety of energy sources to produce food, feed and fibre. Crop farmers use large amounts of liquid fuel for field operations. Larger farmers rely on diesel-fuelled tractors for tilling, planting, cultivating, harvesting and applying chemicals. Petrol is used primarily for smaller and older tractors and smaller trucks.

Which field crops are more energy intensive? We only have data on direct energy expenditures for field preparation, planting, production and harvesting major field crops. Comparable crop drying and indirect energy expenditure data are not available. In Fig. 3.6, per acre expenditures on fuel and electricity are presented for major field crops. Direct energy expenditures per acre are highest for rice, followed by peanuts, cotton, maize, grain sorghum and soybeans (Economic Research Service, 2004c). If real energy costs increase significantly, crops with higher direct energy production costs may be more adversely impacted and ultimately, the mix of crops produced may change as farmers substitute more crops that are less energy-intensive in their rotations. Ultimately though, the profitability of these substitutions are constrained by relative output price adjustments. For example, soybeans may become relatively more profitable (having both lower direct and indirect energy costs as well) and substitute for crops with higher direct energy costs per acre, but with increased supply, the price and profitability of soybeans will decrease relative to maize and other substitute crops. Significant energy expenditures may be incurred for drying some crops (e.g., maize grain, tobacco, peanuts), but we could not obtain comparable energy expenditure data. Yet, considering on-farm energy expenditures on production and drying over time may not be meaningful. Farm energy demands for drying vary from year to year depending on weather conditions at harvest time and crop drying activities have migrated off-farm in recent years similar to processing and other value-adding activities.

Energy Expenditures by Type of Farm

The Agriculture Resource Management Survey (ARMS), conducted by the US Department of Agriculture's National Agricultural Statistics Service (NASS) is the primary source of on-farm energy expenditure data. NASS collects farm level expenditure information on all field crop, farm animal, fruit and nut, vegetable and nursery and greenhouse production activities. These data are not collected by crop or animal enterprise, but rather, for the whole farming operation. A farm is then classified as a particular farm type if over 50% of the value of production (not used on the farm) can be attributed to a particular crop or animal enterprise. For example, a farm with over half of the value of production (not used on the farm) attributable to the maize enterprise is classified as a maize farm. On-farm energy expenditure estimates using unpublished ARMS data were compiled and summarized for 2002. The energy expenditure summaries by farm type are reported in Appendix Table 3.4 on pp. 99-104 (National Agricultural Statistics Service, 2003). It is important to note that the ARMS expenditure estimates are reported by farm type as opposed to by animal or crop enterprise, so they are not comparable with the estimates in Fig. 3.6. Further, it is not possible to report ARMS estimates of fuel and electricity expenditures on a per acre or per animal unit basis because they represent a composite of livestock and crop expenditures for a particular type of farm as opposed to enterprise type. Thus, the ARMS energy expenditure estimates are reported on per dollar value of production and on a share of total cash expenditures for particular types of farms. Fuel and electricity expenditures per unit of output and the share of fuel and electricity expenditures in total cash expenditures for crop and livestock farm types are reported in Appendix Table 3.4 (National Agricultural Statistics Service, 2003).

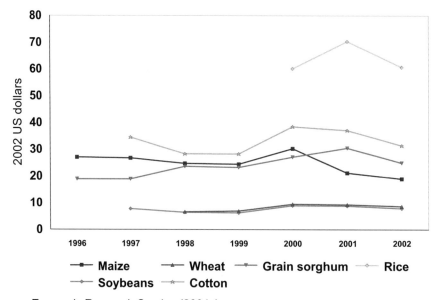

Source: Economic Research Service (2004c).

Figure 3.6. Expenditure on energy (fuel, lubrication and electricity) in major agricultural crops, per acre, 1996-2002.

Primary Uses of Energy on Crop Operations

Peanut farms had the highest energy expenditure per dollar of output at US$0.15 or 11.6% of cash expenses for fuel and electricity, rice farms were second at US$0.149 per dollar of output or 12.8% of cash expenses, cotton farms were third at US$0.11 per dollar of output or 9.5% of cash expenses, tobacco farms were fourth at US$0.10 per dollar of output or 12.8% of cash expenses, followed by wheat farms at US$0.092 per dollar of output or 9.3% of cash expenses, maize farms at US$0.07 per dollar of output or 8.5% of cash expenses and soybean farms at US$0.073 per dollar of output or 7.4% of cash expenses. These results do parallel those reported in Fig. 3.6 above even though each relies on different systems for deriving estimates. It is important to note that none of these estimates included indirect energy expenditures that would have changed the relative ranking of some crop farms, especially maize farms.

Primary Uses of Energy on Animal Operations

On livestock farms, producers relied on electricity, natural and LP gas for heating and cooling poultry and swine facilities. Dairy farms relied heavily on electricity to power their milking, cooling and handling equipment. Livestock farms[1] accounted for US$2.72 billion of fuel and lubricant and US$1.59 billion of electricity expenditures, or 39% of total farm fuel purchases and 47% of electricity purchases (Appendix Table 3.4). In terms of cash expenses per unit of output and share of cash total expenses, beef operations were highest at US$0.062 per dollar of output or 6.6% of cash expenses for fuel and electricity purchases, dairy farms were second with US$0.047 per dollar of output or 5.5% of cash expenses, then poultry farms with US$0.038 per dollar of output or 15.2% of cash expenses and last were pig farms with US$0.032 per dollar of output or 6.2% of total cash expenses.

Primary Uses of Energy on Nursery, Greenhouse, Fruit/Nut and Vegetable Operations

In 2002, fruit and tree nut farms expended US$208 million on fuels and lubricants, US$218 million on electricity and US$0.042 per dollar of output on fuels and electricity (Appendix Table 3.4). Fuels and electricity accounted for 5.7% of their total cash expenses. Vegetable producers purchased US$280 million of fuels and lubricants and US$236 million of electricity and US$0.043 per dollar of output on fuels and electricity. Fuels and electricity accounted for 6.1% of total cash expenses. Fruits and nuts and vegetables spent the most on electricity followed by diesel fuel. Nursery and greenhouse producers spent US$514 million on fuels and lubricants, US$256 million on electricity and US$0.048 per dollar of output on fuels and electricity. These items accounted for 6.0% of cash expenses, with electricity and natural gas being the two key energy inputs. Energy is obviously a significant expense for these specialty crops or outputs but not as important as frequently hypothesized. Although nurseries and greenhouses are more vulnerable to weather, which increases their outlays for natural gas relative to fruits and vegetable, natural gas only accounts for about US$0.01 per dollar of output.

Share of Energy Expenditures in Total Farm Expenditures

Direct and indirect energy expenditures are a significant share of farm expenses accounting for 12% (excluding electricity) of total farm production expenses in 2002 (National Agricultural Statistics Service, 2004; Appendix Table 3.5). On-farm fuel (petrol, diesel, LP

---

[1] Livestock farms are farms on which over 50% of their value of production is accounted for by livestock.

gas and natural gas) expenses were 3.3% of total farm production expenditures. Expenditures on fertilizers, pesticides and other indirect energy consumption accounted for 9.2% of total farm expenditures (Appendix Table 3.5). Since farmers are dependent on direct and indirect energy inputs, energy prices can have a significant effect on farm expenditures. For example, during the energy crises of the 1970s, energy price increases led energy's share of total farm production expenses to increase from 11.2% in 1972 to 15.9% in 1981 (Fig. 3.7). Direct energy costs increased from 3.3% of total farm production expenditures in 1972 to 6.1% in 1981. As fuel supplies stabilized, direct energy costs decreased to 3% to 4% of total farm production expenditures after 1985 and have remained in that range through 2002 (National Agricultural Statistics Service, 2004).

The farm costs of indirect energy increased significantly in the 1970s. The share of indirect energy in total farm production expenditures went from 8.0% in 1972 to 11.3% in 1975. The share of indirect energy expenditures decreased to 7.8% in 1983 and has been in the 9% to 11% range since. In 2002, the share of indirect energy expenditures was 9.2% (National Agricultural Statistics Service, 2004).

In 2002, US farmers spent about US$10.4 billion on direct energy use. Farmers spent about US$7.0 billion on fuels, oils and lubricants and almost US$3.4 billion on electricity (Fig. 3.8). Of the US$7.0 billion spent on fuels, oils and lubricants, crop farms spent US$4.5 billion and livestock operations spent US$2.5 billion. Crop farms spent almost US$1.9 billion on electricity while livestock operations spent US$1.4 billion. As illustrated in Fig. 3.8, farmers' expenditures were highest on diesel fuel, followed by electricity and petrol (National Agricultural Statistics Service, 2003).

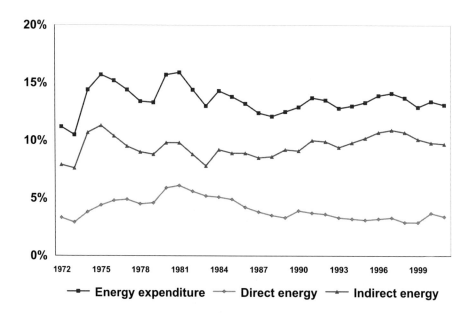

Source: National Agricultural Statistics Service (2004).

Figure 3.7. Energy's share of farm production expenses, 1972-2002.

Irrigation

Irrigation is another significant energy consumer in US agriculture. The amount of land irrigated and water used in the USA varies from year to year, depending on the quantity of rainfall, the price of the commodity and the price of energy. Based on the *1998 Farm and Ranch Irrigation Survey*, 50 million acres or about 28% of the farmland in the USA was irrigated, compared to about 46 million acres irrigated in 1994 (Appendix Table 3.6). The highest irrigated crop acreages were maize, lucerne, cotton, soybeans and orchards (e.g., fruit trees, vineyards and nut trees). These crops accounted for 57% of irrigated cropland, with maize accounting for nearly 21% of irrigated land in 1998 (National Agricultural Statistics Service, 1999a). California had the highest irrigated acreage, irrigating about 8.1 million acres of farmland. Other states that had significant irrigated acreage include Nebraska (5.7 million acres), Texas (5.2 million acres), Arkansas (4 million acres), Idaho (3 million acres) and Colorado (3 million acres). Other states irrigating over a million acres included Florida, Mississippi, Montana, Oregon, Utah, Washington and Wyoming.

Many irrigation systems in the USA are gravity flow systems that require little energy for water distribution. However, about 38 million irrigated acres relied on pumps to provide irrigation water in 1998. Irrigation pumps consume diesel fuel, petrol, natural gas and electricity. In 1998, the energy consumed cost farmers US$1.2 billion, or an average of US$32 per irrigated acre (National Agricultural Statistics Service, 1999b). Electricity was the principal power source to irrigate 20.2 million acres at a cost of US$800 million (Fig. 3.9). Diesel fuel was used to irrigate another 10 million acres and natural gas was used on about 6 million acres (Appendix Table 3.7).

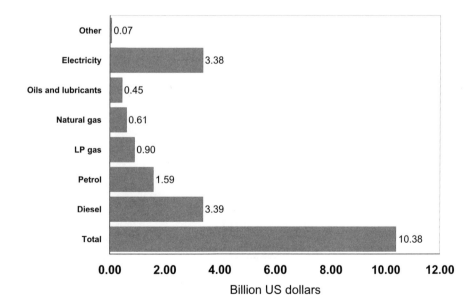

Source: National Agricultural Statistics Service (2003).

Figure 3.8. Energy expenditures on US farms by type of energy, 2002

Farm Energy Consumption by Production Region

Based on energy expenditures and prices reported by farmers, US farmers used approximately 3.4 billion gallons (gal) of diesel fuel, 1.2 billion gal of petrol and 0.9 billion gal of LP in

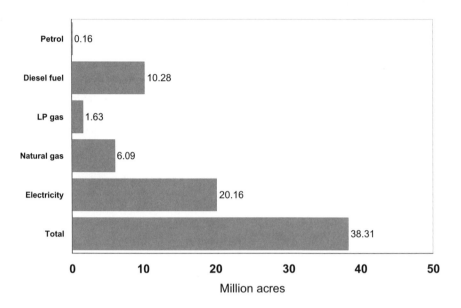

Source: National Agricultural Statistics Service (2003).

Figure 3.9. Energy source for irrigated acreage on US farms, 1998.

Source: Economic Research Service (2004d).

Figure 3.10. Farm production regions.

2002. Individual state data are aggregated into 10 farm production regions (Fig. 3.10). The Corn Belt region used 966 million gal of fuel (petrol, diesel and LP gas), the most fuel used among farm production regions (Fig. 3.11). Farmers in the Northern Plains region were the second largest user of fuels using 832 million gal in 2002. The Lake States, Pacific, Mountain and Southern Plains regions used between 506 and 575 million gal. Fuel use in the Appalachian, Delta and Southeast regions ranged from 370-428 million gal. The Northeast region used 286 million gal of fuel.

Natural gas consumption by farm operations in 2002 was almost 60 billion cubic feet (Appendix Table 3.8). This estimate was derived from ARMS estimates of farm expenditures on natural gas and natural gas prices estimated by the Energy Information Administration (EIA) (Energy Information Administration, 2001). Natural gas use is prominent in regions where farmers use it for drying crops and for irrigation such as the Northern Plains, Southern Plains and the Lake States. Also, a large amount of natural gas is used in the Pacific region for heating greenhouses. Regional use of natural gas can vary significantly, since it is used for functions that are affected by the weather.

Electricity use on farms was estimated by dividing ARMS farm expenditure estimates on electricity by EIA electricity price estimates (Energy Information Administration, 2004f). Farmers consumed about 41 billion kilowatt hours (kWh) of electricity in 2002. The Pacific region is the largest user of electricity consuming about 8.5 billion kWh (Fig. 3.12). The Pacific region uses large amounts of electricity in the dairy industry, which depends on electricity for milking, refrigeration and handling operations. Fruit and tree nut growers depend on electricity for irrigation and frost control and crop farmers use significant amounts on irrigation.

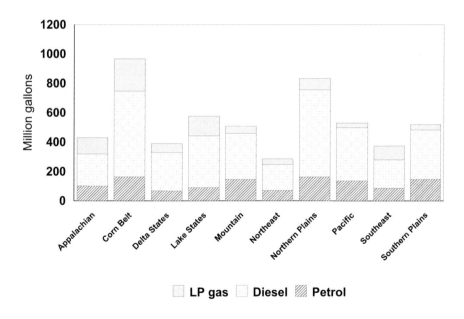

Source: National Agricultural Statistics Service (2003).

Figure 3.11. Fuel consumed by farmers in NASS production regions, 2002.

Which Producers Are Most Affected by an Increase in Energy Prices?

There are some higher energy price implications to take away from the energy expenditures per acre on crop production. First, rice involves the highest direct energy expenditure per acre and soybeans the lowest among the major field crops. Likewise, rice farms are among the highest in terms of energy expenditures per dollar of output and as a share of cash expenses. Soybean farms are among the lowest.

Livestock producers will be impacted by higher real energy costs, but some more than others. First, the energy cost per dollar of output is quite high in beef production. In particularly, the energy cost per dollar of beef output for diesel (US$0.022/US$ output), petrol (US$0.015/US$ output) and electricity (US$0.017/US$ output) are high relative to other livestock enterprises. At the same time, the total cash expenses associated with beef production indicate energy expenses are a smaller share of total cash expenses than for pigs, dairy, or poultry. Second, the opposite is true of poultry where the cost per dollar of output is low, but the energy share of cash expenses is very high. Thus, poultry is more vulnerable or sensitive to energy price increases than beef or dairy and pigs. If we consider all crops and livestock, the direct energy expenditure per dollar of output is US$0.057 and the energy share of cash expenses is 7.3%.

Do Producers Respond to Energy Price Increases?

In the short run, farmers have limited options to mitigate the effects of higher energy prices. Some producers may be able to reduce field operations by switching from conventional tillage practices to reduced till, by adjusting fertilizer application rates when the higher

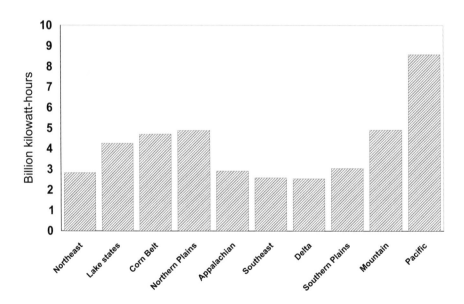

Source: National Agricultural Statistics Service (2003); Energy Information Administration (2004f).

Figure 3.12. Electricity consumed on farms in NASS farm production regions, 2002.

costs of nitrogen exceed the value of the marginal output, or by allowing crops to dry naturally in the fields.

Over the long term, farmers have more flexibility and can utilize more energy-efficient practices and equipment. This would be similar to what happened following the real energy price increases during the 1970s and early 1980s. Currently, more advanced technologies and farming practices such as precision farming (e.g., yield monitoring, global positioning system (GPS), calibrated application of pesticides and fertilizers) are available and can be adopted in response to higher energy prices.

Higher energy costs in the western USA are of particular concern to horticultural producers. Horticulture accounts for 40% of total US crop value and over 20% of all US agricultural exports. California alone produces one-half of all US horticultural products and accounts for over one-half of horticultural exports. Many horticultural producers in California are adjusting to higher costs since over 60% of the California production area is planted to orchards and vineyards. In addition to higher costs for water and fertilizers to grow high-value perennial crops, over one-half of total California horticultural production requires significant energy for post-harvest processing.

Although it is common to talk about higher nominal energy prices at the petrol pump, producers and consumers react primarily to real energy prices (corrected for inflation). Nominal and real energy prices are reported in Figs. 3.13-3.14 for all fuels, diesel, petrol, natural gas and electricity. Nominal prices surged in early 2000, but real energy prices present a different picture. Real prices for major energy inputs peaked in the early 1980s and with the exception of natural gas are currently similar to real prices in 1970. Currently, real energy prices create limited market incentives to encourage conservation of energy inputs in agriculture, industry, services, or consumption. Although there are a number of ways of achieving energy conservation, including moral suasion (e.g., 'don't be fuelish'), energy efficiency ratings, EPA fleet mileage standards and other voluntary and mandatory approaches, market prices are the most effective inducement. Current prices are a small inducement. Other relative prices, including output and input prices faced by the producer, also can provide important inducements that can work against and for energy conservation. For example, guaranteed floor prices for commodities at levels above variable costs can provide incentive to plant additional acres to such commodities and use more energy in crop production.

Based on evidence from several studies, energy use like other input use is responsive to higher real energy prices in agricultural production as well as in all other sectors of the economy (e.g., Berndt and Wood, 1979, 1981; Mensah and Miranowski, 1988). Generally, we are interested in both the reduction in energy demand associated with higher real prices, or the own price elasticity of demand and the substitution effects between energy and other relatively less expensive inputs. Table 3.1 contains the own price elasticities of demand for energy and the other major inputs used in agriculture. In the aggregate, these results indicate that a 10% increase in real energy prices will result in a 6% decrease in energy use in agriculture, assuming all other prices are unchanged. Table 3.1 also provides an indication of the relative energy intensity of indirect energy inputs in production agriculture and why increases in real prices to produce indirect energy inputs, fertilizers and pesticides, will cause a similar reduction in the use of those inputs.

At the same time, we can estimate what quantities of other inputs would substitute for energy as the price of energy becomes relatively more costly, or the Allen elasticities of substitution, which are the most commonly reported measure. During the energy crises of the 1970s, several industry studies found a complementary relationship between capital and energy, implying that a relative increase in energy prices would not only reduce energy use in production but would also reduce capital investment. The anticipated consequence of reduced energy use and energy-capital net complementarity is that the associated reduction in capital investment would deter economic growth (e.g., Berndt and Wood, 1975). The results reported in Table 3.1 indicate that we found all inputs substituting for energy, including capital, unlike the earlier manufacturing sector estimates. In the aggregate, these results

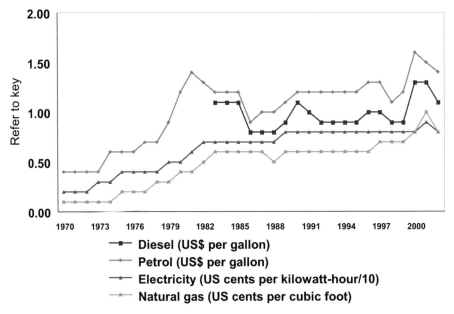

Source: Energy Information Administration (2004b); see Appendix Table 3.9.

Figure 3.13. Nominal prices of major fuel sources, 1970-2002.

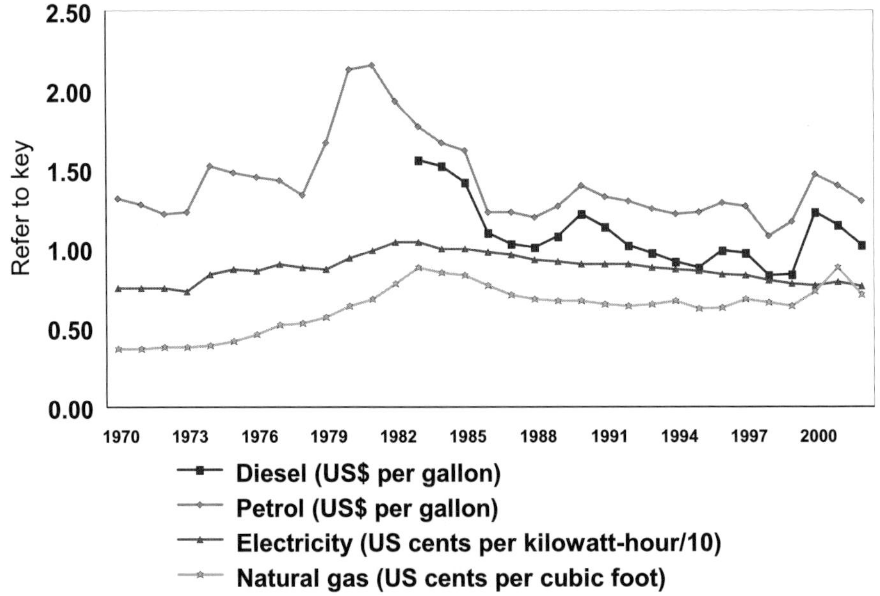

Source: Energy Information Administration (2004b); Executive Office of the President (2004); see Appendix Table 3.10.

Figure 3.14. Real prices of major fuel sources (in 1996 US dollars), 1970-2002.

Table 3.1. Allen full elasticity of substitution.

Input	Land	Labour	Capital	Energy	Fertilizers	Pesticides
Land	-0.28					
Labour	-0.27	-0.39				
Capital	0.73	0.65	-0.86			
Energy	0.35	0.59	1.13	-0.60		
Fertilizers	0.20	0.82	0.97	0.60	-0.66	
Pesticides	0.08	0.66	0.82	0.70	1.04	-0.53

Source: Ball (2004).

indicate that a 10% increase in relative energy prices can be expected to increase the use of other inputs by 4% to 11%. Although not reported in Table 3.1, other specifications of input substitution elasticities were estimated and again, capital and energy were generally found to be substitutes.

We do not normally think of input changes in aggregate agriculture. In more intuitive terms, what does an increase in real energy prices mean for the typical maize-soybean producer? The adoption of conservation tillage provides an intuitive illustration of the important role that energy prices play in agricultural production decisions. Because conservation tillage and no-till reduce soil loss, efforts were made to encourage adoption in the 1970s, but the adoption process was slow. Adoption of conservation-till on major field crops, such as maize and soybeans, began to increase in about 1980, largely in response to higher real and relative energy prices faced by farmers (Miranowski, 1980). Using conventional-till requires several (five to seven) trips over the field which consumes a significant amount of energy and buries almost all the plant residue leaving the soil prone to

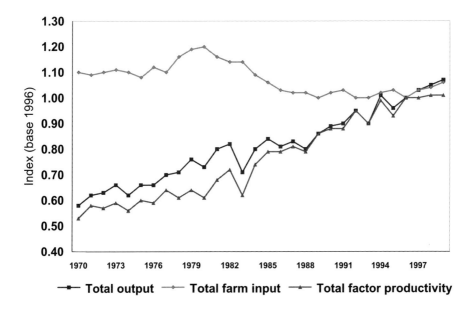

Source: Economic Research Service (2004a): see Appendix Table 3.11.

Figure 3.15. Indices of farm output, input use and productivity in US agriculture, 1968-2000.

erosion. Conservation tillage requires significantly less energy than conventional-till because it requires less field preparation trips and frequently eliminates the energy-inefficient practice of mouldboard ploughing (Zinser *et al.*, 1985). Conservation tillage and no-till practices increased from 41 million acres in 1982 to 103 million acres in 2002, or 37% of total planted cropland.

What is Happening to Farm Energy Efficiency over Time?

Energy use grew during the 1960s and 1970s, peaking at 2.4 quad in 1978. After 1978, a different energy consumption scenario emerged in production agriculture. Total farm energy consumption declined 30% from 1978 to 2002. Indices of other inputs decreased as well, while agricultural output increased 45%. As a result, total factor productivity grew at 2% annually (Fig. 3.15). US agriculture became more energy-efficient. First, the sustained productivity growth in the agricultural sector, combined with reduction in energy and other input use, led to significant improvements in energy efficiency (Fig. 3.16). Second, real energy price increases in the mid-1970s to 1982 provided incentives for farmers to become more energy-efficient, as noted earlier. Farmers reduced direct energy use by 30% from 1978 to 2002 and indirect energy use by 38% from 1980 to 2002. The combined effects of growing agricultural productivity and increasing fuel (and other input) efficiency throughout this period was that energy use relative to agricultural output decreased by 7% between 1978 and 2002 (Fig. 3.17).

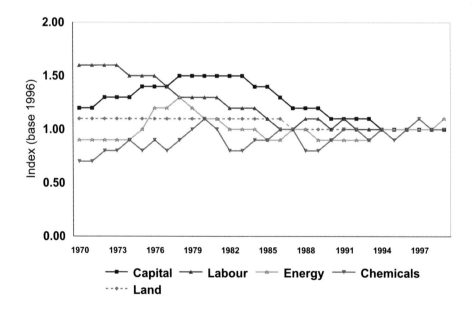

Source: Economic Research Service (2004a): see Appendix Table 3.12.

Figure 3.16. Indices of major farm input use in US agriculture, 1968-2000.

Off-farm Energy Consumed in Agriculture Processing

Energy is a key input in processing food after it leaves the farm or ranch. Food processing consumed over 1 quad Btu in 2000 (US Census Bureau, 2004). Wet maize milling, the major user, consumed about 175 trillion Btu. The primary energy source in food processing is natural gas, accounting for over 50% of energy consumed, followed by electricity accounting for another 20% of energy consumed (US Census Bureau, 2004).

Relative to all manufacturing industries, food processing is more energy-efficient based on the gross value of shipments measure (US Census Bureau, 2004). Within the different categories of food processing, the milling, processing and oil extraction category is the most energy-intensive and has witnessed little or no improvement in energy efficiency over time. The sugar processing and the confectionery category was not energy inefficient in the 1970s but improved substantially in the 1990s. The very limited improvement in food processing energy efficiency over time may be due to very limited price incentives. Natural gas and electricity accounted for over 70% of energy use in the food processing sector and natural gas and electrical prices have been more stable over time than other fuels at least prior to the last couple of years. Without a significant real price increase, there may be little incentive to improve energy efficiency in food processing.

Energy Consumption in Agriculture Compared to US Energy Consumption

US agriculture is a relatively intensive user of energy. The share of energy used in production agriculture is similar to the sector's contribution to gross domestic product (GDP). Both direct and indirect energy consumption for farm production required 1.7 quad in 2002, or about 1.7%

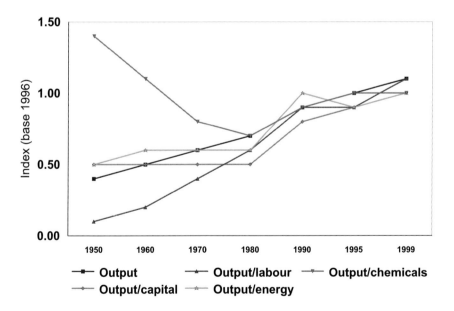

Source: Economic Research Service (2004a): see Appendix Tables 3.11-3.12.

Figure 3.17. Partial productivity indexes in US agriculture, 1950-1999.

Table 3.2. Energy intensity: energy needed for a US dollar output in different sectors of the economy, 1970-2001.

Year	Food manufacturing	Agriculture	Industry[a]	US economy
	(thousand Btu/1996 adjusted US dollars)			
1970-71	10.9	11.8	26.7	18.6
1980-81	7.5	11.7	21.9	15.9
1992	7.8	8.5	21.7	12.5
1995	7.6	8.7	20.5	12.1
2000	6.7	7.3	19.2	10.7
2001	7.9	7.2	N/A[b]	10.4

[a] Construction, mining and manufacturing.
[b] N/A = not available.
Source: US Census Bureau (2004); Executive Office of the President (2004).

of total energy consumed in the USA. US agricultural production accounted for about 1.7% of US GDP (Energy Information Administration, 2004d; Executive Office of the President, 2004).

A more important question may be what is happening to energy efficiency in the agricultural sector relative to other productive sectors in the economy? Table 3.2 contains a comparison of energy intensity per real dollar of output for agriculture, food processing, industry and the entire US economy. Measured in Btu, agriculture's and food manufacturing's energy intensity per dollar of real output are below industry and the US economy over the last three decades. At the same time, agriculture and food manufacturing have realized modest gains in energy efficiency over time (Fig. 3.18), while industry has realized more significant gains and the US economy as a whole has made substantial progress in increasing energy

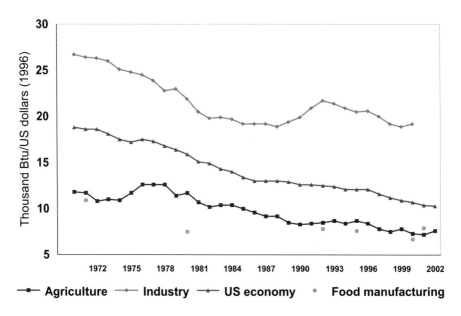

Source: National Agricultural Statistics Service (2003); Energy Information Administration (2004c); Executive Office of the President (2004); US Census Bureau (2004).

Figure 3.18. Energy intensity (Btu consumed per US dollar of output) in US agriculture, food manufacturing and industry, 1970-2002.

efficiency. It is important to note that these estimates use real gross output as the measure of energy intensity and not value added, which may generate different results. Unfortunately, value-added estimates were not available for all the sectors considered. Also, industrial and manufacturing sectors typically provide measures of gross and net energy consumption while agriculture is reported only on a gross basis. As the magnitude of on-farm energy production increases in significance, gross measures may not be appropriate for comparison. With the production of bio-renewable, wind, methane and solar power produced for on-farm and off-farm consumption, net energy consumption estimates may need to be considered.

Even given the limitations of the comparative data, agriculture and food manufacturing are more energy efficient than other segments of the economy and agriculture production has realized improvements over the last two decades. Energy efficiency should continue to improve in production agriculture if the current rate of productivity growth continues, but it will be more difficult for agriculture to realize efficiency gains if agricultural productivity growth slows at current real energy prices.

The Potential Effects of Information and Biotechnology on Agriculture Energy Efficiency

Since 1990, there have been two important revolutions in both biotechnology and information. These revolutions have had and are having a dramatic impact on the US economy and agriculture as well. Biotechnology has permitted important breakthroughs in plant and animal modifications that permit the production of a wide range of new products, including pharmaceuticals and plants resistant to pests, diseases and herbicides. At the same time, the information revolution has speeded the process of mapping plant and animal genomes and manipulating genetic materials improving the efficiency of production and service systems and reducing the cost of information to producers and consumers.

What are some of the implications for energy efficiency in plant production? First, by inserting genes in the plant to control plant pests, the need for pesticides, a major consumer of indirect energy, will be reduced in production agriculture. Second, herbicide-resistant plants will reduce the need for multiple herbicide treatments and will reduce overall herbicide use and save energy as well. Third, research is underway to improve nutrient utilization in plants and to allow the plant itself to fix nitrogen in the soil, reducing commercial fertilizer and indirect energy consumption. Fourth, GPS, yield and soil monitors and geographic information system (GIS) data are key components of precision farming systems. Such systems have the potential to more efficiently use nutrients and control pests and ultimately save on both direct and indirect energy needs in crop production. Several other projects are underway to reduce energy and other crop inputs in production including drought-tolerant crops to reduce irrigation needs.

Likewise, animal agriculture is making important breakthroughs that improve productivity and reduce energy needs in agriculture. Less energy is going to be consumed if productivity can be increased while consuming less energy. For example, if fewer cows can produce the same amount of milk using recombinant bovine somatotropin (rbST), then energy can be saved in dairy production. Further, the information revolution has transitioned meat animal production from an era of 'attentive' husbandry to an era of 'knowledge' or 'informed' husbandry. Computers are now used to monitor health condition of pigs in finishing facilities based on feed and water consumption. This provides a preventive approach to animal health care and reduces overall energy needs.

These are only a few examples of areas where technology and information can be used to improve energy efficiency in crop and animal production. Many more exciting ideas are currently in the development stages.

Opportunities to Integrate Farm Energy Use and Farm Energy Production

Higher energy prices and the desire to reduce the US dependence on imported oil provide incentives for agriculture to move in the direction of integrated farming operations that supply part of their own energy. Crops, crop residues, forest residues and energy crops planted on idle or marginal cropland serve as feedstocks for ethanol, biodiesel and methane production. Ethanol from grains accounts for most of US biofuel production. In 2003, 2.81 billion gallons of ethanol were produced across the US and production capacity is expanding. Yet, the US has been slow to develop integrated farm systems that consume energy from the feedstocks farmers produce. Rather, most of the feedstocks are sold into commercial markets. Small-scale facilities, such as ethanol plants, lack the scale economies realized by commercial plants in the 100 to 200 million gallon range. Wind energy, methane and possibly biodiesel have more potential for on-farm production and consumption. Wind energy possibly offers the greatest potential in farming regions that have appropriate wind conditions and farming systems that are dependent on electricity, e.g., dairy operations. Commercial wind technology has made tremendous strides in recent years and is becoming a competitive source of electric power. Approximately 1 megawatt wind turbines are becoming the norm in the US wind power industry. At the same time, integration into the farm production system will be more successful if producers can tie into the electric grid to obtain energy when they have deficiencies and sell energy when they have surpluses and receive 'fair' terms-of-trade.

Recent advances in solar energy technology can offer important on-farm substitution opportunities, including providing water for livestock, powering electric fencing and providing lighting in more isolated areas. As technology continues to improve, other substitution opportunities are anticipated in farming and ranching operations.

Effects of Policy on Energy Use

There are a number of government proposals and programmes to promote improved efficiency of energy use in agriculture. Although these proposals and programmes have somewhat different goals, they generally are designed to promote biobased energy products and inform the public of the potential benefits, assist farmers and rural people in the purchase of renewable energy systems and procure biobased products for government use (e.g., Energy Information Administration, 2004e). Under the *Farm Security and Rural Investment Act of 2002* (*2002 FSRI Act*), a number of new provisions were introduced and are in various stages of implementation and operation (National Association of Conservation Districts, 2004). Some key provisions are:

- Labelling of 'USDA Certified Biobased Products' that provides financial assistance to support product testing and certification. Infrastructure is under development, including biobased product testing protocols and electronic information dissemination systems;
- Federal procurement of biobased products to provide energy for the federal fleet and in government service delivery;

- Biodiesel fuel education programme that is being implemented through grants to educate governmental and private entities that operate vehicle fleets and the public about the benefits of biodiesel;
- Authorized loans, loan guarantees and grants to farmers, ranchers and rural small businesses to purchase renewable energy systems and to make energy efficiency improvements. The programme covers a portion of the renewable energy system costs;
- The biomass research and development section extends authority of the *Biomass Research and Development Act of 2000* to 2007 and provides funding for biomass research and development (R&D) projects in rural areas with some projects being supported in cooperation with the US Department of Energy.

It is too early to assess the impacts of the provisions of the Energy Title of the *2002 FSRI Act*. These provisions provide incentives for research, development, education and procurement of biobased energy products. The *2002 FSRI Act* creates incentives to invest by labelling and creating product identity, creating market opportunities, subsidizing capital costs and supporting R&D to make biobased products more competitive in the market place (Economic Research Service, 2004b).

Rural Energy Security and Disruption Costs in Crop and Animal Agriculture

Over 31% of total US natural gas consumption and 23% of total US electricity consumption occurs in the months of December, January and February (Energy Information Administration, 2004c). Data on seasonal energy use by agriculture are not available. However, in much of the USA, crop harvest is completed in the autumn months and on-farm energy use in crop production decreases. During the winter period, expenditures for drying and conditioning grains are significant. In states with milder climates, such as Florida and California, winter vegetable and fruit crop production continues. Livestock, poultry and dairy operations, on the other hand, are active year-around and account for most of the on-farm energy consumption during the late autumn and winter months in most states. Derived energy demand from livestock production would likely increase during winter months, especially in the less temperate states.

While seasonal data are not available on energy consumption in the food processing industry, modern livestock production now witnesses only limited seasonal fluctuation and thus harvesting and processing continue with little seasonal fluctuation. Lower temperatures could be expected to increase heating expenditures but reduce cooling expenditures in processing and harvesting.

Even more important than the relative intensity of energy use in the agricultural sector is the impact of energy disruptions at critical points or periods in production and processing. If an energy disruption occurs at harvest time, crops and especially perishable crops may rot in the field or on trees and vines. Likewise, perishable crops must be processed soon after harvest. If energy is not available, such crops will spoil. Energy disruptions in animal production and harvesting may have equal or greater impacts on farmers and consumers. If energy is not available to run blowers on barns, thousands of animals may perish on a hot summer day, or if energy is not available to harvest animals, processing may develop a backlog. If a backlog develops, producer prices will fall especially if processing capacity constraints are reached. Then, consumer prices will rise sharply.

Rural Energy Security and Disruption Costs in Agriculture Processing and Distribution

Processed dairy products, such as cheese, butter, ice cream and yogurt, require energy to maintain proper temperatures during processing. With energy disruptions and without proper refrigeration, a large amount of milk and other dairy products will perish or spoil. Consumers many experience higher prices and possible shortages of milk, cheese, butter and other processed dairy products.

Nitrogen fertilizer production tends to be located near sources of natural gas. Natural gas is the primary feedstock for anhydrous ammonia, the most common form of nitrogen fertilizer as well as a feedstock for other fertilizer products. Fertilizer plants are clustered in the natural gas-producing states, including Texas, Oklahoma and Louisiana. Florida has the largest concentration of phosphorus producers. The USA imports potassium predominantly from Canada (Lin *et al.*, 1995; Economic Research Service, 2003). Fertilizers come primarily in solid and liquid forms and are shipped by trains, trucks, barges and ocean vessels. Liquid fertilizer is also shipped by pipeline. Given the concentrated production of fertilizer in natural gas-producing states and the use of pipelines and major carriers to transport the liquid and gaseous forms, rail, highway and pipeline disruptions could impact fertilizer availability at planting time. Such disruptions could have a significant impact on both fertilizer prices and crop yields.

Costs of Rural Energy Security

Given the seasonal nature of agriculture, a disruption in energy supplies would have limited effect on most field crop production, except during the planting and harvesting times. Energy disruptions could have a significant impact on livestock, poultry and dairy production and processing. To the extent that energy disruptions increase transportation and storage costs, a disruption could raise consumer prices while lowering farm prices in affected areas. The magnitude of these effects would depend on the nature of the energy disruption and the availability of and substitution possibilities for alternative forms of energy. Agricultural production and processing is particularly dependent on electricity and natural gas. A disruption in the supplies of these energy forms would have potentially large effects, particularly for perishable commodities.

Food and energy security is only one aspect of the homeland security issue, but a very important dimension in ensuring a safe, wholesome and moderately priced food supply for the American consumer. Unfortunately, we have relatively little data on the seasonal or real time use of energy in US agriculture, especially on food processing activities. There is almost no information on critical interruption points in the sector and thus limited knowledge of the impact of such disruptions. A better knowledge base would allow the food system to better position itself to deal with energy disruptions at critical disruption points in the food system.

For example, large-confined pig operations have very different energy demands over the production cycle. During the cooler months, farrowing requires supplemental heat to ensure the survival of piglets, but during the heat of the summer months, millions of finishing pigs would not survive without proper ventilation. The producer may be able to stockpile large amounts of LP gas in large storage tanks on-farm to protect piglets in the winter but has more limited options to store electricity to run fans during peak heat periods unless the producer has back-up generators.

There is almost no energy use data on the off-farm food processing and harvesting portion of US agriculture, either annual or otherwise. Yet, the area with the greatest potential for energy disruption impacts may be the processing sector, much of which is a continuous process throughout the year in modern livestock agriculture and in important fruit and vegetable areas. Energy disruption in processing and harvesting would not only limit the supply of product readily available to consumers, but it could lead to major market disruptions. To illustrate this point, consider the December 1998 pork market crisis. The US pork industry operates with limited excess capacity. In December 1998 and January 1999 when the number of pigs sent to slaughter exceeded harvesting capacity, market prices not only decreased significantly, but prices fell to depression-era levels until the excess supply was eliminated. If energy disruptions shut down pork harvesting, even at only a few large plants and for a short period, the backlog could cause pork prices to tank once again. Similar impacts could occur in other livestock species and in perishable fruits and vegetables where widespread spoilage and loss of product could occur. At present, there is little knowledge of these potential impacts, of the critical disruption points, of avoidance costs and of market impacts.

Field crop production is less prone to energy disruptions except during the planting and harvesting periods when direct and indirect energy disruptions could reduce crop yields and increase product loss due to delayed planting or planting without a full complement of inputs. Also crop production could be disrupted during harvesting and drying when delays may lead to increased field loss and increased crop damage from inadequate drying. An even more important energy disruption from an economic perspective may occur in the processing of field crops if major processing facilities are forced to shut down due to energy disruptions. For example, large food and bio-product maize processing plants such as high fructose corn syrup (HFCS) and ethanol operate on a continuous process and consume over a unit train (i.e., 350,000 bushels) of maize per day. If such plants are forced to temporarily shut down for any source of energy disruption from transportation to electricity, it could have major impacts on HFCS, ethanol and feed markets. Some of these facilities have their own energy sources but others are vulnerable. Unfortunately, because of large-scale economies, the US food, feed and biofuels production system is highly concentrated into large-scale plants that are centrally located. Thus, there is a need to develop more complete information and assessments of energy disruption impact for the processing segment of US agriculture.

Conclusions and Implications for Energy Consumption in US Agriculture

Energy conservation should be a goal for agriculture as well as all sectors of the US economy. Improved energy efficiency reduces the vulnerability of producers and consumers to energy price shocks, reduces the adverse impacts of long-term real energy price increases and reduces potential environmental impacts of fossil fuel consumption. As agriculture becomes more technologically sophisticated through the use of precision farming systems, biotechnology crops and enhanced information systems, more energy-efficient production systems in agriculture should result. If real energy prices increase over time, price incentives will speed the adoption of new information and biotechnologies, encourage investment in more energy-saving capital and information and continue the productivity growth and structural transition in American agriculture.

If energy conservation is a good thing, do farmers need more encouragement to increase energy efficiency? Real price incentives are usually the most effective inducement to promote energy conservation and efficiency. Direct regulation and government intervention is typically a highly inefficient approach to achieving energy conservation. Yet, except for the late 1970s and early 1980s, real prices have provided little consistent information and few incentives to producers on the potential economic benefits of energy efficiency. If producers are behaving rationally and maximizing profits for their operations, they currently have limited incentives to improve energy efficiency. Further, commodity programmes provide incentives to produce crops on lands that may otherwise be fallow or in pasture/range. Government incentives to promote adoption of renewable energy and energy-efficient investments are an option, but they run the risk of distorting investment behaviour and not providing longer-term incentives to achieve more energy efficiency.

What areas offer the greatest potential for energy efficiency gains in agriculture? Farm production management for field crops definitely offers opportunities for improved efficiency. Tillage, nutrient, water, drying, pest and precision management have important potential. Over 40% of US cropland is already under conservation-till or no-till. With appropriate market or government energy price incentives, further energy conservation is possible, especially given the magnitude of field crop production. Nutrient management may offer the largest potential energy savings in field crop production. Indirect energy consumption, particularly through nitrogen fertilizer use, is a sizeable component of on-farm energy consumption. Better management of application to reduce runoff, leaching and volatilization of nutrients may save substantial amounts of energy, reduce potential environmental problems and have a modest impact on crop yield. But such an increase in the efficiency of nitrogenous fertilizer nutrients will require price incentives provided by the marketplace, government programmes to improve nutrient management, or possibly environmental measures. Currently, such incentives to encourage more efficient nitrogen use are limited. Crop drying operations are heavily dependent on autumn weather characteristics over which producers have little control. If the real price of fuel for crop drying increased significantly, adjustments in drying practices would occur, including more field drying where possible. Finally, irrigation management provides an opportunity to substitute drip, trickle and other more energy efficient systems for less energy and water efficient systems, but these more efficient systems are more costly and only feasible on certain, high-valued crops.

Potential improvements in energy efficiency in livestock production appear more limited given the nature of production in different regions of the country and the reliance on energy to meet heating and cooling requirements. Broiler and egg production are both sensitive to appropriate climate control facilities. Significant improvements have been made in the efficiency of climate control systems for poultry, dairy and pig systems and significant energy-saving breakthroughs are not anticipated under current real energy prices. Given the share of cash expenditures on energy for poultry production (15%), the industry is quite vulnerable to real price increases.

In greenhouse, horticultural and specialty crop farms, energy accounts for roughly 5% to 6% of cash expenditures, similar to many other crop and livestock activities (National Agricultural Statistics Service, 2003). Given the relative share of cash expenditures on energy and current real energy prices, there do not appear to be significant incentives to improve the energy efficiency of production. An exception is the real price of natural gas, which has been at an all-time high for the last three decades. Higher real natural gas should provide incentives in both specialty crop and poultry production to find more energy-efficient production practices and systems.

Finally, this chapter adheres to market-based premises to achieve energy efficiency. Energy consumption (and efficiency) in agriculture is driven by real and relative energy prices, but also recognizes that energy efficiency improvements may be achieved through agricultural productivity growth, especially in the absence of real and relative price incentives. Government commodity and renewable energy programmes provide incentives and disincentives for energy consumption in agriculture that both serve to encourage and discourage improvement in energy efficiency. The agricultural sector may be more vulnerable to energy disruptions than price shocks and real energy price increases. For most crops and farms, energy expenditures are a small share of the value of output and cash expenses so price shocks and real energy price increases may be able to be absorbed with disruptions. Energy supply disruptions could prove far more costly to field crop, animal, horticultural and greenhouse operation, especially during critical periods in the production process.

Acknowledgements

The author wishes to thank Jim Duffield, Bob Dubman and Eldon Ball, US Department of Agriculture, for sharing ARMS, agriculture productivity series and energy and other farm input consumption data; Subramanian Kumarappan, Iowa State University, for research assistance in data collection and estimation; and Daniel Monchuk, Iowa State University, for comments.

References

Ball, E. (2004) *Agricultural Productivity Internal Database*. Data derived, compiled and updated from US Department of Agriculture, National Agricultural Statistics Service and US Department of Agriculture, Economic Research Service and other sources. US Department of Agriculture, Economic Research Service, Washington, DC. Portions of the database can be accessed at http://www.ers.usda.gov/Data/AgProductivity/

Berndt, E. and Wood, D. (1975) Technology, Prices and the Derived Demand for Energy. *Review of Economics and Statistics* 57: 259-268.

Berndt, E. and Wood, D. (1979) Engineering and Econometric Interpretations of Energy-Capital Complementarity. *American Economic Review* 69: 342-354.

Berndt, E. and Wood, D. (1981) Engineering and Econometric Interpretations of Energy-Capital Complementarity: Reply and Further Results. *American Economic Review* 71: 1105-1110.

Bhat, M., English, B., Turhollow, A. and Nyangito, H. (1994) *Energy in Synthetic Fertilizers and Pesticides: Revisited*. US Department of Energy, Oak Ridge National Laboratory, Oak Ridge, Tennessee.

Conservation Technology Information Center. (2004) National Crop Residue Management Survey. *Conservation Tillage Trends 1989-2002*. [Accessed 2004.] Available from http://www.ctic.purdue.edu/Core4/CT/CTSurvey/NationalData8902.html

Donaldson, D., Kiely, T. and Grube, A. (2002) *Pesticide Industry Sales and Usage, 1998 and 1999 Market Estimates*. US Environmental Protection Agency, Biological and Economic Analysis Division, Office of Prevention, Pesticides and Toxic Substances, Washington, DC.

Economic Research Service. (1994) *Agricultural Resources and Environmental Indicators*. Agricultural Handbook Number 705. US Department of Agriculture, Economic Research Service, Washington, DC.

Economic Research Service. (1997) *Agricultural Resources and Environmental Indicators*, Agricultural Handbook Number 712. US Department of Agriculture, Economic Research Service, Washington, DC.

Economic Research Service. (2003) *Agricultural Resources and Environmental Indicators*, Agricultural Handbook Number 722. US Department of Agriculture, Economic Research Service, Washington, DC. [Accessed 2004.] Available from http://www.ers.usda.gov/publications/arei/ah722/

Economic Research Service. (2004a) *Agricultural Productivity in the United States: Data*. US Department of Agriculture, Economic Research Service, Washington, DC. [Accessed 2004.] Available from http://www.ers.usda.gov/Data/AgProductivity/#data

Economic Research Service. (2004b) *Farm Policy – Title IX Energy*. US Department of Agriculture, Economic Research Service. [Accessed 2004.] Available from http://www.ers.usda.gov/Features/farmbill/titles/titleIXenergy.htm

Economic Research Service. (2004c) *Fuel Consumption Estimates Data: Commodity Costs and Returns*. US Department of Agriculture, Economic Research Service, Washington, DC. [Accessed 2004.] Available from http://ers.usda.gov/data/costsandreturns/Fuelbystate.xls

Economic Research Service. (2004d) *Old Farm Production Regions. Research Emphasis – Harmony between Agriculture and the Environment: Current Issues*. US Department of Agriculture, Economic Research Service, Washington, DC. [Accessed 2004.] Available from http://www.ers.usda.gov/Emphases/Harmony/issues/resourceregions/resourceregions.htm

Energy Information Administration. (2001) *Natural Gas Monthly*. US Department of Energy, Energy Information Administration, Washington, DC.

Energy Information Administration. (2003a) *Electric Power Monthly*. US Department of Energy, Energy Information Administration, Washington, DC.

Energy Information Administration. (2003b) *Natural Gas Monthly*. US Department of Energy, Energy Information Administration, Washington, DC.

Energy Information Administration. (2004a). *Annual Energy Review 2003 with Projections to 2025*. DOE/EIA-0383(2003). US Department of Energy, Energy Information Administration. [Accessed 2004.] Available from http://www.eia.doe.gov/emeu/aer/pdf/pages/sec2.pdf

Energy Information Administration. (2004b). *Energy Prices*. US Department of Energy, Energy Information Administration. [Accessed 2004.] Available from http://www.eia.doe.gov/price.html

Energy Information Administration. (2004c) *Energy Statistics, Data and Analysis*. US Department of Energy, Energy Information Administration. [Accessed 2004.] Available from http://www.eia.doe.gov/emeu/

Energy Information Administration. (2004d) *Monthly Energy Review*. US Department of Energy, Energy Information Administration. [Accessed 2004.] Available from http://www.eia.doe.gov/emeu/mer/consump.html

Energy Information Administration. (2004e) *Summary Impacts of Modeled Provisions of the 2003 Conference Energy Bill*. US Department of Energy, Energy Information Administration, Washington, DC. [Accessed 2004.] Available from http://www.eia.doe.gov/oiaf/servicerpt/pceb/pdf/sroiaf(2004)02.pdf

Energy Information Administration. (2004f) *US Electric Utility Average Revenue per Kilowatt-hour (Retail Price) Data*. [Accessed 2004.] Available from http://www.eia.doe.gov/cneaf/electricity/page/at_a_glance/sales_tabs.html

Executive Office of the President. (2004) *Economic Report of the President*. 108th Congress, 2nd Session. H. Doc. 108-145. US Government Printing Office, Washington, DC. [Accessed 2004.] Available from http://www.gpoaccess.gov/usbudget/fy05/pdf/2004_erp.pdf

The Fertilizer Institute. (2004a) *Production Cost Surveys*. Various issues from 1990-2004. Compiled by International Fertilizer Development Center, Muscle Shoals, Alabama.

The Fertilizer Institute (2004b) *US Fertilizer Use*. The Fertilizer Institute, Washington, DC. [Assessed 2004.] Available from http://www.tfi.org/Statistics/worldfertuse.asp

Lin, B., Padgitt, M., Bull, L., Delvo, H., Shank, D. and Taylor, H. (1995) *Pesticide and Fertilizer Use and Trends in US Agriculture*. AER-717. US Department of Agriculture, Economic Research Service, Washington, DC.

Mensah, E. and Miranowski, J. (1988) *The Energy – Capital Complementarity's Controversy Revisited: Evidence from US Agriculture*. Presented paper, International Commodity Market Modeling, 25th Conference of Applied Econometric Association, October 24-26, Washington, DC.

Miranowski, J. (1980) Estimating the Relationship Between Pest Management and Energy Prices, An Environmental Damage. *American Journal of Agricultural Economics* 62: 995-1000.

National Agricultural Statistics Service. (1999a) *1998 Farm and Ranch Irrigation Survey, 1997 Census of Agriculture. Vol. 3, Special Studies, Part 1*. US Department of Agriculture, National Agricultural Statistics Service, Washington, DC. [Accessed 2004.] Available from http://www.nass.usda.gov/census/census97/fris/fris.htm

National Agricultural Statistics Service. (1999b) *Energy Expenses for On-Farm Pumping of Irrigation Water by Type of Energy: 1998 and 1994, Table 17. Census of Agriculture, 1998 Farm & Ranch Irrigation Survey*. US Department of Agriculture, National Agricultural Statistics Service, Washington, DC. [Accessed 2004.] Available from http://www.nass.usda.gov/census/census97/fris/tbl17.pdf

National Agricultural Statistics Service. (2002) *Agricultural Prices*. (April, Page-B9) US Department of Agriculture, National Agricultural Statistics Service, Washington, DC. [Accessed 2002.] Available from http://usda.mannlib.cornell.edu/reports/nassr/price/pap-bb/2002/

National Agricultural Statistics Service. (2003) Unpublished estimates on farm expenditures on petrol, diesel fuel, liquid petroleum gas, natural gas and electricity from the Agricultural Resource Management Survey. US Department of Agriculture, National Agricultural Statistics Service, Washington, DC.

National Agricultural Statistics Service. (2004) *Farm Production Expenditures*. US Department of Agriculture, National Agricultural Statistics Service, Washington, DC. [Accessed 2004.] Available from http://usda.mannlib.cornell.edu/reports/nassr/price/zpe-bb/

National Association of Conservation Districts. (2004) *Title IX – Energy Joint Explanatory Statement of the Committee of Conference*. [Accessed 2004.] Available from http://nacdnet.org/govtaff/FB/FBmanagersT9.htm

Shapouri, H., Duffield, J. and Wang, M. (2002) *The Energy Balance of Corn Ethanol: An Update*. Agricultural Economic Report 813. US Department of Agriculture, Office of the Chief Economist, Office of Energy Policy and New Uses, Washington, DC.

Taylor, H. (1994) *Fertilizer Use and Price Statistics, 1960-93*. SB-893. US Department of Agriculture, Economic Research Service, Washington, DC.

Uri, N. and Day, K. (1991) Energy Efficiency, Technological Change and the Dieselization of Agriculture in the United States. *Transportation Planning and Technology* 16: 221-231.

US Census Bureau. (1986) *1984 Farm and Ranch Irrigation Survey*. US Department of Commerce, Economics and Statistics Administration, Bureau of the Census, Washington, DC. [Accessed 2004.] Available from http://usda.mannlib.cornell.edu/data-sets/land/87014/

US Census Bureau. (1989) *1988 Farm and Ranch Irrigation Survey, 1987 Census of Agriculture*. Volume 3, Related Surveys, No. 1. US Department of Commerce, Economics and Statistics Administration, Bureau of the Census, Washington, DC.

US Census Bureau. (1995) *1994 Farm and Ranch Irrigation Survey, 1992 Census of Agriculture*. Volume 3, Related Surveys. US Department of Commerce, Economics and Statistics Administration, Bureau of the Census, Washington, DC. [Accessed 2004.] Available from http://www.census.gov/prod/1/agr/92fris/

US Census Bureau. (2004) *Annual Survey of Manufactures*. US Department of Commerce, Economics and Statistics Administration, Bureau of the Census, Washington, DC. [Accessed 2004]. Available from http://www.census.gov/mcd/asmhome.html

Zinser, L., Miranowski, J., Shortle, J. and Monson, M. (1985) Effects of Rising Relative Energy Prices on Soil Erosion and its Control. *American Journal of Agricultural Economics* 67: 558-562.

Appendix

US agriculture energy consumption data

Appendix Table 3.1. Direct energy from major US agricultural inputs (trillion Btu), 1965-2002.

Year	Petrol[a,b]	Diesel[a,b]	LP gas[a,b]	Natural gas[a,c]	Electricity[a,c]	Total Direct
1965	654	203	110	135	93	1195
1966	604	230	111	140	99	1184
1967	570	252	112	143	99	1176
1968	572	265	113	146	106	1202
1969	550	293	114	150	114	1221
1970	541	298	114	153	124	1230
1971	523	328	110	157	134	1252
1972	476	325	123	149	139	1212
1973	590	414	126	159	142	1431
1974	447	345	87	109	154	1142
1975	555	354	91	134	169	1303
1976	482	418	111	168	205	1384
1977	479	434	102	118	250	1383
1978	432	575	112	173	301	1593
1979	431	498	114	125	293	1461
1980	371	468	100	93	278	1310
1981	354	429	94	82	283	1242
1982	288	416	98	97	286	1185
1983	277	452	79	73	263	1144
1984	256	435	86	76	283	1136
1985	232	426	87	54	253	1052
1986	199	379	65	62	243	948
1987	186	425	58	36	291	996
1988	190	404	59	48	308	1009
1989	172	386	60	28	336	982
1990	178	355	58	48	318	957
1991	159	372	54	75	316	976
1992	183	436	78	48	330	1075
1993	187	456	69	59	334	1105
1994	196	482	77	68	334	1157
1995	178	505	70	85	372	1209
1996	178	488	72	75	401	1214
1997	188	499	91	65	304	1147
1998	163	507	79	64	331	1144
1999	164	530	90	87	280	1150
2000	156	469	78	76	313	1092
2001	136	463	66	58	370	1093
2002	146	469	79	62	356	1111

[a] Source: 1965-1992, Economic Research Service (1997).
[b] Source: 1993-2002 derived by dividing farm energy expenditures on petrol, diesel and LP gas by the average price farmers pay for these fuels using National Agricultural Statistics Service (2002) and National Agricultural Statistics Service (2003).
[c] Source: 1993-2002 derived by dividing farm energy expenditures on natural gas and electricity by their respective prices, using the average price residential consumers pay for natural gas and electricity using National Agricultural Statistics Service (2003), Energy Information Administration (2003a,b) and Energy Information Administration (2004e).

Appendix Table 3.2. Indirect energy from major US agricultural inputs (trillion Btu), 1965-2002.

Year	Nitrogen[a]	Phosphate[a]	Potash[a]	Fertilizer total[a]	Pesticides[b]	Total Indirect
1965	285	60	28	373	49	422
1966	329	66	32	427	50	477
1967	372	73	36	481	56	537
1968	422	75	38	534	57	592
1969	428	80	39	547	63	609
1970	465	78	40	583	59	642
1971	502	82	42	626	73	699
1972	496	83	43	622	74	696
1973	515	87	46	647	81	728
1974	570	87	51	708	84	792
1975	533	77	44	654	95	749
1976	645	88	52	785	103	889
1977	657	95	58	810	112	922
1978	620	87	55	762	106	867
1979	663	95	62	821	116	937
1980	707	92	62	861	119	980
1981	738	92	63	893	118	1011
1982	682	82	56	820	114	934
1983	546	73	48	667	106	773
1984	666	64	58	788	112	900
1985	690	61	56	807	109	916
1986	624	55	51	730	105	835
1987	563	49	48	661	96	756
1988	565	48	49	662	98	760
1989	555	45	46	647	99	746
1990	566	45	49	660	100	760
1991	561	41	46	648	96	744
1992	553	38	45	637	100	737
1993	532	38	45	615	98	713
1994	571	36	46	652	106	758
1995	513	32	43	588	106	694
1996	522	28	43	594	108	702
1997	506	27	53	586	108	694
1998	488	25	42	554	106	660
1999	476	20	38	534	108	642
2000	453	18	37	507	108	616
2001	423	17	37	477	108	585
2002	442	19	37	498	108	606

[a] Source: Taylor (1994); The Fertilizer Institute (2004b).
[b] Source: Donaldson *et al.* (2002). Estimates are available from 1965-1999. The 1999 estimate is used for 2000-2002.

Appendix Table 3.3. US commercial fertilizer consumed (million tons), 1965-2002.

Year	Total
1965	10.9
1966	12.4
1967	14.0
1968	15.0
1969	15.5
1970	16.1
1971	17.2
1972	17.2
1973	18.0
1974	19.3
1975	17.6
1976	20.8
1977	22.1
1978	20.6
1979	22.6
1980	23.1
1981	23.7
1982	21.4
1983	18.1
1984	21.8
1985	21.7
1986	19.7
1987	19.1
1988	19.6
1989	19.6
1990	20.6
1991	20.5
1992	20.7
1993	20.9
1994	22.4
1995	21.3
1996	22.1
1997	22.4
1998	22.3
1999	21.7
2000	21.6
2001	20.6
2002	22.5

Source: The Fertilizer Institute (2004b).

Appendix Table 3.4. Farm business fuel expenses, by farm type defined with value of production, 2002.

Item	General cash grain	Wheat	Maize	Soybean	Rice	Tobacco	Cotton	Peanut	General crop[a]
(Farm type defined with value of production)									
Number of farms	75,746	36,300	124,695	68,886	4,321	49,347	12,885	3,382	426,229
Percentage of farms	3.5	1.7	5.8	3.2	0.2	2.3	0.6	0.2	19.8
Sample size	575	406	668	297	56	158	173	37	1,416
Average expansion factor	80	82	125	128	47	254	56	86	280
(Thousand US dollars)									
Total fuel/lubes purchases									
Petrol/gasohol	107,606	54,244	126,500	48,726	10,919	44,293	33,818	9,353	184,732
Diesel	356,547	137,259	485,427	159,098	53,481	65,406	112,006	22,278	409,160
Natural gas	45,996	9,372	148,782	1,964	1,300	4,061	22,661	2,596	28,725
LP gas (propane, butane)	73,996	6,424	134,306	17,792	2,355	72,935	5,134	1,060	40,395
Oils and lubricants	38,658	15,120	44,777	15,972	7,057	12,401	17,643	2,776	57,088
All other fuels	733	605	388	345	NR[b]	1,408	856	0	23,865
All purchased fuels/lubes	623,534	223,024	940,181	243,896	75,466	200,505	192,118	38,064	43,966
Electricity	154,727	41,203	232,350	50,206	20,930	46,962	104,419	22,523	413,171
(US dollars)									
Average fuel/lubes purchases									
Petrol/gasohol	1,421	1,494	1,014	707	2,527	898	2,625	2,766	433
Diesel	4,707	3,781	3,893	2,310	12,376	1,325	8,693	6,588	960
Natural gas	607	258	1,193	29	301	82	1,759	768	67
LP gas (propane, butane)	977	177	1,077	258	545	1,478	398	314	95
Oils and lubricants	510	417	359	232	1,633	251	1,369	821	134
All other fuels	10	17	3	5	NR	29	66	0	56
All purchased fuels/lubes	8,232	6,144	7,540	3,541	17,463	4,063	14,911	11,255	1,745
Electricity	2,043	1,135	1,863	729	4,843	952	8,104	6,660	969

[a] General crop farms are farms on which over 50% of their value of production is accounted for by crop production.
[b] NR = not reported.
Source: National Agricultural Statistics Service (2003).

Appendix Table 3.4 continued. Farm business fuel expenses, by farm type defined with value of production, 2002.

Item	General cash grain	Wheat	Maize	Soybean	Rice	Tobacco	Cotton	Peanut	General crop[a]
	(Percentage)								
Fuel purchases distribution									
Petrol/gasohol distribution	6.78	3.42	7.97	3.07	0.69	2.79	2.13	0.59	11.64
Diesel distribution	10.51	4.05	14.31	4.69	1.58	1.93	3.30	0.66	12.06
Natural gas distribution	7.56	1.54	24.44	0.32	0.21	0.67	3.72	0.43	4.72
LP gas (propane, butane) distribution	8.23	0.71	14.93	1.98	0.26	8.11	0.57	0.12	4.49
Oils and lubricants distribution	8.67	3.39	10.04	3.58	1.58	2.78	3.96	0.62	12.80
All other fuels distribution	1.12	0.93	0.59	0.53	NR[b]	2.16	1.31	0.00	36.53
All purchased fuels distribution	8.91	3.19	13.43	3.48	1.08	2.86	2.74	0.54	10.63
Electricity distribution	4.57	1.22	6.87	1.48	0.62	1.39	3.09	0.67	12.21
	(Farm numbers)								
Number of farms reporting fuel expense									
Petrol/gasohol	59,690	29,096	83,696	43,622	2,951	36,115	9,040	2,824	201,114
Diesel	73,316	33,598	111,634	55,725	4,085	41,235	11,994	3,382	241,103
Natural gas	5,282	2,032	8,129	931	236	1,958	2,596	264	5,587
LP gas (propane, butane)	28,736	7,279	55,223	13,900	681	8,958	3,001	503	30,633
Oils and lubricants	63,798	31,965	87,759	44,567	4,019	42,655	10,919	3,064	201,245
All other fuels	3,534	1,745	2,972	1,329	459	3,843	1,323	0	9,944
All purchased fuels	75,472	34,866	120,996	59,275	4,225	48,791	12,373	3,382	319,716
Electricity	66,792	27,021	102,492	39,759	3,257	38,600	8,563	2,464	171,853

[a] General crop farms are farms on which over 50% of their value of production is accounted for by crop production.
[b] NR = not reported.
Source: National Agricultural Statistics Service (2003).

Appendix Table 3.4 continued. Farm business fuel expenses, by farm type defined with value of production, 2002.

Item	General cash grain	Wheat	Maize	Soybean	Rice	Tobacco	Cotton	Peanut	General crop[a]
					(Fuel use ratios)				
Fuel purchases/value of production[b]									
Petrol/value of production	0.0097	0.0192	0.0076	0.0121	0.0158	0.0183	0.0126	0.0231	0.0099
Diesel/value of production	0.0320	0.0487	0.0290	0.0394	0.0772	0.0270	0.0419	0.0550	0.0218
Natural gas/value of production	0.0041	0.0033	0.0089	0.0005	0.0019	0.0017	0.0085	0.0064	0.0015
LP gas (propane, butane)/ value of production	0.0066	0.0023	0.0080	0.0044	0.0034	0.0302	0.0019	0.0026	0.0022
Oils and lubricants/ value of production	0.0035	0.0054	0.0027	0.0040	0.0102	0.0051	0.0066	0.0068	0.0030
All other fuels/value of production	0.0001	0.0002	0.0000	0.0001	0.0005	0.0006	0.0003	0.0000	0.0013
All purchased fuels/ value of production	0.0560	0.0791	0.0561	0.0603	0.1089	0.0829	0.0718	0.0939	0.0397
Electricity/value of production	0.0139	0.0146	0.0139	0.0124	0.0302	0.0194	0.0390	0.0556	0.0221
Fuel purchases/total cash expenses[c]									
Petrol/total cash expenses	0.0110	0.0194	0.0091	0.0123	0.0111	0.0211	0.0109	0.0180	0.0143
Diesel/total cash expenses	0.0363	0.0491	0.0351	0.0401	0.0541	0.0312	0.0360	0.0430	0.0318
Natural gas/total cash expenses	0.0047	0.0034	0.0108	0.0005	0.0013	0.0019	0.0073	0.0050	0.0022
LP gas (propane, butane)/ total cash expenses	0.0075	0.0023	0.0097	0.0045	0.0024	0.0348	0.0017	0.0020	0.0031
Oils and lubricants/ total cash expenses	0.0039	0.0054	0.0032	0.0040	0.0071	0.0059	0.0057	0.0054	0.0044
All other fuels/ total cash expenses	0.0001	0.0002	0.0000	0.0001	0.0004	0.0007	0.0003	0.0000	0.0019
All purchased fuels/ total cash expenses	0.0635	0.0798	0.0680	0.0614	0.0764	0.0956	0.0618	0.0734	0.0577
Electricity/total cash expenses	0.0158	0.0147	0.0168	0.0126	0.0212	0.0224	0.0336	0.0434	0.0321

[a] General crop farms are farms on which over 50% of their value of production is accounted for by crop production.
[b] Value of production is the value of all crops and livestock produced and not used on the farm.
[c] Cash expenses are all expenses except depreciation and non-cash hired labour expenses.
Source: National Agricultural Statistics Service (2003).

Appendix Table 3.4 continued. Farm business fuel expenses, by farm type defined with value of production, 2002.

Item	Fruits and tree nuts	Vegetables	Nursery and greenhouse	Beef cattle	Pigs	Poultry	Dairy	General livestock	All[a]
Number of farms	61,633	28,145	47,217	704,177	35,287	43,562	76,187	354,413	2,152,411
Percentage of farms	2.9	1.3	2.2	32.7	1.6	2.0	3.5	16.5	100.0
Sample size	502	293	480	2,536	267	665	1,062	773	10,364
Average expansion factor	123	87	97	263	96	61	64	429	178
				(Thousand US dollars)					
Total fuel/lubes purchases									
Petrol/gasohol	69,666	72,109	100,185	428,876	39,634	44,385	101,890	109,955	1,586,892
Diesel	99,600	170,797	128,794	616,315	75,799	51,148	320,220	129,803	3,393,139
Natural gas	NR[b]	5,482	178,692	51,958	13,271	65,446	11,358	5,070	608,785
LP gas (propane, butane)	12,813	14,422	66,214	74,486	75,207	227,528	49,150	25,395	899,611
Oils and lubricants	13,579	16,290	17,392	103,824	9,631	12,053	39,286	22,342	445,888
All other fuels	NR	1,108	22,274	4,300	1,184	470	4,572	2,377	65,325
All purchased fuels/lubes	208,195	280,207	513,551	1,279,758	214,725	401,031	526,476	294,943	6,999,640
Electricity	218,645	235,991	255,891	466,031	139,121	275,657	516,006	189,491	3,383,323
				(US dollars)					
Average fuel/lubes purchases									
Petrol/gasohol	1,130	2,562	2,122	609	1,123	1,019	1,337	310	737
Diesel	1,616	6,068	2,728	875	2,148	1,174	4,203	366	1,576
Natural gas	NR	195	3,784	74	376	1,502	149	14	283
LP gas (propane, butane)	208	512	1,402	106	2,131	5,223	645	72	418
Oils and lubricants	220	579	368	147	273	277	516	63	207
All other fuels	NR	39	472	6	34	11	60	7	30
All purchased fuels/lubes	3,378	9,956	10,876	1,817	6,085	9,206	6,910	832	3,252
Electricity	3,548	8,385	5,419	662	3,943	6,328	6,773	535	1,572

[a] All is an aggregate or average (ratios) over all farm classifications.
[b] NR = not reported.
Source: National Agricultural Statistics Service (2003).

Appendix Table 3.4 continued. Farm business fuel expenses, by farm type defined with value of production, 2002.

Item	Fruits and tree nuts	Vegetables	Nursery and greenhouse	Beef cattle	Pigs	Poultry	Dairy	General livestock	All[a]
Fuel purchases distribution					(Percentage)				
Petrol/gasohol distribution	4.39	4.54	6.31	27.03	2.50	2.80	6.42	6.93	100.00
Diesel distribution	2.94	5.03	3.80	18.16	2.23	1.51	9.44	3.83	100.00
Natural gas distribution	NR[b]	0.90	29.35	8.53	2.18	10.75	1.87	0.83	100.00
LP gas (propane, butane) distribution	1.42	1.60	7.36	8.28	8.36	25.29	5.46	2.82	100.00
Oils and lubricants distribution	3.05	3.65	3.90	23.28	2.16	2.70	8.81	5.01	100.00
All other fuels distribution	NR	1.70	34.10	6.58	1.81	0.72	7.00	3.64	100.00
All purchased fuels distribution	2.97	4.00	7.34	18.28	3.07	5.73	7.52	4.21	100.00
Electricity distribution	6.46	6.98	7.56	13.77	4.11	8.15	15.25	5.60	100.00
Number of farms reporting fuel expense					(Farm numbers)				
Petrol/gasohol	40,299	23,784	33,888	429,721	25,116	31,901	57,512	184,877	1,295,244
Diesel	41,826	17,156	26,152	526,557	24,631	33,874	72,533	159,874	1,478,675
Natural gas	993	911	6,612	11,424	1,162	5,154	4,151	5,033	62,456
LP gas (propane, butane)	10,082	6,127	17,571	79,347	15,559	23,326	26,181	21,824	348,933
Oils and lubricants	38,966	20,929	27,311	484,099	20,687	32,062	59,320	164,992	1,338,357
All other fuels	806	2,485	5,135	16,620	2,255	651	6,520	11,050	70,671
All purchased fuels	53,568	27,216	45,634	638,629	32,393	42,536	75,457	269,823	1,864,351
Electricity	44,482	19,273	39,150	445,439	30,335	41,789	70,508	204,019	1,355,795

[a] All is an aggregate or average (ratios) over all farm classifications.
[b] NR = not reported.
Source: National Agricultural Statistics Service (2003).

Appendix Table 3.4 continued. Farm business fuel expenses, by farm type defined with value of production, 2002.

Item	Fruits and tree nuts	Vegetables	Nursery and greenhouse	Beef cattle	Pigs	Poultry	Dairy	General livestock	All[a]
				(Fuel use ratios)					
Fuel purchases/value of production[b]									
Petrol/value of production	0.0068	0.0059	0.0063	0.0152	0.0036	0.0025	0.0046	0.0233	0.0087
Diesel/value of production	0.0098	0.0141	0.0080	0.0219	0.0069	0.0029	0.0143	0.0275	0.0187
Natural gas/value of production	0.0012	0.0005	0.0112	0.0018	0.0012	0.0037	0.0005	0.0011	0.0033
LP gas (propane, butane)/value of production	0.0013	0.0012	0.0041	0.0026	0.0068	0.0129	0.0022	0.0054	0.0049
Oils and lubricants/value of production	0.0013	0.0013	0.0011	0.0037	0.0009	0.0007	0.0018	0.0047	0.0025
All other fuels/value of production	0.0000	0.0001	0.0014	0.0002	0.0001	0.0000	0.0002	0.0005	0.0004
All purchased fuels/value of production	0.0204	0.0231	0.0321	0.0455	0.0195	0.0228	0.0236	0.0624	0.0385
Electricity/value of production	0.0214	0.0195	0.0160	0.0166	0.0126	0.0157	0.0231	0.0401	0.0186
Fuel purchases/total cash expenses[c]									
Petrol/total cash expenses	0.0093	0.0085	0.0079	0.0162	0.0069	0.0099	0.0054	0.0129	0.0111
Diesel/total cash expenses	0.0133	0.0202	0.0102	0.0233	0.0132	0.0115	0.0170	0.0153	0.0238
Natural gas/total cash expenses	0.0016	0.0006	0.0141	0.0020	0.0023	0.0147	0.0006	0.0006	0.0043
LP gas (propane, butane)/total cash expenses	0.0017	0.0017	0.0052	0.0028	0.0131	0.0510	0.0026	0.0030	0.0063
Oils and lubricants/total cash expenses	0.0018	0.0019	0.0014	0.0039	0.0017	0.0027	0.0021	0.0026	0.0031
All other fuels/total cash expenses	0.0001	0.0001	0.0018	0.0002	0.0002	0.0001	0.0002	0.0003	0.0005
All purchased fuels/total cash expenses	0.0279	0.0332	0.0405	0.0484	0.0375	0.0899	0.0279	0.0347	0.0491
Electricity/total cash expenses	0.0293	0.0280	0.0202	0.0176	0.0243	0.0618	0.0273	0.0223	0.0237

[a] All is an aggregate or average (ratios) over all farm classifications.
[b] Value of production is the value of all crops and livestock produced and not used on the farm.
[c] Cash expenses are all expenses except depreciation and non-cash hired labour expenses.
Source: National Agricultural Statistics Service (2003).

Appendix Table 3.5. Expenditure on major US farm inputs (billion 1996 US dollars).

Year	Land rent	Labour	Capital	Fuels	Fertilzer, lime and soil conditioners	Agricultural chemicals	Others	Total
1980	20.6	17.4	34.4	16.9	17.7	5.3	121.9	234.1
1981	17.5	15.8	30.7	16.3	15.8	5.7	108.5	210.2
1982	16.9	17.4	22.4	14.8	14.3	5.3	105.8	196.9
1983	18.0	15.9	20.6	13.8	12.1	4.9	104.0	189.4
1984	18.0	14.6	25.0	9.7	12.3	5.4	93.8	178.8
1985	17.1	14.8	20.2	8.7	11.6	5.3	92.3	170.0
1986	12.6	13.1	19.0	6.2	9.1	4.7	75.4	140.1
1987	13.1	12.9	20.7	5.6	8.2	4.8	76.3	141.5
1988	12.5	12.7	20.8	5.5	8.6	4.8	82.2	147.1
1989	13.6	13.6	21.8	5.3	8.6	5.4	80.3	148.5
1990	13.7	15.4	21.8	6.0	8.2	5.5	81.0	151.5
1991	13.5	14.6	19.8	5.5	8.3	6.2	76.7	144.6
1992	15.9	14.9	19.1	6.0	9.1	7.0	85.9	157.9
1993	15.7	15.6	17.2	5.4	8.9	7.1	94.9	164.8
1994	16.0	15.6	16.6	5.3	9.5	7.5	96.6	167.1
1995	16.3	16.3	16.1	5.6	10.2	7.8	99.2	171.6
1996	18.3	17.1	17.6	5.8	10.9	8.5	96.8	175.0
1997	18.2	18.0	18.4	5.9	10.7	8.9	100.1	180.2
1998	16.8	18.5	19.4	5.2	10.3	8.7	99.0	178.1
1999	15.2	18.9	19.1	5.1	9.5	8.2	100.4	176.3
2000	15.1	19.4	20.1	6.6	9.4	8.0	99.4	177.9
2001	15.0	19.9	20.2	6.1	9.4	7.9	100.4	179.0
2002	14.4	19.4	19.9	5.7	8.6	7.4	97.4	172.8

Source: National Agricultural Statistics Service (2004); Executive Office of the President (2004).

Appendix Table 3.6. Irrigated crop area by method (million acres), 1984-1998.

	Sprinkler systems	Gravity flow systems	Drip or trickle systems	Sub irrigation	Total irrigated area
1984	16.88	27.46	0.84	0.62	45.80
1988	18.42	27.41	0.87	0.58	47.29
1994	21.46	25.09	1.75	0.36	46.42
1998	22.96	25.08	2.10	0.55	50.03

Source: US Census Bureau (1986, 1989, 1995); National Agricultural Statistics Service (1999a).

Appendix Table 3.7. US on-farm pumping of irrigation water (million acres), 1998.

Electricity	20.16
Natural gas	6.09
LP gas	1.63
Diesel fuel	10.28
Petrol	0.16
Total	38.31

Source: National Agricultural Statistics Service (1999b).

Appendix Table 3.8. US direct and indirect energy consumption, 1965-2002.

Year	Petrol (billion gallons)	Diesel (billion gallons)	LP gas (billion gallons)	Natural gas (billion cubic feet)	Nitrogen (tons)	Phosphate (tons)	Potash (tons)	Pesticides (million pounds)
1965	5.5	1.6	N/A	131.5	46	35	28	435
1966	5	1.8	N/A	136.3	53	39	32	446
1967	4.8	1.9	N/A	139.2	60	43	39	496
1968	4.8	2	N/A	142.2	68	44	38	505
1969	4.6	2.3	N/A	149	69	47	39	553
1970	4.5	2.3	N/A	151.8	75	46	40	520
1971	4.4	2.5	N/A	155.8	81	48	42	643
1972	3.9	2.5	N/A	147.8	80	49	43	650
1973	4.9	3.2	N/A	157.7	83	51	46	715
1974	3.7	2.6	1.4	108.1	92	51	51	744
1975	4.6	2.4	1	132.9	86	45	44	841
1976	4	2.8	1.2	166.7	104	52	52	913
1977	4	2.9	1.1	117.1	106	56	58	989
1978	3.6	3.2	1.3	171.6	100	51	55	932
1979	3.6	3.2	1.1	124	107	56	62	1025
1980	3.1	3.2	1.1	92.3	114	54	62	1053
1981	2.9	3.1	1	81.4	119	54	63	1046
1982	2.4	2.9	1.1	96.2	110	48	56	1011
1983	2.3	3	0.9	72.4	91	41	48	941
1984	2.1	3	0.9	75.4	111	49	58	988
1985	1.9	2.9	0.9	53.6	115	47	56	961
1986	1.7	2.9	0.7	61.5	104	42	51	927
1987	1.6	3	0.6	35.7	102	40	48	846
1988	1.6	2.8	0.6	47.6	105	41	50	867
1989	1.4	2.5	0.7	27.8	106	41	48	873
1990	1.5	2.7	0.6	47.6	111	43	52	884
1991	1.3	2.8	0.6	74.4	113	42	50	848
1992	1.6	3.1	0.9	46.3	115	42	50	884
1993	1.4	3.3	0.7	57.2	114	44	51	864
1994	1.4	3.5	0.9	66.1	126	45	53	939
1995	1.4	3.6	0.7	82	117	44	51	933
1996	1.3	3.2	0.8	72.4	123	43	53	955
1997	1.5	3.6	1	63.4	123	46	54	955
1998	1.3	3.7	0.9	62.2	123	47	53	956
1999	1.3	3.8	1	84	125	43	50	956
2000	1.3	3.4	0.9	73.6	123	42	50	956
2001	1.1	3.3	0.7	55.9	115	42	49	956
2002	1.2	3.4	0.9	59.9	129	46	50	956

N/A = not available.
Source: National Agricultural Statistics Service (2003).

Appendix Table 3.9. Nominal prices of major energy inputs in US agriculture, 1970-2002.

Year	Diesel (US cents/ gallon)	Petrol (US cents/ gallon)	Electricity (US cents/ kilowatt-hour)	Natural gas (US cents/ cubic foot)
1970	N/A	38.7	2.2	0.109
1971	N/A	39.4	2.3	0.115
1972	N/A	39.1	2.4	0.121
1973	N/A	41.8	2.5	0.129
1974	N/A	56.2	3.1	0.143
1975	N/A	59.7	3.5	0.171
1976	N/A	62	3.7	0.198
1977	N/A	65	4.1	0.235
1978	N/A	65.2	4.3	0.256
1979	N/A	88.2	4.6	0.298
1980	N/A	122.1	5.4	0.368
1981	N/A	135.3	6.2	0.429
1982	N/A	128.1	6.9	0.517
1983	107.8	122.5	7.2	0.606
1984	109.1	119.8	7.15	0.612
1985	105.3	119.6	7.39	0.612
1986	83.6	93.1	7.42	0.583
1987	80.3	95.7	7.45	0.554
1988	81.3	96.3	7.48	0.547
1989	90	106	7.65	0.564
1990	106.3	121.7	7.83	0.58
1991	101.9	119.6	8.04	0.582
1992	93.4	119	8.21	0.589
1993	91.1	117.3	8.32	0.616
1994	88.4	117.4	8.38	0.641
1995	86.7	120.5	8.4	0.606
1996	98.9	128.8	8.36	0.634
1997	98.4	129.1	8.43	0.694
1998	85.2	111.5	8.26	0.682
1999	87.6	122.1	8.16	0.669
2000	131.1	156.3	8.24	0.776
2001	125	153.1	8.62	0.964
2002	112.9	144.1	8.45	0.785

N/A = not available.
Source: Energy Information Administration (2004a).

Appendix Table 3.10. Real prices of major energy inputs in US agriculture (1996 dollars), 1970-2002.

Year	Diesel (US cents/ gallon)	Petrol (US cents/ gallon)	Electricity (US cents/ kilowatt-hour)	Natural gas (US cents/ cubic foot)
1970	N/A	132.1	7.51	0.372
1971	N/A	127.8	7.46	0.373
1972	N/A	121.5	7.46	0.376
1973	N/A	122.9	7.35	0.379
1974	N/A	152.1	8.39	0.387
1975	N/A	147.9	8.67	0.424
1976	N/A	144.9	8.65	0.463
1977	N/A	142.6	8.99	0.515
1978	N/A	133.7	8.82	0.525
1979	N/A	167.4	8.73	0.565
1980	N/A	212.8	9.41	0.641
1981	N/A	215.8	9.89	0.684
1982	N/A	192.6	10.37	0.777
1983	155.7	176.9	10.40	0.875
1984	151.9	166.9	9.96	0.852
1985	142.2	161.5	9.98	0.826
1986	110.5	123.1	9.81	0.771
1987	103.2	122.9	9.57	0.712
1988	101.0	119.7	9.30	0.680
1989	107.7	126.9	9.16	0.675
1990	122.4	140.2	9.02	0.668
1991	113.5	133.2	8.96	0.648
1992	101.9	129.8	8.95	0.642
1993	96.8	124.6	8.84	0.654
1994	91.9	122.1	8.72	0.667
1995	88.3	122.7	8.55	0.617
1996	98.9	128.8	8.36	0.634
1997	96.8	127.0	8.29	0.683
1998	82.8	108.4	8.03	0.663
1999	83.9	117.0	7.82	0.641
2000	123.0	146.7	7.73	0.728
2001	114.6	140.4	7.90	0.884
2002	101.9	130.9	7.63	0.709

N/A = not available.
Source: Energy Information Administration (2004b); Executive Office of the President (2004).

Appendix Table 3.11. Indexes of farm output, input use, and productivity (base year 1996).

Year	Total farm output	Total farm input	Total factor productivity
1970	0.585	1.098	0.533
1971	0.625	1.086	0.575
1972	0.630	1.097	0.574
1973	0.655	1.109	0.591
1974	0.618	1.101	0.562
1975	0.656	1.084	0.605
1976	0.664	1.122	0.592
1977	0.701	1.097	0.639
1978	0.711	1.158	0.614
1979	0.759	1.186	0.640
1980	0.731	1.202	0.608
1981	0.796	1.163	0.684
1982	0.820	1.138	0.720
1983	0.706	1.137	0.621
1984	0.805	1.089	0.739
1985	0.842	1.064	0.791
1986	0.814	1.033	0.788
1987	0.831	1.023	0.813
1988	0.797	1.015	0.785
1989	0.860	1.001	0.860
1990	0.894	1.017	0.879
1991	0.902	1.027	0.879
1992	0.948	0.999	0.949
1993	0.902	1.003	0.899
1994	1.006	1.017	0.989
1995	0.959	1.034	0.927
1996	1.000	1.000	1.000
1997	1.035	1.031	1.004
1998	1.049	1.038	1.011
1999	1.070	1.063	1.007

Source: Economic Research Service (2004a).

Appendix Table 3.12. Index numbers of consumption of major outputs and inputs in agriculture (base year 1996).

Year	Output		Input				
	Crops	Livestock	Capital	Land	Labour	Energy	Chemicals
1970	0.544	0.729	1.235	1.146	1.632	0.937	0.721
1971	0.606	0.739	1.243	1.127	1.601	0.919	0.723
1972	0.610	0.748	1.264	1.111	1.581	0.909	0.772
1973	0.648	0.756	1.283	1.099	1.588	0.922	0.796
1974	0.595	0.744	1.337	1.093	1.470	0.877	0.860
1975	0.674	0.703	1.371	1.094	1.473	1.032	0.774
1976	0.668	0.738	1.403	1.099	1.453	1.159	0.911
1977	0.724	0.752	1.429	1.105	1.403	1.216	0.805
1978	0.743	0.751	1.463	1.107	1.333	1.272	0.871
1979	0.812	0.774	1.490	1.103	1.305	1.158	0.952
1980	0.744	0.803	1.549	1.094	1.264	1.128	1.118
1981	0.850	0.816	1.529	1.084	1.276	1.082	1.007
1982	0.857	0.807	1.513	1.073	1.220	1.021	0.825
1983	0.656	0.826	1.478	1.065	1.205	0.989	0.808
1984	0.833	0.815	1.410	1.059	1.189	1.025	0.883
1985	0.872	0.837	1.384	1.055	1.107	0.920	0.903
1986	0.820	0.843	1.305	1.050	1.033	0.855	1.049
1987	0.826	0.861	1.233	1.046	1.033	0.951	0.959
1988	0.730	0.879	1.188	1.040	1.064	0.950	0.800
1989	0.834	0.883	1.153	1.034	1.053	0.943	0.819
1990	0.894	0.895	1.131	1.028	1.046	0.940	0.949
1991	0.886	0.918	1.120	1.021	1.078	0.944	0.963
1992	0.964	0.941	1.092	1.016	1.018	0.936	0.978
1993	0.869	0.947	1.071	1.011	0.978	0.939	0.944
1994	1.031	0.985	1.038	1.007	0.990	0.967	1.008
1995	0.920	1.005	1.029	1.003	1.027	1.015	0.923
1996	1.000	1.000	1.000	1.000	1.000	1.000	1.000
1997	1.043	1.011	0.996	0.997	1.006	1.022	1.085
1998	1.031	1.039	0.989	0.994	0.978	1.037	1.047
1999	1.035	1.067	0.985	0.990	1.008	1.054	1.038

Source: Economic Research Service (2004a).

Appendix Table 3.13. US energy intensity in agriculture, industry and the economy.

Year	Output (1996 US dollars, in billions)			Energy Consumption (quadrillion units)			Energy Intensity (thousand BTU/1996 US dollars)		
	Agriculture[a]	Industry[bc]	GDP	Agriculture	Industry[b]	Economy	Agriculture	Industry[b]	Economy
1970	159.17	1110.9	3609.8	1.6	29.6	54.0	11.8	26.7	18.8
1971	167.01	1119.6	3731.2	1.7	29.6	57.0	11.7	26.4	18.6
1972	177.10	1177.9	3914.5	1.7	31.0	58.9	10.8	26.3	18.6
1973	195.47	1259.8	4186.4	1.8	32.7	62.4	11.0	26.0	18.1
1974	177.75	1268.5	4235.8	1.8	31.9	65.6	10.9	25.1	17.5
1975	174.83	1187.1	4190.7	1.9	29.5	67.8	11.7	24.8	17.2
1976	180.07	1286.1	4354.5	2.0	31.5	69.3	12.6	24.5	17.5
1977	182.75	1357.4	4507.7	1.9	32.4	72.7	12.6	23.9	17.3
1978	195.98	1437.4	4771.8	2.2	32.8	75.7	12.6	22.8	16.8
1979	209.86	1479.9	4932.9	1.9	34.0	74.0	11.4	23.0	16.4
1980	195.27	1469.2	4935.5	2.1	32.2	72.0	11.7	21.9	15.9
1981	210.89	1508.8	5041.5	2.3	31.0	76.0	10.7	20.5	15.1
1982	208.37	1405.6	4909.5	2.3	27.8	78.0	10.2	19.8	14.9
1983	183.82	1389.9	5108.7	2.5	27.6	80.0	10.4	19.9	14.3
1984	196.15	1510.8	5479.3	2.4	29.7	80.9	10.4	19.7	14.0
1985	197.53	1519.3	5694.4	2.3	29.1	78.3	10.0	19.2	13.4
1986	184.95	1484.2	5884.7	2.3	28.5	76.3	9.6	19.2	13.0
1987	189.70	1542.1	6090.2	2.1	29.6	73.2	9.2	19.2	13.0
1988	192.58	1640.0	6358.8	1.9	30.9	73.1	9.2	18.9	13.0
1989	203.13	1631.8	6577.4	2.0	31.6	76.7	8.5	19.4	12.9
1990	206.23	1620.2	6710.1	2.0	32.2	76.4	8.3	19.9	12.6
1991	204.77	1530.1	6682.5	1.8	31.9	76.7	8.4	20.9	12.6
1992	214.31	1526.3	6889.8	1.8	33.1	79.2	8.5	21.7	12.5
1993	209.95	1561.2	7076.8	1.8	33.5	82.8	8.7	21.4	12.4
1994	226.84	1653.4	7360.1	1.7	34.5	84.9	8.4	20.9	12.1
1995	219.20	1707.3	7539.7	1.7	34.9	84.6	8.7	20.5	12.1
1996	228.55	1745.4	7816.9	1.7	35.9	84.5	8.4	20.6	12.1
1997	235.72	1805.1	8161.5	1.8	36.1	85.9	7.8	20.0	11.6
1998	232.42	1857.2	8494.2	1.8	35.6	87.6	7.5	19.2	11.2
1999	229.66	1927.4	8883.8	1.9	36.4	89.2	7.8	18.9	10.9
2000	232.34	1986.8	9223.0	1.9	38.2	91.2	7.3	19.2	10.7
2001	233.99	1874.6	9272.5	1.9	N/A[d]	94.2	7.2	N/A	10.4
2002	227.36	1860.0	9454.7	1.8	N/A	94.7	7.6	N/A	10.3

[a] Deflated using US Bureau of Labor Statistics farm products, processed foods and feeds index (PFFPLUS02).
[b] Construction, mining and manufacturing.
[c] Deflated using gross domestic product (GDP) deflator.
[d] N/A = not available.
Source: Ball (2004); Energy Information Administration (2004a); Executive Office of the President (2004).

Chapter 4

Energy Systems Integration: Fitting Biomass Energy from Agriculture into US Energy Systems

Otto C. Doering III

Introduction

The USA is the world's dominant consumer of energy. Highly capitalized and integrated supply systems are currently in place that are matched with similarly capitalized and integrated utilization systems. The systems themselves determine which fuels will be used. Any change often requires that the systems be modified or changed. To some extent, the complex and interdependent characteristics of this energy system inhibits change in fuel source, delivery, or use. Change from the current state requires that one has some idea of why things are the way they are today. Without understanding the reasons for today's energy structure and logistics, new approaches face the danger of being irrelevant at best or disruptive at worst. Even under the best conditions, change may be marginal at best because of the scale and scope of existing systems, let alone the presence of other rigidities in the system or vested interests maintaining the status quo. The pace of change accelerates only when there is stress on the system or some breakdown or emergency.

In the 1970s, during the Organization of the Petroleum Exporting Countries (OPEC) oil embargo, a radical vision flourished of energy-independent farms, homes and businesses each having local biomass, wind, or solar energy systems. This was not only a fuel change vision but also one of moving from centralized to distributed power. These concepts were articulated by Amory Lovins in a *Foreign Affairs* article, though he was by no means the first to express them (Lovins, 1976). This notion of energy self-sufficiency probably gained more adherents on farms with the visions of ethanol-powered machinery and methane gas heat sources derived from the farm than it did in many other places. Part of the lack of progress for this vision was the fact that existing energy systems provided for relatively reliable low cost energy with low levels of expenditure of capital and management time. Biomass, solar and wind were not in the running given the logistics and economics of the systems in place other than during the disruptive conditions of the oil embargo and the price spikes that followed.

Alternative energy sources and systems, in this case, those that would stem from the development of agricultural biomass, do not and will not operate in a vacuum disconnected from what else goes on. Consumers have cost and convenience decisions about which energy source to use in terms of the utilization of conventional energy sources and systems now in place. An alternative energy source with different cost, convenience, or other attributes may not provide a perfect or even an acceptable substitute in important respects to the consumer. If this is so, changes will likely be marginal.

© CAB International 2005. *Agriculture as a Producer and Consumer of Energy* (eds J.L. Outlaw, K.J. Collins and J.A. Duffield)

In the absence of an emergency that forces the abandonment of current energy sources and infrastructure, we have to start with existing systems and ask whether a particular biomass energy source integrates well in existing systems or what alternative systems that accommodate biomass energy might have to look like to meet or exceed current consumer expectations. As we think about an increased biomass component of national energy use, we need to ask questions like:

- To what extent can biomass sources and systems meet the form, time and place determinants of energy utility that consumers obtain from current sources and systems?
- Are the form, time and place determinants the right benchmark?
- What are the compelling reasons for current energy sources and systems to predominate that tie them to consumers and discourage their replacement by other systems and sources?
- How might biomass energy successfully integrate into current systems?
- What are some of the critical challenges in integrating biomass systems?
- What policy issues will have to be dealt with for biomass sources to integrate or successfully create replacements for existing energy systems?

Some years ago Kenneth Boulding, an economist and sage, wrote a pointed and humorous poem poking fun at different academic disciplines and how they viewed the solutions to all problems falling within each of their own narrow domains. His experience was that there was great danger in these narrow views. The intent here is to not fall into that trap and instead take a broad view spanning different expertize and concerns.

The Utility of Energy Use

There are several aspects of energy consumption that are critical to its usefulness. For the most part, energy use today is paid for and consumed on the basis of specific requirements of form, time and place. We want a specific kind of energy whose form accomplishes what we desire at a particular time and at a specific place. Over time, the energy sources and logistics we employ today have been chosen and designed to meet these requirements. These are so ingrained in our current utilization that we take them for granted. Unless we are forced into change by external events, any new source or systems will likely have to meet these existing requirements. Part of this is because we think in terms of specific sources or forms of energy as being essential to meeting particular needs or services we desire.

There are important characteristics of our energy use that indicate where we are today in terms of our desire for energy services. Any new energy sources or systems would have to conform to the extent of meeting the perceived needs that are already met by the existing systems. As an example, the USA uses vast amounts of liquid fuel and much of this for transportation. It is not only because we have lower vehicle fuel efficiency (which economists would ascribe to lower petrol prices than other countries), but also the fact that we have developed as a continental economy. Public transportation 'works' in Europe to some extent because of shorter distances and much higher population density. Thus, Australia and Canada have liquid fuel uses that are similar to ours. Our liquid fuel system is currently set up to provide such vast amounts of liquid fuel because of the need for individual car transportation across the country. The large public subsidy of the interstate highway system locks in place a truck transportation system that competes with rail at a different level than would be the case if the subsidies for the interstate highway system were not there or if the public helped provide the road bed for rail transportation.

The form, time and place requirements for specific activities can be compelling in locking users into a fuel type. As an example, the State of Indiana tried to make extra propane available to its farmers for crop drying during the propane gas shortage of the 1970s. The state located propane for purchase on the Gulf Coast. This propane (form) had to be available at maize harvest (time) in Indiana (place) to dry the grain to prevent spoilage. A clean gas fuel was essential because most grain dryers vent the heat of combustion through the grain, which is much more efficient than using a heat exchanger. The grain drying infrastructure was thus in place and demanded a specific form of fuel that would not pollute the grain. The fuel also had to be *liquefied* petroleum gas so that it could be delivered to farms far from pipelines and stored in on-farm tanks, the existing infrastructure. The right form was located, just in time before harvest, but it turned out to be impossible to deliver it to the place it was needed. First, pipeline capacity was already booked. In addition, the state found it was competing with another user for whom the form, time and place were of even higher value. Rail cars for transporting the propane from the Gulf Coast were not available. A major automobile company in the Upper Midwest used gas for heat treating metal transmission gears. Using coal to heat treat metals adds carbon and changes the characteristics of the metal; gas does not. The car company had hundreds of tank cars filled with propane in their yards on demurrage for heat treating transmission parts in case of natural gas shortages that were threatened. This effectively used up the available tank car transportation capacity for bringing additional propane to Indiana. The value in use for this propane gas for heat treating was so high that it was worth paying for the propane and to hold propane in reserve even though it might not be necessary.

The critical question is, when and under what conditions might we be able to break some of the form, time and place bounds and be more flexible about fuels and energy use? When Roger Sant went into the new Federal Energy Agency, he brought a new line of thought about energy that for a short time changed the thinking in a few corners of the federal government (Sant, 1979; Sant *et al.*, 1984). His notion was that energy needed to be viewed as a provider of services as Amory Lovins had suggested (Lovins, 1976). Beyond that, what society desired was the service from an energy input and the specific service was not necessarily tied to a particular fuel form and maybe not even on the existing specific time and place fuel requirements. Thus, planning and policy should proceed starting with the service desired, not with the current fuel required. This heresy was promulgated not too long after the nation was seized with the phobia over petrol shortages and prices induced by the OPEC oil embargo. Sant's response was that the desired service is transportation and petrol is only one way, or just the current way, that we get this service. What would be the alternative ways we might provide the service of transportation? One example of Sant's approach was the rail transportation study commissioned by the US Department of Transportation. It demonstrated that if we were short of liquid fuel, one viable response would be to electrify the rail system – a very different response from developing oil shale or biodiesel to expand the sources of liquid fuel (Whitford *et al.*, 1981). This electrification approach was also more cost effective.

Why We Are Where We Are Today

The nature of our continental geography and the continental economy are certainly important factors that have shaped our energy use in form, time and place. There is a whole complexity of decisions that have been made over time interacting with technology and the extent to which various activities respond to the market, consumers' preferences and perceptions and to political decisions. New York only became the gateway to America when the Erie Canal was built allowing access to the Great Lakes and the upper Midwest – a political decision by an individual governor. Without it, New Orleans might well have been the gateway, but this would have depended upon the rapid development of steam boat technology for traffic on the Mississippi as well as navigation improvements in the river. The rail system initially followed the canal routes, but went beyond them as the nineteenth century progressed. The rail connections through Chicago, Illinois, helped move the centre of grain production and the grain business west from Buffalo, New York, to Chicago while the centre of wheat production moved west as well because of transportation and the *Homestead Act of 1862*. Even with good rail connections, the meat packing industry did not centralize in Chicago until Gustavius Swift invented the further technological refinement of the refrigerated railcar – saving on transport and allowing large-scale centralized slaughtering for the finished carcasses that were then shipped east.

Wood and coal were the primary energy sources of the nineteenth century. Coal became the preferred fuel for both mobile and stationary power because of its caloric density as well as its availability. The caloric density also made it more practicable for high thermal demand situations. The great development of oil follows the invention of the automobile and extensive natural gas utilization was spurred by special needs for industrial use and low-cost availability for home heating – competing with both coal and oil. All of these fuels had characteristics, price advantages or locational advantages that brought about their adoption for one purpose or another. Many of the conditions that led to fuel use today in a particular form and at a particular place have not changed, so there is not an obvious incentive to think in terms of fuel switching, let alone switch the focus of thinking to focus on the services delivered by a particular fuel.

How natural gas entered the energy scene is a good example of how we got to where we are today. Natural gas was initially used for lighting. Natural gas was often a waste product found on the way to oil that had to be efficiently discarded (i.e., flamed off). The natural gas associated with oil discoveries in Pennsylvania was utilized for industrial purposes. In the upper Midwest, natural gas from the Gulf Coast or Southwest made its entry as a home heating fuel after World War II. It was only after World War II that welding techniques and other advances in pipe construction allowed the extension of a national gas pipeline system.

As an example of the switch to gas, in some Chicago suburbs, there were central heating companies that burned coal and provided hot water heat to entire neighbourhoods. When pipeline natural gas arrived, these central heating plants faded away and homes installed their own individual gas furnaces. Initially the central heating plant had replaced each home burning its own coal. Then, natural gas at each home replaced the central heating plant. Part of this related to inexpensive natural gas, part to the wearing out of a capital stock of central heating plants and their delivery infrastructure, part to convenience in use and part to coal's negative attributes as a 'dirty' fuel and natural gas's attributes as a clean fuel. Cheap natural gas began to replace coal, gas from coke, and even oil in industrial use. Ultimately, natural gas began to replace coal in electric generation. Coincidently, the federal government began regulating the price of natural gas. In 1938, the price of natural gas was regulated in the belief that through the pipeline it was a natural monopoly like the railroads or electric power. Strict price regulation was dropped in 1989 after the natural gas shortages in the 1970s and 1980s. What followed was a natural gas bubble as high prices of the 1980s encouraged exploration, increased production and drove prices down again. Natural gas use was also increasingly encouraged by environmental/clean air regulations.

What we have in place today is an extensive mature natural gas infrastructure that is tuned to deliver gas from major sources, the US Gulf Coast and Southwest, Canada, Mexico and some limited liquefied natural gas from outside the USA that comes into a small number of East and West Coast ports. Natural gas users are tied as much to the distribution system as to the fuel itself.

Much of current fuel use relates to perceptions and experiences about the reliability of supply. One of the early marketing points for natural gas was that it was a reliable fuel. It did not depend upon the United Mine Workers (who might strike and not dig coal), or the weather, or anything else that might disrupt an oil delivery truck. Natural gas users perceive it as an absolutely reliable fuel or energy source that requires no on-site storage. In many instances, such as home heating, wood, coal or oil require a storage activity by the end user. Residential use of energy, whether it is gas, oil, or electricity, aligns with a perception of reliability that has become an imbedded attribute of the fuel or energy source – much like the old AT&T telephone system reliability standard with its extensive built-in redundancies. One of the important questions is what trade-offs are residential users of energy willing to make between reliability and other factors? Since World War II, the expectation of reliability has been and is still virtually 100% for home fuels and energy used in the home – California and Northeast US electricity blackouts not withstanding.

Industrial and commercial users of fuels and energy have had to modify their reliability expectations since the OPEC oil embargo of the 1970s. Suppliers of energy or fuels to commercial and industrial firms now offer a reliability menu where different levels of reliability relate to different prices paid for the fuels or to different fuel options. This is very much the case today with interruptible natural gas contracts and similar interruptible contracts from electricity suppliers. After the OPEC oil embargo, the commercial and industrial sectors invested in dual fuel capacity when possible. In practice, this was a natural gas/oil trade-off that was first envisioned to provide back-up if there were disruptions in oil supplies. As this capacity became widespread it has become an equilibrating factor between fuel oil and natural gas prices as boilers are switched between fuels on the basis of price sensitivity. What this indicates is that the commercial and industrial sectors have a different sense of reliability from their experience since the embargo as compared to the previous 25 years after World War II when reliability was assured once infrastructure was in place. Prior to the embargo, prices between fuels were more stable for those fuels that could be substituted one for another. There were not repeated economic incentives for fuel switching.

Liquid fuel for transportation was perceived as extremely reliable in the post-World War II period, until it was interrupted by the OPEC oil embargo. The perception is still one of reliability of supply, if not of price. The distribution system is extensive and functions smoothly. It has also consolidated in terms of a reduction in the number of suppliers and retailers, but this likely affects pricing power more than it does the physical availability of fuel for mobile transportation. Liquid petroleum fuels are also extremely reliable in terms of quality – thus presenting a uniform form, time and place set of characteristics to the consumer.

Change From Where We Are Today

Economists and engineers like to believe that changes in prices and changes in technology, respectively, are the primary movers of change. After the OPEC oil embargo, many economists believed, and many policy makers made policy as if, the demand for energy was inelastic. In other words, even if energy prices increased tremendously we would not see much in the way of reduced demand. The policy implications of this were that fuels in their historical form, time and place had to be provided and that there was little adjustment capability. A counter argument was that there was some elasticity in energy demand over the short and the long term (Peck and Doering, 1976). The implications of this argument for national policy and for the behaviour of energy suppliers are most important.

Resource fixity can also be a potential cause of change. The 'limits to growth' work in the early 1970s is a prime example (Meadows *et al*., 1972). The argument was that a fixed resource base, a growing population and increased per capita consumption would require very different energy consumption, systems and sources. The models used in the analysis did not allow for the impacts of price increases that might reduce demand or the development of new technologies in reaction to resource scarcity that provided technology stimulating price increases. An economic historian made an effective critique for this fixed resource/ growing population approach looking at past experiences with resource scarcity (Hueckel, 1975). What was shown from the historical experience was similar to the notion of induced technological innovation developed earlier for agricultural innovation (Hayami and Ruttan, 1985). Resource scarcity would bring about resource price increases leading to new technology for more frugal use of the resource or the development of a substitute for the resource. The critical question here is – will such induced technological innovation work with respect to our current energy resources and delivery systems? Then, will energy from agricultural biomass be one of the resource substitutes, a part of the new technology?

Integrating Agricultural Biomass

Initially, and for some time thereafter, agricultural biomass energy will have to integrate into current energy delivery and utilization systems and meet existing form, time and place criteria. It will also have to meet reliability and quality expectations and environmental standards. From experience so far, it also faces a relative cost hurdle that is not inconsequential. The policy debate centres on the extent of government involvement in encouraging such new energy sources. The provision of public funds for such an effort also raises the issue of the cost of such efforts as compared with other approaches that might be undertaken.

Liquid Fuels from Biomass

Existing form, time and place criteria have already strongly influenced the development of agricultural biomass. Ethanol and biodiesel are viable because they have been able to fit into the form, time and place requirements already in place. As an example, one of the initial concerns about ethanol was its water-absorbing characteristic which caused difficulties with gasohol transportation and storage. What occurred over time were adjustments in handling, blending and the purging of existing transportation and storage systems so that this is no longer a serious problem. System modifications and modified operational protocols have made this water-absorbing characteristic a non-issue through modification of existing delivery systems. Ethanol is available in a timely fashion. Its form, given the nature of the blend, is for practical purposes the same as the existing petroleum-based petrol. Additionally, it is distributed across the same geography as its counterpart petrol. Utilization technology has been adjusted so that initial problems with gasket failure or fuel system fouling from loosened sludge and varnish have been all but eliminated through equipment modifications. Reliability of product supply, performance and quality are no longer seriously questioned by users. One would expect that biodiesel can follow a similar path.

Biomass energy does have its own set of negative concerns. While ethanol is associated with improved air quality under some conditions, this does not necessarily hold across all conditions. The economics of ethanol require that the protein co-product be fed to livestock. However, the waste from animals fed such a distiller's grain diet is higher in phosphorus which is often the limiting environmental factor in the absorptive capacity of animal systems (Knowlton, 2003; Koelsch and Lesoing, 1999). As we increasingly have to control phosphates out of concern for water quality, increased ethanol production (where its co-products are fed to livestock) will present greater challenges until technology is developed to remove the phosphates. Similar concerns are expressed about harvesting biomass residues (e.g., maize stover) with respect to impacts on soil loss and nutrient removal. This can be managed. Biomass production and utilization is similar to most other resource system development in that it is not necessarily environmentally friendly 'green' – just by definition.

Integrating Biomass Based on National Needs and Energy Services Concepts

For something like ethanol fuel, the key question is what are the criteria for moving in the direction of not only integrating with existing systems (which it now does quite well) but also in the context of our overall energy needs. Additionally, we might consider the opportunity cost of the expansion of ethanol production. If we were to spend the same resources on something other than ethanol, would we be better off (and by what criteria – time, form, or place)? Because we have made the commitment to pursue ethanol, this may be an academic question. It is not an academic question for other biomass energy sources as yet less developed, such as biodiesel or burning biomass for electricity generation.

Ethanol development and utilization has been largely dependent upon government subsidies. There are two important political rationales for these. One is to increase farm income through increasing the price of ethanol feedstocks – today, primarily maize. The second rationale is to reduce our dependence on foreign energy sources, particularly petroleum. It is this latter rationale that will be considered here. To put this criteria into perspective, the USA annual petrol usage approached 140 billion gallons (gal) a year in 2002 (British Petroleum, 2003). Total gross crude and product imports into the US were 62% of our total consumption in 2003 (Energy Information Administration, 2004). How much do we accomplish if we convert a 10 billion bushel (bu) maize crop (a rough approximation of our annual production over the last decade) to ethanol. At a conversion ratio of 2.7 gal/bu, this would yield 27 billion gal of ethanol, roughly 19% of our petrol use. Conversion of the whole maize crop to ethanol would replace a bit more than 14% of our total oil imports and there would be no maize for our livestock industry or for maize exports, but there would be more distillers dried grains with solubles (DDGS) and other feed co-products.

Starting from where we are today, increasing ethanol production to 5 billion gal annually, as envisioned in the pending energy bill, will provide 3.5% of our current petrol needs. Current ethanol production of about 3 billion gal meets about 2.0% of our petrol needs at a subsidy cost of US$0.52/gal of ethanol produced. Thus far, what has been purchased with ethanol subsidies are a modest diversification of fuel supply and suppliers, as well as some potential reduction in liquid fuel price volatility. We could import ethanol from Brazil to increase the diversity of our imported fuel suppliers. Another potentially important consideration is whether ethanol will eventually be made economically from cellulose.

One alternative would be to increase the efficiency of our petrol use by 2.0% which would be the equivalent reduction in imports from the production of a like proportion of our petrol use in terms of ethanol. If fleet car and light truck mileage averages 22 miles per gallon (mpg), then the equivalent mpg saving to substitute conservation for ethanol production would be nearly 0.5 mpg (Heavenrich and Hellman, 2000). The Congressional Budget Office (2003) estimates that raising the Corporate Average Fuel Economy (CAFE) standards by 3.8 mpg, to 31.3 mpg for cars and 24.5 mpg for trucks, would result in an overall reduction in petrol consumption of 10%. The total cost of this would be between a US$3 billion and US$3.6 billion reduction of producers and consumers welfare.

In the discussion of ethanol production and its subsidies, the issue of the ethanol production energy balance has been a long-term running dispute. For over 30 years there has been controversy around the question of whether ethanol took more energy to produce than its caloric yield as a fuel. To some extent this has been an argument of accountancy, depending upon where the envelope was drawn or what specifications and production systems were adopted in counting energy for ethanol production. The initial analysis that raised the issue depended upon an example which was neither typical nor representative in many important respects (Pimentel, 1976). This study resulted in immediate and numerous counter accountancies. This argument continues to this day (Shapouri *et al.*, 2002; Pimentel, 2003).

If one takes a form, time and place energy systems perspective on this argument over ethanol's energy balance, it becomes a much less important dispute. The reason we would want to produce ethanol is because we have a scarcity of liquid fuel, on which much of our transportation system depends. If we can take energy that we do have in abundance that is not in liquid form and convert it to a liquid form we have improved our situation – even if it took more energy in solid or gas form than we ended up with in the liquid form. When the ethanol energy accounting dispute first surfaced, if we recognized coal as the primary energy embodied in machinery and equipment, assumed coal would power the distillation process and assumed abundant natural gas for the fertilizer for the maize, we then were successfully creating a scarce energy form (a liquid) from other solid and gas forms that we had in abundance.

If we are bound by caloric accounting, we get one answer. If we consider form, time and place requirements, especially form, we get a different one. Dollar accounting will often be somewhere else. The value judgment expressed here is that for subsidizing ethanol production as a policy decision – either dollar-based economic accounting that considers opportunity cost or form, or time and place accounting that considers national security and import dependency – make more sense than energy accounting.

What judgments might be drawn if we put ethanol into the broad form, time and place accounting emphasizing the broader energy system and national needs? On the basis of this approach one would consider the forms of fuel that go into ethanol production. Fuels that are scarce or involve import dependency would be a negative factor – irrespective of the caloric balance question. Two such fuels that are major inputs for ethanol production are petroleum and natural gas. The ethanol distillation process is fuelled primarily by natural gas and new capacity appears to be predominately natural gas. Natural gas is the feedstock for anhydrous ammonia fertilizer used to grow the maize input for ethanol and petroleum fuels the machinery

Table 4.1. Resource cost of transportation petroleum equivalents at 5% discount rate (US$/barrel), 1981.

Area of development	Cost
Shale oil	32.21-35.73
Coal liquids	40.69-46.08
Car fuel economy	25.37-38.23
Biomass	59.00-52.10[a]
Railroad electrification	22.82-24.88

[a] Maximum feasible development for biomass would involve decreased costs given the greater opportunity for new technology which would not hold for the others.
Source: Whitford et al. (1981).

that produces the maize. Strategically, does it make sense to use large amounts of natural gas and petroleum, which we import, to produce a liquid fuel, with the purpose of cutting liquid fuel imports?

What if we set liquid biofuels in an even broader context of transportation services and look at their production in an opportunity cost context. A 1981 study sponsored by the Department of Transportation compared the discounted cost per barrel of a range of transportation options (Whitford et al., 1981). The costs shown in Table 4.1 have certainly changed since then. Petroleum was about US$35/barrel (bbl) during the period. The price range reflects development of transportation petroleum equivalents either on a 'business as usual basis' (listed first) or 'maximum feasible development' (listed second). The range is presented in Table 4.1.

Certainly these numbers cannot be said to hold today. However, they do characterize the energy dilemma in a different way – how do we get more transportation bang for the buck, not just how do we pump more petroleum into transportation?

A version of this approach was taken by the American Council for an Energy-Efficient Economy (Geller, 2001). Here the comparison was made between the added supply of oil from the US Arctic National Wildlife Refuge (US Geological Survey estimate with oil at an inflation-adjusted US$34/bbl) as compared to increased fuel efficiency in cars and light trucks. The added supply from Alaska is estimated at 0.58 million barrels per day (MMbbl/d). The US Energy Information Administration (EIA)'s current estimate is a peak production of 1 to 1.35 MMbbl/d 25 years after initial production (Energy Information Administration, 2004). The fuel that might be saved by increasing a CAFE standard for cars and light trucks to 28.1 mpg would be 0.28 MMbbl/d. If the standard were increased to 35.2 mpg, the savings would be 1.47 MMbbl/d. The policy question then becomes the consequences of increasing the CAFE standard versus the development of the existing resource. While different estimates show different savings, what is being put forward here is comparing the costs and magnitudes of increases in production with potential reductions in demand. Supply enhancement and demand reduction are complements not substitutes. Economic sense indicates that policy should encourage both alternatives based on their relative cost effectiveness.

Integration of Solid Biofuels into Electricity Production

Conversion of agricultural biomass into electricity was widely promoted during the Clinton administration to reduce carbon accumulation in the atmosphere. There are a variety of ways this is being explored. These include utilizing methane or gasified biomass solids, burning biomass solids directly and co-firing with traditional (boiler) fuels such as coal. Co-firing wood waste, sawdust, switchgrass (*Panicum virgatum*), or harvested wood chips with coal has already been tried in a variety of configurations (Tillman, 2000). Biomass production of electricity has a number of positive attributes. The energy source is renewable, it is low in sulphur and it represents a cycling activity that does not add a great deal of net carbon dioxide to the atmosphere. In co-firing situations, there is a loss of boiler efficiency as increasing amounts of biomass materials are co-fired. There are a myriad of other technical issues that relate to fuel volatility, boiler design, fuel feeding systems, optimal proportions of biomass, ash volumes and contents, potash contents, high reactive alkali contents, etc. The Chariton Valley, Iowa, switchgrass co-fire testing is often referred to for technical information for this approach (Amos, 2002; Antares Group, 2002; Hintz et al., 2002).

For many, the key to biomass energy for electricity is biomass fuel availability. There is an extensive agronomic literature on various specific sources like switchgrass (Lewandowski et al., 2003, as an example). Any number of extensive studies have been made on biomass availability regionally and nationally – one of the earliest of these was conducted for the Congressional Office of Technology Assessment (Purdue University, 1979). These studies have generally assessed the energy content available, harvesting costs, production and harvesting technologies, environmental trade-offs and total costs of bio-energy materials. This has been accomplished looking at the possible trade-offs for agriculture with the use of existing crop and Conservation Reserve Program (CRP) land and changes in potential crop prices and subsidy payments (Walsh et al., 2000; Atchison and Hettenhaus, 2003; Gallagher et al., 2003; De La Torre Ugarte et al., 2003).

If one is considering utilizing large amounts of biomass to generate electricity, the issue is not one of biomass availability per se, but one of available electric plant and/or transmission capacity location with respect to biomass location. It is of little direct help if we want to deliver electricity to consumers from biomass just to know that there is 'x' amount of biomass out there in the country. A national projected supply curve for biomass at different price levels, or regional supply curves do not necessarily help that much when we need to integrate that with electricity production, transmission and distribution. In spite of that, there has been extensive data gathering and modelling undertaken to attempt to make such projections (for example, Haq, 2002).

If delivery of biomass-based electricity to consumers is to be accomplished, there is a need to work on a specific locational basis in terms of generating plant location, biomass supply location and in terms of transmission and delivery of electricity to users. One example of how to approach this is a study assessing co-firing with existing electricity plants in the Northeast USA (Antares Group and Parson's Power, 1996). This work included a screening process examining existing coal-fired plants and identifying those that might be suitable candidates for co-firing on the basis of a large number of criteria, including everything from biomass availability to emissions, necessary system modifications, potential concerns of plant neighbours, etc. There are other studies that have looked at new plant siting issues (Bluhm et al., 1995; Fruin, 1998; Rose and Husain, 1998) but the tendency to examine biomass availability without considering the additional parameters necessary for actual electric power generation and delivery continues.

Confounding Characteristics of Generation Today

There are characteristics of the current generating stock that will partially determine how much penetration there will be by biomass in electric generation. A first one is scale. There are real economies of scale for coal-fired generating plants. Different fuels also embody different capital requirements. Looking across the conventional generation spectrum, nuclear power entails the greatest capital cost and the lowest fuel cost. Safety issues aside, nuclear is

Table 4.2. Approximate capital costs for different power systems.

Type of system	(US$ cost per kW capacity)
Peaking gas turbine	400-500
Combined cycle gas turbine	650-800
Conventional coal	1,000-1,300
Coal integrated gasification – combined cycle	1,200-1,500
Nuclear	2,000-3,000

Source: Utility industry staff, Power Systems Engineering Research Center, Cornell University, Ithaca, New York, 2004, personal communication.

not feasible without extremely low capital costs – maybe even lower than the couple of percentage points that Commonwealth Edison paid its bond holders in the 1950s when it built its large nuclear capacity. At the other end of the spectrum is natural gas-fired generation. This has the lowest capital cost and usually the highest fuel cost. Coal is in between and generates most of the electricity used today. However, no baseload coal plant has been built in the last decade and few are in the works. Coal has traditionally been the baseload power source and will likely continue to be so. Until recently, natural gas turbines were largely the solution for peaking requirements. These relative capital costs are shown in Table 4.2.

A number of factors occurred that halted the expansion of baseload coal generation. One factor has been the increased costs of meeting clean air regulations. Another was a collapse in natural gas prices during a natural gas glut in the 1990s (following the shortages and high prices in the 1980s). This made natural gas systems all that more attractive. Related to this was the fact that gas supply companies began to offer low-priced long-term delivery contracts again – a practice that had been discontinued during the earlier 1980s gas shortages. Another important factor was the uncertainty of capital recovery for new capacity construction under deregulation. A firm wanting to build a new power plant, but facing regulatory uncertain capital cost recovery would minimize capital expenditure risk and opt for low capital cost natural gas systems given the greater regulatory certainty of recovering operating costs (fuel). Even if total electricity costs were higher with gas, recovery is what mattered for investors – so gas plants were built.

With the shortages of electrical capacity a few years ago, natural gas turbines were the wise investors' solution. General Electric built and sold record numbers of gas turbine generating systems. Production increased from 50 to 100 units annually to 500 to 600 units before the 2002 recession and high gas prices significantly reduced demand. The amount of natural gas used in electricity generation expanded quickly. Figure 4.1 shows recent EIA projections that indicated a tremendous expected increase in generation by natural gas given the low prices at that time.

As CO_2 has become a pollution issue, natural gas is even more desirable because it releases less carbon into the atmosphere per heat unit provided than coal. The generation issue, not enough installed capacity to meet electricity demands and the concurrent blackouts ultimately became a fuel issue. Natural gas, supposedly at the peaking end of the generation spectrum, also ended up being a baseline fuel. Natural gas supplies became extremely tight and in California natural gas shortages limited electricity production. Figure 4.2 shows recent EIA projections of natural gas use in electricity generation accounting for the new environment of natural gas price and availability. These tables also reflect a dampening in the overall demand projection from 5295 billion kilowatts for 2020 in EIA's 2002 projection to 4872 billion kilowatts in EIA's 2004 projection.

In 2004, we now have an excess of generating capacity. A large number of the gas peaking and larger more efficient combined cycle gas generators produced in the 2000 to 2003 period have been mothballed or are being used at less than capacity. Industry estimates indicate substantial excess capacity is waiting to come on line. No new capacity of any sort

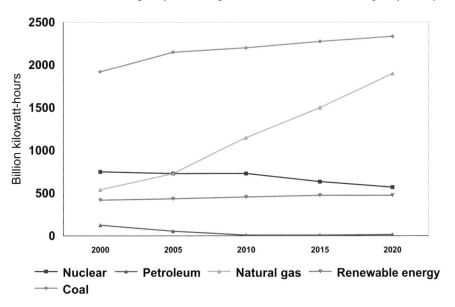

Source: Energy Information Administration (2001).

Figure 4.1. Projections of electricity generation by fuel, 2000-2020.

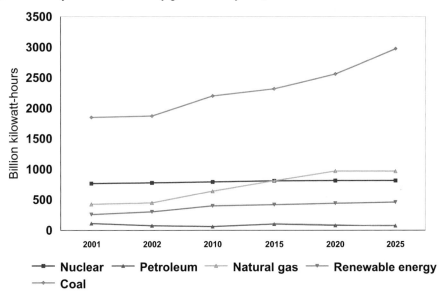

Source: Energy Information Administration (2003).

Figure 4.2. Projections of electricity generation by fuel, 2001-2025.

is likely to be installed until this excess capacity, which has already been paid for, is utilized or assigned as part of the reserve margin. It will take several years to absorb this excess capacity. Thus, currently there is no real economic need for new biomass-utilizing capacity nationally. However, some new baseload coal capacity will likely have to be built. Natural gas is now priced out of baseload.

At the same time we have a transmission system that is severely stressed. Starting a decade ago with deregulation, investment in transmission became extremely unattractive. In the deregulation environment, with the rules then in place, a traditional electric utility was driven to do everything possible to squeeze as much revenue as possible out of the existing capital stock. This meant running equipment and transmission as close to capacity as possible and testing reliability limits as never before. This is important to biomass electricity generation because the flexibility and robustness of the electric infrastructure is substantially reduced from where it was a decade or so ago. If one is planning to add modest sized 100 megawatt (mW) generators across the countryside, in out of the way places near to biomass, there is substantially less flexibility in the existing generation and transmission system to utilize this. That is why the approach of the Northeast study screening those existing plants that might co-fire 5% to 10% biomass appears to be by far the most realistic option today. Rules that have been under discussion by the US Federal Energy Regulatory Commission (FERC) are geared at keeping the current system afloat. These rules are not geared to go back to the level of robustness and redundancy of the 1970s or 1980s. Where the cardinal rule of real estate is location, location, location, the rule for the electric grid today is infrastructure, infrastructure and infrastructure. What those interested in biomass have been concentrating on is biomass supply, supply, supply. These two trains are passing in the night!

As a complex system, the electric grid and the necessary generation factors to keep it up and running has become more brittle. With this comes much longer lead times to make changes. When a system is pushed hard and one attempts to introduce changes, the system also may be more prone to instability or failure. The extent of the stress on the system is great enough now that many reconfiguration options are closed. At the same time, the cost and difficulty of siting new transmission lines is very real. One large new transmission line that is likely to be constructed has been over a decade in planning and siting. Costs incurred so far are somewhere around US$2 million a mile, before construction has even begun.

Contrast with Ethanol

The contrast with ethanol is stark. The reason the biomass community has been able to devote their energy to the production of ethanol is that the systems logistics for almost everything else beyond production was in place. Initial technical bottlenecks in blending, transportation and utilization have been overcome and there was enough flexibility in the existing delivery and utilization system to do so. Lead time to make these changes was not inconsequential in spite of the fact that the adjustments required were not as severe as those possible in the current vision of how biomass will fit into the electric grid.

What is being proposed for other biomass development reflects the ethanol experience – supply is critical and logistics, system concerns and utilization will be solved over time if we just force the product into the system. The *Interagency Strategic Plan of 2001* has as its technology goals: to reduce costs of technologies for supply, conversion, manufacturing and application systems for biomass; to accelerate commercial readiness and acceptance of biomass fuels, products and systems for fuels, heat, power, etc.; to assess the environmental and ecosystem impacts; and to foster innovation-driven science and quickly incorporate results (Biomass Research and Development Board, 2001, p. 3). This document reflects the desire to introduce products, not to fix the grid or greatly change an existing system so that something like electricity from biomass could access easily from points most economic for biomass electricity production. The encouragement for commercialization that is part of this interagency plan does not call for subsidies at any level, let alone at the level currently in place for ethanol.

The 2002 *Roadmap for Biomass Technologies in the United States* takes a similar product development focus (Biomass Research and Development Technical Advisory Committee, 2002). For product uses and distribution, the main concerns are bio-refinery development and demonstration projects. Public policy measures are incentives, education, resource supply, environmental measures, standards and codes and financial assistance. All of these have been important to the success of ethanol. The plea by this group for life cycle assessment to demonstrate the advantages of bio-based products would change the metric from the current Office of Management and Budget (OMB) benefit/cost guidelines.

Energy Services and Time, Form and Place

While we may view electrical energy from biomass in the context of comparison with other means of generating electricity, we do not usually view it more broadly in the context of opportunity costs and the best way to provide electrical services. These would be services such as 'comfort and convenience in buildings' which would include heating, cooling, light, etc. (Sant *et al.*, 1984, pp. 49-72). The same exercise can be undertaken with respect to energy used in industrial production (Sant *et al.*, 1984, pp. 73-92). Under this approach, one would try to improve our situation by doing the most cost-effective and energy-efficient job possible in the provision of energy services. This would be within the context of the different fuel alternatives and form, time and place requirements for process heat, motor power, etc.

Natural Gas

Marginal change tends to be the norm in politically stable democracies and in large complex technically based economies. Natural gas pricing and availability is going to have a lot to do with biomass development over the next decade. Being locked into a new scenario of lower availability and higher prices for this fuel may reduce or enhance the opportunities to do other things, like expand ethanol production and produce electricity from biomass.

In simple terms, for the moment, natural gas supplies have hit the wall relative to demand. 'North America is moving to a period in its history in which it will no longer be self-reliant in meeting its growing natural gas needs; production from traditional US and Canadian basins has plateaued' (National Petroleum Council, 2003, p. 7). Those associated with the natural gas industry point to several factors that have tipped usage over the edge of supply in North America. During the natural gas shortages of the 1980s there was a moratorium on new electric capacity that used natural gas. When the resultant bubble of natural gas occurred and prices declined, natural gas became the obvious fuel for new electric generation and industrial uses. After the OPEC oil embargo, industrial firms installed dual fuel systems to the extent possible. This meant that they used whichever was less expensive, oil or natural gas. As oil prices increase, this industrial dual capacity tends to bring natural gas prices along with oil as less oil and more gas is used. North American natural gas prices were previously somewhat insulated from oil prices. Natural gas also is the premium environmental fuel, leaving products of combustion that can even be vented through food. It yields the least amount of CO_2 of the basic fossil fuels. Many of the applications of natural gas relate to the particular form and characteristics of the fuel – as in the heat treating of metals or making of glass. It is also a premium chemical feedstock – still the best way to produce anhydrous ammonia. However, because of high domestic gas prices, anhydrous ammonia is no longer the major user. Imports are a growing source of anhydrous ammonia. The much touted hydrogen economy is likely to be fuelled initially by natural gas – the least expensive source for large amounts of hydrogen today.

Currently, natural gas is the dominant fuel used residentially and commercially, even though these account for only 21% and 14% of the total natural gas use. Figure 4.3 shows that some 34% of the natural gas is used in the industrial sector, accounting for about 40% of the energy use in that sector. Electric generation now consumes 24% of our natural gas and is the fastest growing use for natural gas. However, natural gas is still just over 15% of the fuels used to generate electricity.

The critical question is: what happens to natural gas as a fuel for electricity generation? Gas prices over the long term above US$4.50 per million Btu coupled with the less expensive gas combined cycle generation units would produce electricity above the electricity costs estimated to make possible heavily scrubbed coal systems with the attendant high capital costs and lower coal costs. This does not eliminate the CO_2 differences, but does eliminate many of the other pollutants as major concerns between gas and coal. Figure 4.2 illustrates this point.

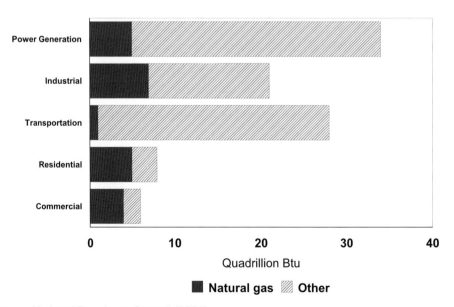

Source: National Petroleum Council (2003).

Figure 4.3. US primary energy use by sector, 2002.

Some key questions that flow from this new price and availability for natural gas include:

- How will this affect production costs for ethanol?
- Will gasification of biomass be a more attractive option for biomass utilization? Gasification would meet form requirements of natural gas and have a transportation and utilization system in place.
- Will natural gas receive different policy treatment because its increase in use will be based on imports? How will this affect biomass development?
- Will there be perceived changes in the reliability of natural gas that would allow other 'less reliable' energy sources to compete?

Some Summary Thoughts on Existing Energy Systems and Biomass Integration

To some extent, we can take this discussion and reassess the initial question about the prospects of biomass as an increased component of national energy use.

- 'To what extent can biomass sources and systems meet the form, time and place determinants of energy utility that consumers obtain from current sources and systems?'

The first observation is that existing energy sources may increasingly be stretched in their ability to meet these determinants. However, are time, form and place the right benchmarks? They are certainly important, but ultimately, over a long time period, we may also have to consider these in the context of alternative ways to provide energy services in a cost-effective way. If we just concentrate on form, time and place, we are asking only a part of the question.

- 'What are the compelling reasons for current energy sources and systems to predominate that tie them to consumers and may discourage their replacement by other systems and sources?'

There are a number of reasons: reliability, cost, ease of access, etc. that go beyond form, time and place. The lesson from this question is that it is a question that needs to be answered specifically for any energy source that is to replace another. Insofar as a biomass energy source can fit into current systems and maintain reliability at competitive cost, a substitution is seamless.

- 'How might biomass energy successfully integrate into current systems and what are some of the critical challenges in doing this?'

The response here is that integration into existing energy systems needs to be a primary consideration in biomass energy development. The challenges of form, time and place for utilization are equally important to the challenges of development. Ignoring them only compounds the integration difficulty. Policy makers have to be willing to commit resources to utilization and integration, not just to development of sources.

- 'What policy issues will have to be dealt with for biomass sources to integrate or successfully create replacements for existing energy systems?'

The issues of natural gas availability and price, gas pipeline capacity and utilization and the issues surrounding the robustness and flexibility of the national electric grid are examples of issues critical for the successful integration of biomass energy into the existing energy system.

Additional Policy Concerns

At some point, we might expect that energy costs, shortages, or strategic vulnerability will get to the point where even a stable democracy will have to change its mode of marginal adjustments and engage in major structural changes. Taking an approach like Roger Sant's or like the US Department of Transportation study of transportation alternatives (Whitford *et al.*, 1981) that shifts from a fuel to energy service focus is a direction that we may be forced into at some point. Options under such a view of energy services are very different. From a policy standpoint, at what moment does one actually set the development of the US Arctic National Wildlife Refuge against something like increased CAFE standards and potential biomass sources? Such a trade-off would lead to a very different set of options and consequences and biomass would either compete effectively or not.

References

Amos, W. (2002) *Summary of Chariton Valley Switchgrass Co-Fire Testing at the Ottumwa Generating Station in Chillicothe, Iowa*. Draft Version. National Renewable Energy Laboratory, Golden, Colorado.

Antares Group. (2002) *Chariton Valley Biomass Project; Environmental Strategies Plan*. Antares Group, Inc., Landover, Maryland.

Antares Group and Parson's Power. (1996) *Utility Coal-Biomass Co-firing Plant Opportunities and Conceptual Assessments*. US Department of Energy, Northeast Regional Biomass Program, Washington, DC.

Atchison, J. and Hettenhaus, J. (2003) *Innovative Methods for Corn Stover Collecting, Handling, Storing and Transporting*. US Department of Energy, National Renewable Energy Laboratory, Golden, Colorado.

Bluhm, G., Conway, R., Roningen, V. and Shapouri, H. (1995) The Economics of Biomass-Derived Energy and Fuels. *Proceedings of the Second Biomass Conference of the Americas*. Portland, Oregon.

British Petroleum. (2003) *BP Statistical Review of World Energy 2003*. BP p.l.c., London.

Biomass Research and Development Board. (2001) *Fostering the Bioeconomic Revolution in Biobased Products and Bioenergy*. National Renewable Energy Laboratory, Golden, Colorado.

Biomass Research and Development Technical Advisory Committee. (2002) *Roadmap for Biomass Technologies in the United States*. The committee was established by the *Biomass R&D Act of 2000* (P.L. 106-224). Biomass Research and Development Technical Advisory Committee, Washington, DC.

Congressional Budget Office. (2003) *The Economic Costs of Fuel Economy Standards Versus a Gasoline Tax*. Congressional Budget Office, Washington, DC.

De La Torre Ugarte, D., Walsh, M., Shapouri, H. and Slinsky, S. (2003) *The Economic Impacts of Bioenergy Crop Production on U.S. Agriculture*. US Department of Agriculture, Office of the Chief Economist, Office of Energy Policy and New Uses, Washington, DC.

Energy Information Administration. (2001) *Annual Energy Outlook 2002 with Projections to 2020.* DOE/EIA-0383(2002). US Department of Energy, Energy Information Administration, Washington, DC.

Energy Information Administration. (2003) *Annual Energy Outlook 2003 with Projections to 2025.* DOE/EIA-0383(2003). US Department of Energy, Energy Information Administration, Washington, DC.

Energy Information Administration. (2004) *United States of America, Country Analysis Brief.* US Department of Energy, Energy Information Administration, Washington, DC.

Fruin, J. (1998) *An Application of Geographic Information Systems to Alfalfa Bio-Mass Energy and Marketing Coops.* Presented paper, Sixth Joint Conference on Food Agriculture and the Environment, University of Minnesota, Center for International Food and Agricultural Policy, Minneapolis, Minnesota.

Gallagher, P., Dikeman, M., Fritz, J., Wailes, E., Gauthier, W. and Shapour, H. (2003) Supply and Social Cost Estimates for Biomass from Crop Residues in the United States. *Environmental and Resource Economics* 24: 335-358.

Geller, H. (2001) *Strategies for Reducing Oil Imports: Expanding Oil Production vs Increasing Vehicle Efficiency.* Report # E011. American Council for an Energy-Efficient Economy, Washington, DC.

Haq, Z. (2002) *Biomass for Electricity Generation.* US Department of Energy, Energy Information Administration, Washington, DC.

Hayami, Y. and Ruttan, V. (1985) *Agricultural Development: An International Perspective.* Johns Hopkins Press, Baltimore, Maryland.

Heavenrich, R. and Hellman, K. (2000) *Light-Duty Automotive Technology and Fuel Economy Trends 1975 Through 2000.* Report EPA420-S-00-003. US Environmental Protection Agency, Office of Transportation and Air Quality, Advanced Technology Division, Washington, DC.

Hintz, R., Moore, K. and Tarr, A. (2002) *Cropping Systems Research for Biomass Energy Production: A Final Report Prepared for the Chariton Valley Resource Conservation and Development, Inc.* Iowa State University, Department of Agronomy, Ames, Iowa.

Hueckel, G. (1975) A Historical Approach to Future Economic Growth. *Science* 187 (14): 925-931.

Knowlton, K. (2003) Opportunities to Reduce Phosphorous Losses from Dairy Operations. Presented paper, Meeting of the American Registry of Professional Animal Scientists, California Chapter, October 30-31, Coalinga, California.

Koelsch, R. and Lesoing, G. (1999) Nutrient Balance on Nebraska Livestock Confinement Systems. *Journal of Animal Sciences* 77 (Suppl.2): 63-71.

Lewandowski, I., Scurlock, J., Lindvall, E. and Christou, M. (2003) The Development and Current Status of Perennial Rhizomatous Grasses as Energy Crops in the US and Europe. *Biomass & Bioenergy* 25: 335-361.

Lovins, A. (1976) Energy Strategy: The Road Not Taken. *Foreign Affairs* 55: 65-96.

Meadows, D., Meadows, D., Rangers, J. and Behrens II, W. (1972). *The Limits to Growth; A Report for the Club of Rome's Project on the Predicament of Mankind.* Universe Books, New York, New York.

National Petroleum Council. (2003) *Balancing Natural Gas Policy – Fueling the Demands of a Growing Economy; Volume 1. Summary of Findings and Recommendations.* National Petroleum Council, Washington, DC.

Peck, A. and Doering, O. (1976) Voluntarism and Price Response: Consumer Reaction to the Energy Shortage. *The Bell Journal of Economics* 7 (1): 287-292.

Pimentel, D. (1976) The Energy Crisis: Its Impact on Agriculture. *Scienza & Tecnica 76.* Mondadori, Milan, Italy.

Pimentel, D. (2003) Ethanol Fuels: Energy Balance, Economics, and Environmental Impacts are Negative. *Natural Resources Research* 12(2): 127-134.

Purdue University. (1979) *The Potential for Producing Energy from Agriculture, A Report to the Office of Technology Assessment.* Purdue University, School of Agriculture, Lafayette, Indiana.

Rose, D. and Husain, S. (1998) *Biomass Electric Power Plants: Land Use Impacts for Forestry and Agriculture*. Presented paper, Sixth Joint Conference on Food Agriculture and the Environment, University of Minnesota, Center for International Food and Agricultural Policy, Minneapolis, Minnesota.

Sant, R. (1979) *The Least Cost Energy Strategy*. Carnegie-Mellon University Press, Pittsburgh, Pennsylvania.

Sant, R., Bakke, D. and Naill, R. (1984) *Creating Abundance: America's Least Cost Energy Strategy*. McGraw-Hill, New York, New York.

Shapouri, H., Duffield, J. and Wang, M. (2002) *The Energy Balance of Corn Ethanol: An Update*. Agricultural Economic Report 813. US Department of Agriculture, Office of the Chief Economist, Office of Energy Policy and New Uses, Washington, DC.

Tillman, D. (2000) Biomass Cofiring: The Technology, the Experience, the Combustion Consequences. *Biomass Energy* 19: 365-384.

Walsh, M., Perlack, R., Turhollow, A., De La Torre Ugarte, D., Becker, D., Slinsky, S. and Ray, D. (2000) *Biomass Feedstock Availability in the United States: 1999 State Level Analysis*. US Department of Energy, Oak Ridge National Laboratory, Oak Ridge, Tennessee.

Whitford, R., Fraser, J., Toft, G. and Tyner, W. (1981) *Transportation Energy Futures: Paths of Transition, Volume 1: Issues and Analysis*. Purdue University, Automotive Transportation Center, West Lafayette, Indiana.

Chapter 5

US Oil and Gas Markets: A Scenario for Future Strong Inter-fuel Competition

Kevin J. Lindemer

Introduction

Energy markets are going through a period of unprecedented volatility which has its roots as far back as the Asia financial crisis. This volatility has been characteristic of both world oil and North American natural gas markets, but is the product of a different set of forces for the two markets. The price of oil has been driven by Organization of the Petroleum Exporting Countries (OPEC) discipline, rising demand and fears of supply disruption. Natural gas prices have been driven by rising demand in North America and stable to declining regional production levels.

Despite the greater degree of volatility, both the world oil and the North American gas markets appear set to remain at higher price levels for the foreseeable future. While oil prices are likely to recede from the recent very high levels experienced in 2004, North American natural gas prices are likely to remain much higher than the prices experienced prior to 2000 and higher relative to oil prices. The emerging price dynamic between oil and gas in North America will reshape the demand for oil and gas and potentially have an important impact on the future energy market in the USA.

The Volatile Crude Oil Market

The crude oil market has been very volatile over the past few years. Real (inflation-adjusted) prices were the lowest since the early 1970s in 1999 and reached the highest price since the late 1970s early in 2004 (Fig. 5.1). Several factors caused this volatility:

- OPEC overproduction in late 1997 and early 1998 as oil demand in Asia decreased and drove prices lower (see Grier, 2004).
- Greater OPEC production discipline since 1999 (see Grier, 2004).
- Oil production growth in non-OPEC countries slowed in 1999 after the price collapse in 1998.
- Since 2002, the rate of global oil demand growth has recovered and is gaining at a relatively robust pace, particularly driven by the recent strength in Asia (see Grier, 2004).
- Rising anxieties in the markets stemming from political uncertainties in some major producing areas and the threat of terrorism (see Blustein, 2004; Yetiv, 2004).

It is clear that these factors will continue to influence prices in the years to come as they have in the past few years and continue to cause oil price volatility with prices ranging from US$25/barrel to US$40/barrel. While higher than the average price level during most of the 1990s, this price level is still below the levels of much of 2004.

A Shift in the North American Natural Gas Market

Up until the late 1990s, natural gas production capacity in North America exceeded demand. As a result, the industry experienced several periods of very low natural gas prices with production occasionally being shut down to balance the market, particularly during low demand years, such as warm winters when there were large amounts of gas left in storage after the peak season. However, production has levelled off and demand growth led by the rise in natural gas demand for electric power generation absorbed essentially all of the spare capacity by the late 1990s. Historically, seasonal production fell short of the peak seasonal demand requirement, with storage additions during low demand periods augmenting supply to meet peak demand. Recently, total annual North American production has been unable to meet total annual demand – a situation that will continue well into the future. The growing imbalance will be met with liquefied natural gas (LNG) imports and demand destruction. With the end of spare annual average capacity, North American gas prices have been fundamentally shifted to a much higher price level than experienced for much of the past 25 years (Fig. 5.2). Over the next several years, natural gas prices in the USA are expected to average in the US$5.00/million British thermal units (Btu) to US$7.00/million Btu range at the Henry Hub in southern Louisiana, the principal pricing point for US natural gas markets. Other expectations include:

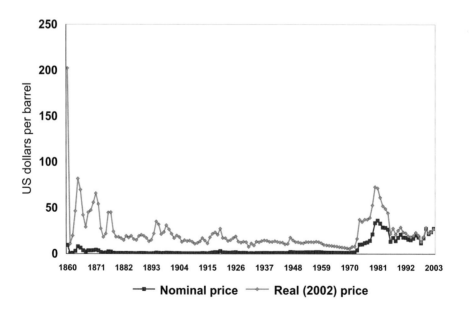

Source: Irving Oil Ltd. (2004).

Figure 5.1. World crude oil prices, 1860-2003.

- Natural gas demand in North America will continue to grow, but growth may be slow as a result of higher prices.
- Incremental natural gas supplies will have high cost with long lead times and strong political challenges.
- LNG faces the issue of siting of import facilities (see Parfomak and Flynn, 2004).
- Arctic gas requires new and long pipelines to bring gas to market (see Brooke, 2000).
- Mature domestic natural gas production has levelled off and is not expected to keep pace with demand growth (see Francis, 2004).

Historically, gas was inexpensive in North America because the potential to produce was greater than demand. In 1999, North America began consuming more gas than could be produced, a trend that is expected to continue. As a result, prices have risen to reflect the alternate supply cost and the price level at which demand switches to an alternative source of energy or is shut down.

What has happened in the natural gas market in North America is similar to the oil market of the early 1970s. Thirty-five years ago, the US was a net exporter of oil. As demand continued to grow, domestic production did not keep pace. This set the stage for oil prices in the USA to be set by imported crude oil rather than domestically produced crude oil and the first real price increase since the end of World War II. Since then, with the advent of OPEC controlling marginal world oil supply and continued robust world demand growth, oil prices have been sustained at a higher price level than that of the 1930s through the 1960s.

In nominal terms, North American oil and gas prices are at historic high levels. In real dollars, oil prices are still well below the levels of the late 1970s and 1980. Gas prices are at record high levels in real dollar terms. The markets and industry and consumer behaviour at present were shaped by the historic relationship between oil and gas prices. As relative prices

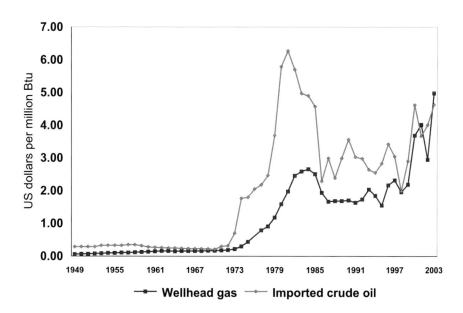

Source: Energy Information Administration (2004a,b).

Figure 5.2. US hydrocarbon prices, 1949-2003.

between oil and gas change, it should be expected that the markets and industry and consumer and behaviour would also change in response to the new price relationship.

Sustained Conversion of Oil and Gas Prices

The premise of this chapter is based on a scenario of a sustained conversion of oil and gas prices. Shaping the market environment of the next several years will be the following key factors and assumptions:

- Today's higher oil and gas prices are the result of factors other than the global resource base limits. At play are strong demand and limited regional supply growth, global politics, regulation, policies, and psychology.
- In addition, the resource base remains abundant globally – resource scarcity is not on the horizon.
- Oil and gas demand will continue to grow in the USA and worldwide. The rate of growth will be more affected by economic growth than by prices.
- North America will become increasingly dependent on imported oil and natural gas.

There is no inherent reason why gas should be priced cheaper than oil – from a quality perspective it should be priced at a premium. Gas burns cleaner, costs less to transform from its raw material state to a saleable product and burns more efficiently in many applications today.

The View Forward – The Effect of the New Oil/Gas Price Relationship

Given the possibility that natural gas prices could be above oil prices more frequently and for longer periods of time, the markets are potentially creating a growth opportunity for oil. If gas prices remain above oil, oil demand will grow faster in sectors where it has been stagnant or declining for much of the last couple of decades.

Historically, oil and gas have been in competition with each other at all levels of the demand chain or more precisely, natural gas has displaced oil in most bulk energy uses due to its lower cost (Fig. 5.3). In general, oil has lost ground to natural gas in all sectors except transportation over the last 25 to 30 years: residential lost about 10% share, commercial lost about 13%, industrial lost about 3% and power generation lost about 14%. While oil lost share to natural gas in stationary markets in the USA, oil's share of total energy demand has stayed relatively constant at about 27%. Growth in transportation demand has been offset by the decline in other sectors.

The focus of the downstream oil industry has been to make transport fuels because there have been no viable substitutes for oil and are no discernible ones on the horizon. While the refining industry has made huge investments to continue to meet environmental regulations associated with fuel quality and emissions, it has not invested in improving the means to supply the stationary fuels markets as a result of historically lower-cost natural gas and the higher value of transportation fuels. The petroleum fuels supplied to stationary markets can generally be characterized as either transport fuels used in stationary applications or as by-product fuels such as residual fuel oil and heating oil.

With the oil/gas price inversion a possible frequent or sustained feature of the North American markets in the future, the question is: Will the downstream oil industry begin to turn to the stationary markets as a growth business? And will customers begin to look to oil as part of their long-term fuel strategy? Competitive markets and rational economic behaviour suggest that the answer to these questions is 'yes'.

Regaining a measurable share of the stationary market may not be as easy for the oil industry as it was to lose it. Recent high gas prices have caused the North America gas market to begin reflecting the cost of imported replacement gas and the cost of shutting down consumption. The new relative price level will allow oil to compete with natural gas rather than retreat from it. Thus far, the existing economic fuel switching has been brought on line. However, there are limits to the amount of natural gas demand that can be switched to oil without further investment on the part of consumers or the oil industry. Investments will be needed in infrastructure, refining, emissions technologies, new product development and market-based solutions.

Changes in Energy Industry Structure Ahead

As the energy industry responds to the new pricing dynamic, the North American energy industry should be expected to change and adapt to the new environment. In the past decade, market forces have transformed the energy industry:

- Super-majors (large oil companies) have formed from the merger of former competitors.
- International power companies are emerging from de-regulation in Europe and the USA.

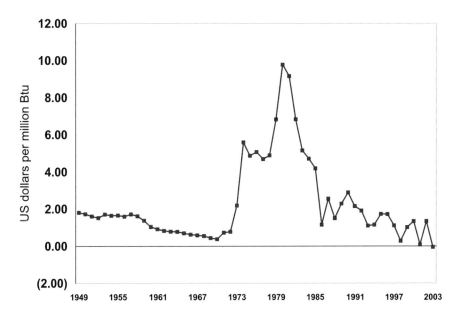

Source: Energy Information Administration (2004a,b).

Figure 5.3. US imported crude oil minus wellhead gas prices, 1949-2003.

- US refining has been transformed by the emergence of large independent refining companies.
- Master limited partnerships, a form of business organization where limited partner units are traded on public exchanges, dominate US midstream industry.

The current structure of the energy industry represents a transformation from what existed a decade ago and is the product of the market environment of the last 20 years. It is likely that what is now underway in the market will have similar or greater implications for the future structure of the industry than past forces.

While it is unclear what the structure of the industry will be a decade from now, there are some things that can be predicted based on historical experience. The need to remain competitive will continue to drive the industry and, thus, an overall reduction in costs is likely. Companies will continue to drive for economies of scale, the number of players and facilities in the USA will continue to contract and those that remain will be larger and more efficient. However, historically, cost reductions have been passed through to the consumer very quickly. Figure 5.4 shows how in the US downstream oil industry reductions in operating costs have led to a parallel reduction in gross margins. This has resulted in relatively constant net margins in real dollars.

While traditional players will likely be driven to rationalize, the emerging market dynamics around the oil and gas price interface and the likely creation of new technology will cause the entrance of new market players. These new entrants increasingly centre their operations on providing energy and related services to broaden their customer base and leverage expertize and infrastructure.

The changing oil and gas price dynamic in the USA over the next several years should be expected to result in adjustments in consumer demand for energy goods and services. As in the past, the industry will respond by delivering these products and services at low cost.

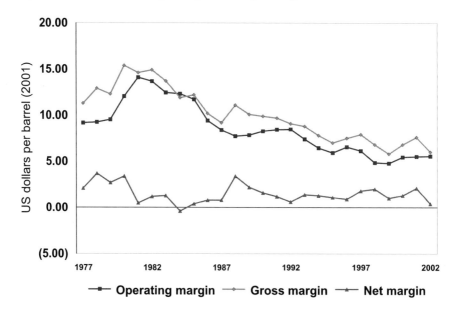

Source: Energy Information Administration (2004c).

Figure 5.4. US refining and marketing margins, 1977-2002.

Global factors will likely result in continued volatility in the prices US consumers pay for oil and gas. However, the emerging trend of relative prices for oil and gas, whereby oil and gas are more competitive on an energy equivalent basis will cause consumers to demand – and receive – new services and sources of energy from the industry.

Implications for the Agricultural Industry

The upward shift in natural gas prices has impacted the agricultural industry through higher fuel, electricity and fertilizer costs. In addition, domestic fertilizer and chemical industries that are natural gas intensive are finding it very difficult if not impossible to compete with supplies from countries that have abundant natural gas resources with little to no local market.

In the scenario outlined in this chapter, the costs of natural gas price-based agricultural inputs, such as ammonia and electricity, should not be expected to decline to the levels of the late 1990s. The net effect on agricultural producers will be to adjust for these higher costs by either passing them on to consumers, reducing demand for these products, or reducing the unit costs of other inputs. Since agriculture, like the oil and natural gas businesses is a commodity business with limited ability to pass on higher costs to the end user, it is likely producers will either reduce their demand or reduce other input costs by improving efficiency. In either case, producers need to lower cost to remain competitive on the global market.

References

Blustein, P. (2004) Oil Prices Reach New Peak As Terrorism Anxieties Jump. *The Washington Post* (June 2): p. A10. [Accessed 2004.] Available from http http://www.washingtonpost.com/wp-dyn/articles/A7943-2004Jun1.html

Brooke, J. (2000) *A Big Push is on for Natural Gas under the Arctic*. The New York Times (September 28): p. A1. [Accessed 2004.] Article text available from http://seattlepi.nwsource.com/national/pipe281.shtml

Energy Information Administration. (2004a) *Annual Energy Review*. US Department of Energy, Energy Information Administration, Washington, DC. [Accessed 2004.] Available from http://www.eia.doe.gov/emeu/aer/contents.html

Energy Information Administration. (2004b) *Monthly Energy Review*. US Department of Energy, Energy Information Administration, Washington, DC. [Accessed 2004.] Available from http://www.eia.doe.gov/emeu/mer/contents.html

Energy Information Administration. (2004c) *Financial Reporting System Public Data*. US Department of Energy, Energy Information Administration, Washington, DC. [Accessed 2004.] Available from http://www.eia.doe.gov/emeu/finance/frsdata.html

Francis, D. (2004) The Escaping Price of Natural Gas. *The Christian Science Monitor* (February 19): p. 11. [Accessed 2004]. Available from http://www.csmonitor.com/2004/0219/p11s02-usec.html

Grier, P. (2004) Why OPEC is Straining to Reassert its Authority. *The Christian Science Monitor* (April 5): p. 2. [Accessed 2004]. Available from http://www.csmonitor.com/2004/0405/p02s01-usfp.html

Irving Oil Ltd. (2004) Internal oil industry statistics database. Data derived from various industry sources including: British Petroleum, *Statistical Review of World Energy*, various, British Petroleum, London, UK; Kemp, P. (ed.) *Petroleum Intelligence Weekly*, various, Energy Intelligence, New York, NY and Energy Information Administration, various, United States Department of Energy, Energy Information Administration, Washington, DC. Irving Oil Ltd., Saint John, New Brunswick, Canada.

Parfomak, P. and Flynn, A. (2004) *Liquefied Natural Gas (LNG) Import Terminals: Siting, Safety and Regulation.* Congressional Research Service Report for Congress RL32205. The Library of Congress, Congressional Research Service, Washington, DC. [Accessed 2004.] Available from http://www.fas.org/spp/civil/crs/RL32205.pdf

Yetiv, S. (2004) Oil Prices Caught in a Global Storm of Angst. *The International Herald Tribune* (August 14): p. 5. [Accessed 2004.] Available from http://www.iht.com/articles/533854.html

Chapter 6

Dry-mill Ethanol Plant Economics and Sensitivity

Douglas G. Tiffany and Vernon R. Eidman

Background

Fuel ethanol is most commonly produced in the USA from maize utilizing two different production processes. Wet-milling involves using the starch of the kernel. Dry-milling (or dry-grind) utilizes the entire kernel. Ethanol plants using dry-milling technology now dominate in satisfying US fuel ethanol demand.

Dry-mill ethanol processing is a commodity-based, processing business that takes maize, which has a certain price volatility, and natural gas, which has its own level of volatility, to produce ethanol, which has a niche in the large market for transportation fuels. Petrol prices in the USA have been low in real terms with some volatility in the last decade. Relatively small movements in maize, ethanol and natural gas prices can result in dramatic shifts in net returns. Gross margins set the economic stage for the ethanol production industry; however, the disciplines of engineering, biochemistry and business management must act in concert in plants of proper scale to execute business plans for success. Plant managers must control important cost categories such as energy for heating, electricity, interest, depreciation, enzymes, repairs, maintenance and other items.

Success in dry-mill ethanol processing requires selecting plant locations with good access to cheap utilities and with good access to transportation and natural gas pipelines. Dry-mill ethanol plants typically locate in areas with historically low maize prices and plentiful maize supplies. Many plants have recently been built in areas on the edge of the traditional Corn Belt and construction of others is underway. Several years of poor maize production, resulting in expensive maize supplies, could severely test the staying power of ethanol plant investors and their bankers. On the other hand, dry-mill plants in non-traditional areas may succeed by emphasizing marketing of wet distillers grains to nearby livestock producers. Dry-mill ethanol plants operate in an industry dominated by large firms. Archer Daniels Midland (ADM), Cargill and Williams Bio-Energy represent over 46% of total national capacity in ethanol production (Renewable Fuels Association, 2004).

Access to funding for expansion of existing firms or for start-ups has become difficult because many major lenders are reaching their internal capacity to finance activity in this industry and are requiring higher levels of equity. By advancing borrowers to 50% equity or greater, it will be possible for lenders to originate ethanol-processing loans and then sell levels of participation to other lenders (J. Thompson, Agribank, St. Paul, Minnesota, 2003, personal communication). Returns in ethanol processing can be very volatile. Staying power and risk management strategies are critical for survival of ethanol plants, particularly in times of high prices for maize and natural gas.

The historic volatility of ethanol and maize prices make this business quite risky unless firms take action to manage marketing risk of ethanol, which dominates the producer's revenue stream, and maize, which dominates expenses. Figure 6.1 shows US average ethanol and maize prices and their respective price volatilities over the period 1994 to 2003.

An important co-product of dry-mill ethanol production is distillers dried grains with solubles (DDGS). DDGS is an economical, partial replacement for maize, soybean meal and dicalcium phosphate in livestock and poultry feeds. Expansion of South American soybean production and ample quantities of cheap soybean meal may temper the opportunity to successfully market DDGS. Some observers expect DDGS to approach the price of maize on a per ton basis. To improve the market for DDGS, ethanol plants have funded research to improve its nutritional value for livestock, with poultry and swine receiving emphasis. Efforts are underway to standardize the quantity of solubles used and the methods of drying the DDGS to retain adequate amino acid levels. In addition, the higher availability of phosphorus in DDGS must be recognized by swine feeders to avoid phosphorus overfeeding, which may contribute to phosphorus pollution. Research is being considered to evaluate the effect of yeast cell wall constituents extracted from DDGS to reduce *Salmonella* levels endemic in poultry.

In an attempt to address the numerous factors that influence the operational and financial success of dry-mill fuel ethanol plants, a spreadsheet was developed that can be used by potential investors in ethanol production, those planning new operations, or managers evaluating the financial impacts of improving an existing plant's performance (Tiffany and Eidman, 2004).

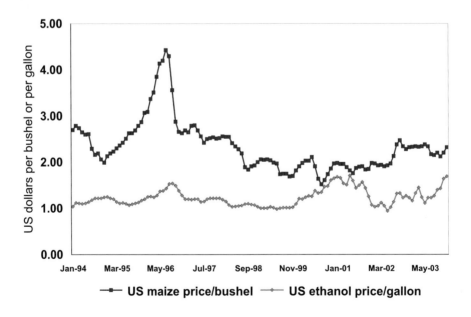

Source: Hart Energy Publishing (2004); National Agricultural Statistics Service (2004).

Figure 6.1. US maize and ethanol prices per bushel and per gallon, 1994-2003.

Project Scope and Selection of Analytical Tool

This study focuses on dry-mill ethanol production because 1) the relatively small plants, most likely to benefit from this type of analysis, tend to use this technology; 2) substantial future expansion is expected in dry-mill capacity; and 3) many dry-mill plants are owned by cooperatives or by limited liability companies (LLCs) comprised of cooperatives, which are willing to share financial and technical data. The spreadsheet developed seeks to accurately represent the 'state of the art' in dry-mill ethanol plant operations and profitability. To this end, ethanol plant managers and bankers with financing experience in ethanol operations were interviewed. A substantial body of literature on this topic was also reviewed. The next section reviews the design features of the spreadsheet and the rationale for that design.

Review of Spreadsheet Features and Underlying Assumptions

Plant Capacity and Debt-Equity Assumptions

Figure 6.2 portrays the scenario of a 40 million gallon (gal) capacity plant that has an investment cost of US$1.50/gal of nameplate capacity (the initial production capacity guaranteed by the facility manufacturer). Subsequent plant improvements raise production 20% above nameplate capacity to 48 million gal of denatured ethanol/year. The scenario assumes 40% initial equity and 7% interest charged on debt. A target rate of return of 12% on equity is based on the rate of return on farm assets recorded by the top 20% of 200 farmers in a farm business analysis group (Southwest Minnesota Farm Business Management Association, 2003). The assumption of a higher expected return on equity is justified because investors will likely face greater problems recovering their funds than will bankers in the event of a business failure.

The spreadsheet offers flexibility in establishing operating conditions of dry-mill plants (see *Column C* of Fig. 6.2), such as nameplate capacity, investment per nameplate gallon, factor of nameplate capacity utilized and factors for equity, debt, interest rate charged and rate of return required by investors. Conversion efficiency assumptions are established in the spreadsheet in *Column C* along with figures representing the amount of anhydrous ethanol extracted per bushel. Pounds of DDGS per bushel and pounds of carbon dioxide (CO_2) extracted per bushel are also listed in *Column C*.

Establishment of Gross Margin

Revenues per bushel are determined by the spreadsheet after entering the revenue sources, including the price for denatured ethanol, the price/ton of DDGS, the price/ton of liquid CO_2 and state and/or federal subsidies paid per gallon of denatured ethanol. The sum of revenue from sales of products and subsidies, on a per bushel basis, minus the price of maize equals the gross margin per bushel. The conversion efficiency default assumptions in Fig. 6.2 are set at 2.75 gal of anhydrous ethanol/bushel (bu), 18 pounds of DDGS/bu and 18 pounds of liquid CO_2 collected per bushel of maize ground and processed. Prices used to establish the gross margin include ethanol at US$1.15/denatured gal, a plant gate price, and US$80.00/ton for DDGS, which is a slightly higher price than the per ton price of maize (US$78.57) when priced at US$2.20/bu. Carbon dioxide is sold as a liquid at US$6.00/ton. Few plants

Ethanol Dry-mill Spreadsheet

Row	B	C	D	E	F	G	H	I
3	7/23/2003 20:30	Cost/Denat. gal Ethanol	Ranges for Column C				Plant Totals	
4	Nameplate Ethanol Prod. (denat. gal)	40,000,000						
5	Investment per Nameplate Gallon	$1.5000	$1.00- $2.00				Plant Cost	$60,000,000
6	Factor of Nameplate Capacity	1.2000	80%- 150%					
7	Debt-Equity Assumptions							
8	Factor of Equity	0.40						
9	Factor of Debt	0.60					Initial Debt	$36,000,000
10	Interest Rate Charged on Debt	0.07						
11	Rate of Return Reqd. by Investors on Equity	0.12						
13	Conversion Efficiency Assumptions				Annual Production			
14	Anhydrous Ethanol Extracted (gal per bu)	2.750	2.5-2.85 gal/bu		Bushels Ground		Denat. Gallons	
15	DDGS per Bushel (lb per bu)	18	15-22 lb/bu		16,581,843		48,000,000	
16	CO_2 extracted per Bushel (lb per bu)	18	15-22 lb/bu					

Row	B	Price per Unit			Revenue/bu Ground		Revenue/Gal. Denatured Sold	Plant Totals
18	Establishment of Gross Margin							
19	Ethanol Price (denatured price) $/gal	$1.15	$.80 to $1.60		$3.3289		$1.1500	$55,200,000
20	DDGS Price $/ton	$80.00	$60-$120		$0.7200		$0.2487	$11,938,927
21	CO_2 Price ($ per ton liq. CO_2)	$6.00	$2- $12 / liq.ton		$0.0540		$0.0187	$ 895,420
22	MN Prod. Subsidy/gal Denat. Ethanol	$0.00			$0.0000		$0.0000	$ -
23	Federal Small Producer Subsidy							$ -
24	CCC Bioenergy Credit							$ -
25	Revenue per Unit				$4.1029		$1.4174	$68,034,347
26	Maize Price Paid by Processor ($ per bu)	$2.20	$1.70-$3.25		$2.2000		$0.7600	$36,480,055
27	Gross Margin				$1.9029		$0.6574	$31,554,292

Row	B	Price per Unit			Cost /Bushel Ground		Cost /gal Denatured Sold	Plant Totals
29	Operating Expenses Per Bushel							
30	Natural Gas Price ($/1,000,000 Btu)	$4.50	$1.50-$9.00/Dtherm					
31	LP (Propane) Price ($ per gallon)	$0.70	$.55-$.72 / gal					
32	Factor of Time Operating on Propane	0.02	0-.12					
33	BTU's of Heat fr Fuel Req./ Denat. gal	35,000	28,500-55,000					
34	Combined Heating Cost				$0.4623		$0.1597	$ 7,665,569
35	Electricity Price ($ per kWh)	$0.05	$.025-$.090/kwh					
36	Kilowatt Hours Required per Denat. gal	1.090	.85-1.2 kWh/denat. gal					
37	Electrical Cost				$0.1578		$0.0545	$ 2,616,000
38	Total BTU's of Fuel and Electricity	45,900						
39	Total Energy Cost				$0.6200		$0.2142	$10,281,569
40		Cost/Denat. gal Ethanol						
41	Enzymes	$0.0480			$0.1389		$0.0480	$ 2,304,000
42	Yeasts	$0.0220			$0.0637		$0.0220	$ 1,056,000
43	Other Proc.Chemicals & Antibiotics	$0.0200			$0.0579		$0.0200	$ 960,000
44	Boiler & Cooling Tower Chemicals	$0.0050			$0.0145		$0.0050	$ 240,000
45	Water	$0.0060	$.005-.010		$0.0174		$0.0060	$ 288,000
46	Denaturant Price per gal	$0.7000			$0.1013		$0.0350	$ 1,679,952
47	Total Chemical Cost				$0.3937		$0.1360	$ 6,527,952
49	Depreciation based on C49 asset life	15 Years			$0.2412		$0.0833	$ 4,000,000
50	Maintenance & Repairs	$0.0125			$0.0362		$0.0125	$ 600,000
51	Interest Expense				$0.1520		$0.0525	$ 2,520,000
52	Labour	$0.0450	$.04-$.06		$0.1303		$0.0450	$ 2,160,000
53	Management & Quality Control	$0.0136	$.010-$.022		$0.0394		$0.0136	$ 652,800
54	Real Estate Taxes	$0.0020			$0.0058		$0.0020	$ 96,000
55	Licences, Fees & Insurance	$0.0040	.0030-.0050		$0.0116		$0.0040	$ 192,000
56	Miscellaneous Expenses	$0.0135	$.01-$.03		$0.0391		$0.0135	$ 648,000
57	Total of Other Processing Costs				$0.6555		$0.2264	$10,868,800
58	Total Processing Costs				$1.6692		$0.5766	$27,678,321
59	Net Margin Achieved Per Unit				$0.2337		$0.0807	$ 3,875,971
60	Farmer-Investor Reqd. Return on Equity	12.00%			$0.1737		$0.0600	$ 2,880,000
61	Increment of Success/Failure to Meet Required Return				$0.0601		$0.0207	$ 995,971
63	Ethanol Plant Profits for Shareholders and Principal Reduction				$3,875,971		$3,875,971	$ 3,875,971

Figure 6.2. Ethanol dry-mill computer spreadsheet.

with less than 40 million gal capacity capture and refine CO_2 due to the high fixed costs of equipment. Generally, CO_2 sales are based on 'net-back' arrangements, in which an outside firm buys the CO_2 and assumes the investment, maintenance and operating expenses for the equipment to clean and liquefy the gas. No state or federal subsidies are assumed in the baseline scenario.

Operating Expense Assumptions

After the dominant expense of maize, the key operating expense in dry-mill plants is natural gas, which is shown priced at US$4.50/decatherm or 1,000,000 British thermal units (Btu). The plant portrayed in Fig. 6.2 requires 35,000 Btu/gal of denatured ethanol produced and sold, similar to the 36,000 Btu reported by the US Department of Agriculture (USDA) for 2001 production (Shapouri et al., 2002). Propane is assumed to be the fuel source 2% of the time. Electricity is priced at US$0.050/kilowatt-hour (kWh) and the assumed use is 1.09 kWh/gal of denatured ethanol produced. Yeast costs are 31% of enzyme costs and are similar to costs for other processing chemicals and antibiotics on a per-bushel-processed basis. The scenario assumes no propagation of yeast on-site, a more costly approach, but one that may reduce opportunities for bacterial infection of the yeast. Denaturant price is US$0.70/gal wholesale and without tax. Approximately US$0.10 worth of denaturant is required per bushel of maize processed. The plant is depreciated over 15 years on a straight-line basis, resulting in US$0.08 of depreciation per denatured gallon produced, or US$0.24/bu processed. Interest expense is US$0.152/bu, which is very similar to the cost for production labour, which is US$0.1303/bu processed. The expenses for management and quality control are approximately one-third of those for production labour. Other expenses are minor.

Results Shown on Spreadsheet and Baseline Conditions

The spreadsheet (Fig. 6.2) displays results in three forms: 1) per bushel of maize ground; 2) per gallon of denatured ethanol; and 3) plant totals in *Columns F, H* and *I*, respectively. Farmer-investors generally prefer the per bushel analysis because their investment in ethanol plants and their value-added dividends are based on shares of stock denominated in bushels of maize to be delivered. In contrast, plant managers generally prefer the per denatured gallon analysis to compare plant efficiency, costs and revenues.

Listed under *Annual Production* is the number of bushels of maize ground to produce the total gallons of denatured ethanol produced (48 million gal for baseline conditions). The operating costs are divided into categories for energy, chemical, depreciation, maintenance, interest, labour, management and other expenses. The primary energy sources are natural gas and electricity. Most plants purchase an interruptible supply of natural gas as the primary fuel for heating and substitute propane during any interruptions in natural gas supply.

In Fig. 6.2 the net profit is US$0.2337/bu of maize processed, or US$3,875,971/year for the plant. If a 12% rate of return on equity were paid to investors, net profits would be reduced to US$0.0601/bu of maize processed. Net profits are US$0.0807/denatured gal produced (including revenue and expenses from DDGS and CO_2) or just US$0.0207/denatured gal when requiring a 12.0% rate of return on equity.

Revenue and expenses from the baseline scenario are shown in Fig. 6.3. The difference in total revenue and total expense appears small for this capital-intensive, high-volume business. In the revenue category, ethanol sales dominate; however, DDGS sales are significant, especially when net margins are so small. In the expenses category, maize dominates, followed by natural gas and depreciation. Electricity, interest expense, enzymes and labour each cost more than US$2 million/year. Denaturant and yeast expenses are each between US$1 million/year and US$2 million/year. Antibiotics and other chemicals approach US$1 million in cost. The summed expenses for water, taxes, licences, fees, insurance, miscellaneous, management and quality and repair total US$2.5 million, nearly equal to the amount of interest charged on the debt.

Key Factors for Profitability of Dry-mill Plants

Five factors (shown below with their baseline levels) dramatically affect profitability of dry-mill ethanol plants:

Maize price	US$2.20/bu
Ethanol price	US$1.15/denatured gal
Natural gas price	US$4.50/decatherm
Ethanol yield	2.75 gal anhydrous ethanol/bu
Capacity factor	Nameplate capacity multiplied by 1.20

The discussion and graphs that follow demonstrate how each of the factors can influence the profitability of a dry-mill ethanol plant under baseline conditions. The analysis was completed by setting baseline conditions and then varying one of the five key factors at a time, holding all other factors constant.

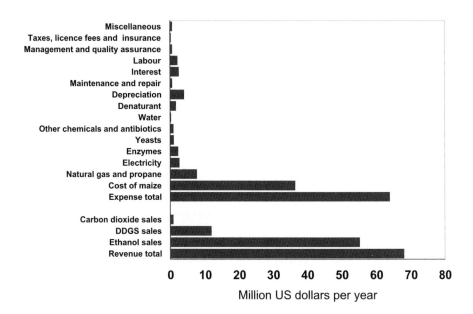

Figure 6.3. Revenue and expenses for a 40 million gallon dry-mill plant under baseline conditions.

Maize Price

Figure 6.4 demonstrates the sensitivity of profits to maize price. Iterative runs revealed that the profits of a 40 million gal/year plant are increased US$165,818 for each US$0.01 decline in maize price. This illustrates the importance of locating ethanol plants in areas with relatively low maize prices. Plant profits drop to zero when maize price rises to US$2.4337/bu.

Ethanol Price

Profits are very sensitive to changes in the ethanol price (Fig. 6.5). Ethanol is a plant's most valuable product and often generates 80% of the total revenue stream. Profits are increased by US$480,000/year for each US$0.01 increase in fuel ethanol price. Under baseline conditions, profits drop to zero when ethanol prices drop to US$1.0693. 'Rack' (bulk wholesale price for tanker quantity) ethanol prices below this level were reached during periods of 1998, all of 1999 and periods in 2002. The net plant-gate ethanol price actually received is often US$0.10 less than the 'rack' price quoted at the nearby refineries due to costs for freight, short-term storage, commissions and other charges.

Because the ethanol market involves fewer transactions than the markets for many farm commodities or agricultural products, price discovery is difficult. Prices for ethanol have locational and seasonal dimensions that correspond to the areas of the country that are required to use ethanol as an oxygenate in petrol, particularly in the winter months. For example, petrol prices undergo locational and seasonal price changes when petrol usage changes and when refineries switch their output mix to produce more heating oil and less petrol. The marketing year for fuel ethanol is commonly divided into two 6-month contracting periods.

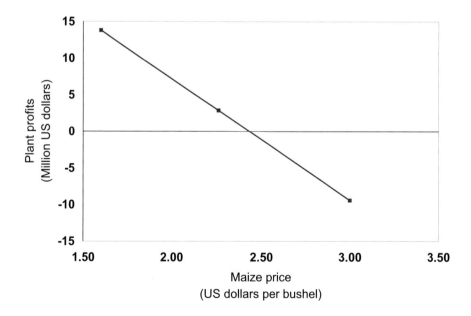

Figure 6.4. Sensitivity of 40 million gallon dry-mill profits to maize price.

October 1 through March 31 is the 'oxygenate season,' when many urban areas are forced by local and national clean air standards to include ethanol in their petrol. April 1 through September 30 is the 'non-oxygenate season'. Fuel ethanol buyers are most interested in securing adequate supplies of ethanol during the oxygenate season, therefore prices tend to be higher during this time. During the 'non-oxygenate season', prices behave based on ethanol's role as an octane enhancer (Redding, 2003).

Dry-mill plants typically employ a combination of three pricing strategies for fuel ethanol. The simplest is to sell at the rack price at a nearby refinery or fuel blending site. This allows for daily pricing and transactions between a plant (or group of plants) and the refiner or blender. The second alternative is fixed-contract pricing for future delivery, in which the ethanol plants agree to deliver a certain quantity of ethanol to the refiner or blender at specified times for an agreed price. The third alternative is gas-plus contracting, in which the price of ethanol is based on prices negotiated for monthly futures contracts for wholesale lead-free petrol listed on the New York Mercantile Exchange (NYMEX). For example, ethanol may be priced at the NYMEX contract price plus US$0.40 or US$0.20/gal. Most marketing directors for ethanol plants develop marketing plans that balance the three ethanol pricing strategies (Todd Kruggel, Renewable Products Marketing Group, Winthrop, Minnesota, 2003, personal communication).

Because it is rare and difficult to stockpile ethanol for later marketing opportunities, marketing managers attempt to price all of their expected production over the next 6 months and actively contract production anticipated for the interval 6 to 12 months out. Some contracts are based on expected production 12 to 24 months in the future. Typical minimum quantities of ethanol sold range from 1 million gal to 2 million gal for delivery in a particular month

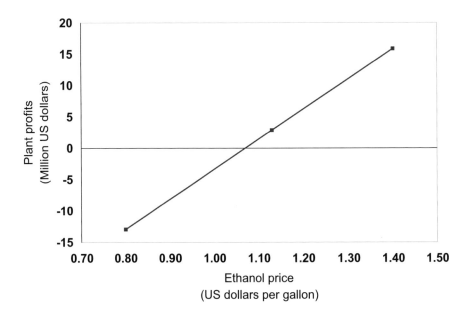

Figure 6.5. Sensitivity of 40 million gallon dry-mill profits to ethanol price.

(Todd Kruggel, Renewable Products Marketing Group, Winthrop, Minnesota, 2003, personal communication).

Determining the effective market price of ethanol for a particular plant is difficult because it involves weighting pricing decisions for various portions of the plant's production through the year. Marketing groups and ethanol cooperatives regularly calculate the net weighted price of ethanol received at the plant gate. A related problem is the difficulty of determining an industry-wide price for ethanol over any period.

Natural Gas Price

Natural gas prices are a constant concern for ethanol plant managers, particularly in the autumn and winter months. Most plants use natural gas to heat the mash, distill the ethanol and dry the DDGS. Natural gas expenditures frequently represent 30% of the operating budget. Many plants have interruptible service arrangements with their natural gas suppliers, necessitating that propane, which is more expensive, be used as a stand-by fuel. Ethanol plant managers cite the importance of plant location on natural gas distribution lines with adequate capacity to reduce the likelihood of interruptions (Joe Johansen, Minnesota Energy, Buffalo Lake, Minnesota, 2003, personal communication). The sensitivity of profits to natural gas prices is shown in Fig. 6.6. A price decrease of US$0.01/decatherm (1,000,000 Btu) from the baseline price of US$4.50/decatherm increases profits by US$16,464 over 1 year. A US$1.00 increase in natural gas price reduces profits by US$1,646,400/year. Plant profits fall to zero when natural gas prices rise to US$6.8542/decatherm. A review of natural gas prices for Minnesota industrial users over the past decade reveals that prices exceeded US$6.00/decatherm only during the period December 2000 to April 2001 (Minnesota Department of Commerce, 2002).

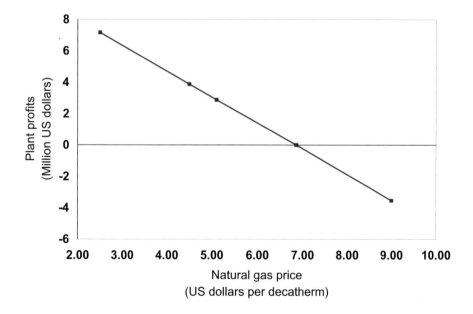

Figure 6.6. Sensitivity of 40 million gallon dry-mill profits to natural gas price.

Ethanol Yield per Bushel

The baseline assumption of 2.75 gal of anhydrous ethanol derived per bushel represents substantial progress over the 2.50 gal/bu often reported for the industry just 5 years ago. More effective enzymes, improved process controls designed to foster yeast activity and better yeast strains have contributed to this improvement. Ironically, improvements in ethanol yield for a fixed capacity to distill product will require smaller quantities of maize to reach the established maximum number of denatured gallons produced. Many plant designs may be unable to utilize higher yield technologies due to limitations in distillation capacity.

Figure 6.7 represents the sensitivity of profits to ethanol yield. An increase in the annual ethanol yield of 0.10 gal/bu increases profits by US$829,674/year. When ethanol yield drops to 2.3627 gal/bu, profits drop to zero. This sensitivity explains the economic interest of yeast companies, enzyme companies, engineering companies and seed maize companies in selling products with the potential to improve ethanol yield. Good ethanol yields result from successful management of numerous subtle factors in plant operations. Disruptions in maize quality, bacterial infections, poor sanitation, mechanical failures and failures of monitoring equipment can, singly or in concert, reduce ethanol yields.

Capacity Factor

Capacity factor refers to the expansion of plant capacity beyond the bonded, warranted nameplate capacity cited by the contracting engineers at construction. Gains in capacity, such as the 1.20 factor used in the baseline scenario, are captured by de-bottlenecking and optimizing the operation of existing plants. As mentioned previously, potential gains cannot always be fully realized because capacity in another area of the plant may be limiting.

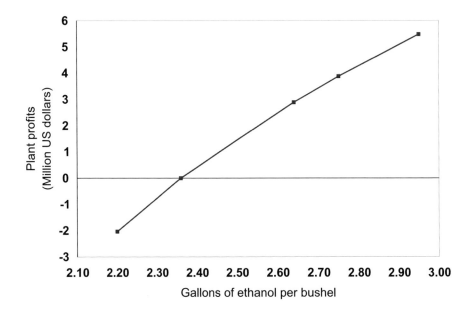

Figure 6.7. Sensitivity of 40 million gallon dry-mill profits to ethanol yield per bushel.

Increasing the capacity factor has favourable impacts on interest charges per bushel processed as well as on depreciation, labour and management charges. Figure 6.8 shows the sensitivity of profits to the capacity factor. Each 0.01 increase in capacity factor increases annual profits by US$86,633. After a plant is thoroughly tested during its shakedown period, decisions are often made concerning how best to expand capacity above nameplate. Figure 6.8 offers some assurance to managers and boards of directors that investments resulting in a higher capacity factor have excellent payoffs. If the capacity factor drops to 0.7526 with all other factors at baseline levels, profits are reduced to zero.

Multiple-factor Interactions

Simultaneous changes in several of the preceding factors also significantly affect profitability. Figure 6.9 demonstrates the effects that changes in maize and ethanol prices have on net profits. Ethanol yield is held at 2.75 gal/bu of maize, natural gas price is held at US$4.50/decatherm and all other conditions are at the established baseline levels. With ethanol at US$1.00/gal, profits are most vulnerable to increases in maize prices. Note that dry-mill plants operating at break-even conditions can tolerate higher maize prices when ethanol prices are higher. Under these conditions, break-even maize prices are US$2.00, US$2.43 and US$2.87/bu at ethanol prices of US$1.00/gal, US$1.15/gal and US$1.30/gal, respectively. Figure 6.10 demonstrates the effects on net profits of changes in maize prices at three natural gas prices – US$4.50/decatherm, US$5.50/decatherm and US$6.50/decatherm when fuel ethanol price is held at US$1.15/gal. Higher natural gas prices dramatically change the price of maize needed to maintain a given level of profit. The break-even maize price of US$2.43/bu at US$4.50/decatherm of natural gas at baseline falls to US$2.33/bu and US$2.24/bu for natural gas priced at US$5.50/decatherm and US$6.50/decatherm, respectively.

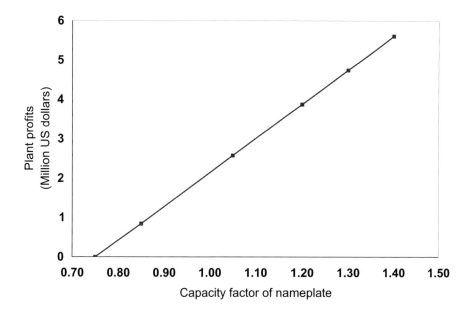

Figure 6.8. Sensitivity of 40 million gallon dry-mill profits to capacity factor of nameplate.

Factors of Lesser Importance to Profitability

Unlike the preceding five key factors, which affect the overall profitability of virtually all dry-mill ethanol plants, the following six factors – capital costs, percentage of debt, interest

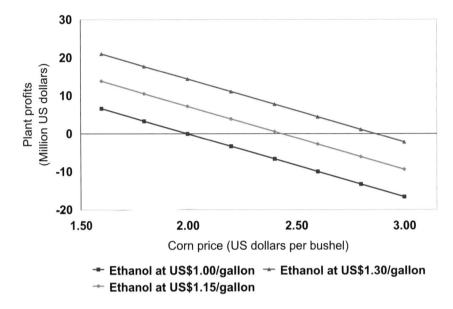

Figure 6.9. Profits for a 40 million gallon dry-mill plant at various maize and ethanol prices.

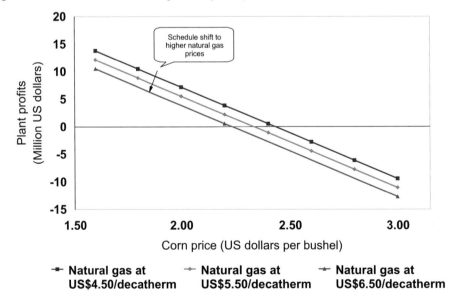

Figure 6.10. Profits for a 40 million gallon dry-mill plant at various maize and natural gas prices.

Capital Costs

Baseline capital costs used in the model plant, i.e., US$1.50/denatured gal of annual nameplate ethanol capacity, are widely reported for modern 40 million gal capacity. Few ethanol plants smaller than 40 million gal of capacity have been built or are planned in the near term. Many of the older, smaller plants, initially built at 15 million gal annual capacity in Minnesota in the mid-1990s were expanded over the last 2 years. Some smaller plants were originally built for US$2.00/denatured gal of annual capacity. Forty million gallon plants have emerged as the modal economy of scale for this technology. For the 40 million gal plant scenario used in this research, additional capital costs of US$0.01/gal of capacity reduce annual net returns by US$43,467. Thus, increasing the investment from US$1.50 to US$1.60/nameplate gal reduces net returns US$434,670/year.

Percentage of Debt

Percentages of debt carried by new ethanol plants typically range from 50% to 60%. Currently, the loan portfolios of senior lenders are heavy in the ethanol-processing category, so higher levels of equity are demanded now than in previous years (Jim Jones, Agribank, St. Paul, Minnesota, 2003, personal communication). In addition, many bankers utilize loan provisions called 'cash flow sweeps,' which allow them to claim higher levels of principal retirement, especially in the early years of a loan. This activity effectively reduces the period of loan repayment from 10 years to 7 years or 8 years. For senior lenders to maintain appropriate diversification of portfolios, they must sell portions of ethanol plant loans to other bankers, who, due to unfamiliarity and discomfort with ethanol plant loans, may demand equity levels 50% or greater. The spreadsheet model reveals that profits increase US$42,000 for each 1% reduction in the level of debt as debt decreases from 60% to 0%. Reducing debt from 60% to 50% adds US$420,000 to net margins. Reducing the debt from 60% to 0% increases profits by US$2,520,000/year.

Interest Rates

The current baseline blended interest rate suggested by lenders and used in the spreadsheet model is 7% (Jim Jones, Agribank, St. Paul, Minnesota, 2003, personal communication). The banking industry is split in its use of rates for financing ethanol plants. Approximately half of the lenders tie their interest rates to the London Interbank Overnight Rates (LIBOR), while approximately half base their loans on the prime rate plus 75 basis points (Jim Jones, Agribank, St. Paul, Minnesota, 2003, personal communication). In recent years, commercial interest rates declined to historical lows, but higher interest rates may return. The model can help predict the effect on profits of higher rates. A 100 basis point increase in interest from 7% (at 60% debt) reduces profits by US$360,000/year.

Debt Interest Rate Interactions

Figure 6.11 demonstrates the effect that changes in interest rates at various levels of debt have on profits for the 40 million gal/year plant in the baseline scenario. At 60% debt, profits are reduced by US$1,080,000/year, over those achieved at 40% debt when interest rates are 9%. This graph helps explain why lenders are eager to implement 'cash flow sweeps' in order to rapidly advance the plants they finance to lower levels of debt and to reduce their vulnerability to periods of lower net returns.

DDGS Prices

The sale of DDGS typically represents 15% to 20% of total plant revenue. Some industry observers worry that the price for DDGS (set at US$80.00/ton in the spreadsheet) might fall as more and more ethanol plants are built over the next few years. However, few individuals think that the price of DDGS will drop below the per ton price of maize even with greater supplies of cheap soybean meal. At US$2.20/bu, maize price on a per ton basis is US$78.57. If DDGS were to be priced as low as US$60.00/ton (because of oversupply of this feed), net profit margins would decline by US$2,984,732. A decline in DDGS price to US$70.00/ton would reduce net returns by US$1,492,366, or US$149,237 for each one-dollar decline in the price per ton of DDGS.

Electrical Prices and Usage

Electricity prices in the USA have generally trended downward over the last decade with the development of additional generation capacity. However, virtually all of this additional capacity has arisen from natural gas-burning generators (Energy Information Administration, 2004). Therefore, increases in natural gas prices will increase electrical prices, the level depending upon the fuel mix of electrical generation capacity in a particular region. In the model used here, a rise in electrical prices from US$0.05/kWh to US$0.08/kWh decreases net profits by US$1,569,600/year, representing a reduction in net profits of US$532,200 for

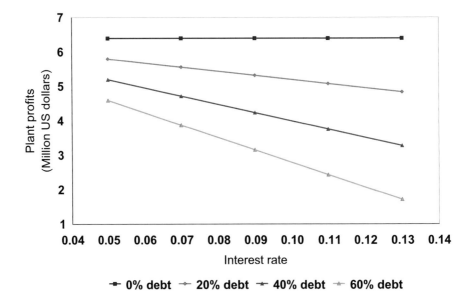

Figure 6.11. Effect of interest rates on 40 million gallon dry-mill plant profits for various debt percentages under baseline conditions.

each increase in electrical prices of US$0.01/kWh. Electrical requirements increase when newer, more robust strains of yeast are used in fermentation. The new strains produce a greater amount of heat, which needs to be controlled because temperatures above 90°F cause yeast cells to make less ethanol or to die. Therefore, the cooling of fermentation chambers, particularly during hot summer days, can significantly increase electricity consumption per gallon of ethanol produced.

Federal, State, or Local Production Subsidies or Incentives

Various state subsidies or incentives may increase ethanol plant profits. The federal government offers credits for purchases of feedstock through the Commodity Credit Corporation Bioenergy Program, which encourages expansion in production and in use of US agricultural commodities. An ethanol plant typically qualifies for this credit only in the year in which expanded production occurs.

Retrospective Analysis of Plant Profits

The spreadsheet was used to determine economic returns for a modern 40 million gal/year plant with current scale economies, investment costs and enhanced ethanol yields in its first year of operation by re-playing a decade of historic prices for products and inputs.

Monthly gross margins, plotted in Fig. 6.12, were obtained by subtracting monthly Minnesota maize prices from the value of dry-mill co-products. Minnesota ethanol prices at the refinery rack were reduced by US$0.10/gal for freight and storage in order to reflect 'plant gate' prices. Prices of DDGS were based on the monthly USDA series for Lawrenceburg,

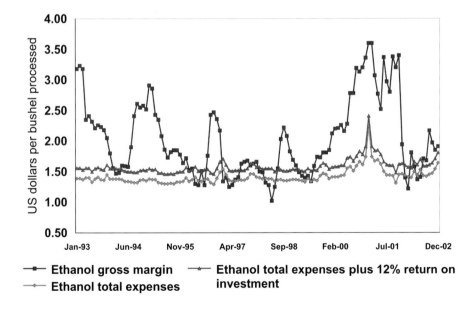

Figure 6.12. Retrospective per bushel dry-mill gross margins, total expenses, and total expenses plus a 12% rate of return on investment for a 40 million gallon dry-mill plant from 1/93-12/02.

Indiana, from January 1993-April 1999 (Economic Research Service, 1999) and Minneapolis, Minnesota, prices provided by Land O' Lakes from May 1999-December 2002 (Land O' Lakes Feed, 2003). Carbon dioxide was assumed to be sold for US$6.00/ton on a net-back basis. Natural gas prices, electrical prices and interest rates were allowed to vary each month. However, chemical prices, depreciation, labour and management, repairs and maintenance and miscellaneous expenses were fixed at the baseline levels for the entire period. Finally, a 12% imputed return on investor equity was added to the total plant expenses to analyse how satisfied investors would be with a plant using current technology and encountering a replay of the past decade's prices and economic conditions.

Figure 6.12 reveals a tumultuous pattern in ethanol gross margins (value of plant products minus maize price) over the decade 1993-2002. Ethanol total expenses and ethanol total expenses plus a 12% return on equity reveal less dramatic patterns. A 12% return on investor equity was not always achieved, particularly in 1995, 1996, 1998, 1999 and 2002. Profits were good during 1993, the second half of 1994 and the second half of 1996 and were exceptional during 2000 and 2001. In 10 of 120 months (8.3% of the time) losses occurred. When a 12% rate of return on investor equity is imposed, deficiencies occur in 20 out of the 120 months, suggesting investor discontent 16.7% of the time. Despite the periods of negative net returns, overall profitability for a modern, 40 million gal plant would occur in a replay of the last decade's product prices, energy costs and interest rates.

Summary and Discussion

This model demonstrates that even a well-located plant will experience times of difficult operating conditions and gross margins. Plants with high debt, high capital costs per unit of capacity, in high-priced maize locations (Gallagher et al., 2000), with poor access to transportation or with inadequate supplies of natural gas or water will be even more vulnerable, financially. Many ethanol plants are sponsored by firms or form associations that market ethanol, DDGS and CO_2 or to purchase enzymes and yeasts. Most plants operating today or in the future will be in alliances or contracted with firms to market their products and purchase inputs.

Investors in new dry-mill ethanol plants should realize that experienced managers and technicians with the skills needed to manage the plants and yeast may be difficult to find. Without key personnel, a plant cannot be profitable, even under times of high gross margins and favourable operating and capital costs. Successful managers and boards of directors must seek strategies to control costs while adopting new technologies and upgrading plants in terms of throughput and ethanol yield.

Risk management strategies are critical in pricing products and inputs. Analysing profitability of a modern plant under the economic conditions that occurred over a decade suggests that all managers will have the challenge of riding-out some times of low or negative margins due to a variety of drivers. Some managers may find opportunities to manage dry-mill plants that have failed. If failed plants can be purchased cheaply enough and the plants lack 'fatal' flaws in engineering or siting, some will return to profitable operations. With such rapid expansion of ethanol production capacity occurring in a business that betrays great volatility in returns, some financial calamities will certainly occur, particularly among firms carrying higher percentages of debt.

Expansion of US production by dry-mill ethanol plants is likely to continue until technology advances occur in lignocellulosic processing permit the usage of fibre for ethanol production. Fuel ethanol will continue to represent a small share of domestic fuel supply. In 2003, the 2.8 billion gal of fuel ethanol produced in the USA represented just 2.11% of the volume of petrol consumed (Energy Information Administration, 2003; Renewable Fuels Association, 2003).

References

Economic Research Service. (1999) *Feed Situation and Outlook Yearbook.* FDS-1999. Appendix Table 16. US Department of Agriculture, Economic Research Service, Washington, DC, pp. 78.

Energy Information Administration. (2003) Finished Motor Gasoline Supply and Disposition, 1988-Present, Table S4. *Petroleum Supply Monthly.* US Department of Energy, Energy Information Administration, Washington, D.C. [Accessed 2004.] Available from http://www.eia.doe.gov/oil_gas/petroleum/data_publications/petroleum_supply_monthly/psm.html

Energy Information Administration. (2004) Capacity Additions and Retirements at U.S. Electric Utilities by Energy Source, 2000. *Inventory of Electric Utility Power Plants in the United States 2000.* US Department of Energy, Energy Information Administration, Washington, DC. [Accessed 2004.] Available from http://www.eia.doe.gov/cneaf/electricity/ipp/html1/t2p01.html

Gallagher, P., Otto, D. and Dikeman, M. (2000) Effects of an Oxygen Requirement for Fuel in Midwest Ethanol Markets and Local Economies. *Review of Agricultural Economics* 22(2):292-311.

Hart Energy Publishing. (2004) *Renewable Fuel News.* (various) Hart Energy Publishing, Rockville, Maryland.

Land O' Lakes Feed. (2003) *Distillers Dried Grains and Solubles Prices Based at Minneapolis.* Unpublished data. Land O' Lakes Feed, Shoreview, Minnesota.

Minnesota Department of Commerce. (2002) Minnesota Statewide Electric Revenue and Consumption by Class of User. *Minnesota Utility Data Book 2002: A Reference Guide to Minnesota Electric and Natural-Gas Utilities, 1965-2000.* Minnesota Department of Commerce, St. Paul, Minnesota. [Accessed 2004.] Available from http://www.state.mn.us/mn/externalDocs/Utility_Data_Book,_1965-2000__030603120425_2002DataBook.pdf

National Agricultural Statistics Service. (2004) Monthly U.S. Corn Prices 1994-2003. *Agricultural Statistics Data Base.* US Department of Agriculture, National Agricultural Statistics Service, Washington, D.C. [Accessed 2004.] Available from http://www.nass.usda.gov:81/ipedb

Redding, J. (2003) *Marketing Ethanol in a Post-MTBE World.* Presented paper, 8th Annual Renewable Fuels Association National Ethanol Conference: Building a Secure Energy Future, February 17-19, Scottsdale, Arizona.

Renewable Fuels Association. (2003) *U.S. Fuel Ethanol Production-Capacity.* Renewable Fuels Association, Washington, DC. [Accessed 2003.] Available from http://www.ethanolrfa.org/eth_prod_fac.html

Renewable Fuels Association. (2004) *Ethanol Industry Outlook 2004.* Renewable Fuels Association, Washington, DC. [Accessed 2003.] Available from http://www.ethanolrfa.org/outlook2004.html

Shapouri, H., Duffield, J. and Wang, M. (2002) *The Energy Balance of Corn Ethanol: An Update.* Economic Report No. 813. US Department of Agriculture, Office of the Chief Economist, Office of Energy Policy and New Uses, Washington, DC., pp.2.

Southwest Minnesota Farm Business Management Association. (2003) *Annual Reports 1982-2002.* Staff Paper Series. University of Minnesota, Department of Applied Economics, St. Paul, Minnesota. [Accessed 2003.] Available from http://agecon.lib.umn.edu

Tiffany, D. and Eidman, V. (2004) *Factors Associated with Success of Fuel Ethanol Producers.* Staff Paper P03-7. University of Minnesota, Department of Applied Economics, St. Paul, Minnesota. [Accessed 2004.] Available from http://www.apec.umn.edu/staff/dtiffany/staffpaperp03-7.pdf

Chapter 7

An Econometric Analysis of the Impact of the Expansion in the US Production of Ethanol from Maize and Biodiesel from Soybeans on Major Agricultural Variables, 2005-2015

John N. (Jake) Ferris and Satish V. Joshi

Background

Since the *Clean Air Act of 1990*, efforts have accelerated through federal and state environmental and energy policies to encourage the use of renewable fuels. While the main purpose has been to improve air and water quality, other reasons cited include reducing the US dependence on foreign oil, promoting economic growth in rural areas, enhancing farm prices and incomes and saving expenditures on federal farm programmes. This chapter provides an analysis of the possible effects on agriculture of alternative scenarios for programmes which encourage the expansion of ethanol and biodiesel production. The analysis focuses on comparisons of scenarios with baselines over the 2006 to 2015 projection period.

Alternative Projections for Renewable Fuels

Projections to 2015 from the US Department of Energy (US DOE) in their *Annual Energy Outlook 2004* and in their analysis of the conference committee report on the proposed *Energy Policy Act of 2003* (EPACT03) were employed in this study with some modifications (US Department of Energy, 2004a,b). The *Annual Energy Outlook 2004* was released in January 2004 and was followed by US DOE's evaluation of EPACT03 in February of that year. In *Annual Energy Outlook 2004,* ethanol consumption was projected to increase from 2.04 billion gallons (gal)/year in 2002 to 3.76 billion gal in 2015 (US Department of Energy, 2004a). These projections serve as the baseline for ethanol in this analysis. The alternative is represented by US DOE's analysis that EPACT03 would require 5.57 billion gal of ethanol by 2015 (US Department of Energy, 2004b). The provision that a Renewable Fuels Standard would require 5.0 billion gal of ethanol and biodiesel by 2012 underlies the US DOE estimates. However, explicit in these projections is a rather small expansion being forecast for biodiesel. Neither the *Annual Energy Outlook 2004* nor the EPACT03 assessments provided any substantial projections for biodiesel to 2015. In essence, *Annual Energy Outlook 2004* projected biodiesel to increase from 0.02 billion gal in 2002 to 0.05 billion gal in 2015. EPACT03 established a target of 0.04 billion gal in 2015.

© CAB International 2005. *Agriculture as a Producer and Consumer of Energy* (eds J.L. Outlaw, K.J. Collins and J.A. Duffield)

Table 7.1. US production of fats and oils as potential biodiesel feedstock, average for 2000-2002.

Source	Million Pounds
Vegetable Oils[a]	
Soybeans	18,585
Maize	2,439
Sunflowers	816
Oilseed rape	622
Other	526
Total	22,988
Animal fats[b]	
Inedible tallow	3,604
Edible tallow	1,864
Lard	309
Total	5,777
Grease	
Yellow[b]	2,612
Brown[c]	3,800
Total	6,412

[a] Source: Economic Research Service (2003).
[b] Source: US Department of Commerce (2003a,b).
[c] Source: Tyson (2002).

The EPACT03 projection is used as the baseline in this study for biodiesel. The alternative scenario for biodiesel selected for this analysis is for all diesel fuel to be a mixture of 2% biodiesel by 2015, which would require 1.09 billion gal of 100% biodiesel labelled B100. However, biodiesel from soybean oil would supply just one-half of the 1.09 billion gal or 545 million gal. The remainder would be derived from other vegetable oils, yellow grease as a residue from away-from-home food institutions and from imports.

These projections differ somewhat from a 2001 study with a target for renewable fuels at 4% of highway use by 2016, which, by 2015 would involve 6,410 million gal of ethanol from maize and 432 million gal of B100 from soybeans (Urbanchuk, 2001). Urbanchuk (2001) projected that 88% of ethanol production in 2015 would have maize as a feedstock with soybeans furnishing 58% of the production of biodiesel. Another 2001 analysis with a target of 3% renewable fuels in US motor fuel consumption by about 2010 used 713 million bushels of maize in ethanol production as a baseline and 1,775 million bushels as the scenario (Food and Agricultural Policy Research Institute, 2001). These projections are equivalent to about 1,977 million gal and 4,922 million gal of ethanol, respectively. The Food and Agricultural Policy Research Institute study projected soybean oil used in biodiesel to be 380 million pounds in their baseline for 2010 and 2,472 million pounds in their scenario. With a conversion rate of 7.68 pounds of soybean oil required to produce one gallon of B100, this amounts to 49 million gal of B100 in the baseline and 322 million gal in the scenario. In early 2004, the US Department of Agriculture (USDA) baseline projected that 1,360 million bushels of maize would be used for fuel alcohol in 2013. This is equivalent to 3,540 million gal of ethanol (US Department of Agriculture, 2004b).

The selection of soybean oil as a feedstock is appropriate since this source has represented over 80% of the US production of vegetable oils in recent years (Table 7.1). A distant second to soybean oil is maize oil at about 10% of the total followed by sunflower and canola (genetically altered rapeplant). Animal fats are another potential feedstock for biodiesel production, which in total represent about one-fourth of the total for vegetable oil. A third category is grease derived from the spent use of vegetable oils and animal fats. Yellow grease consists of fats and oils from cooking in restaurants and could also be derived from rendering plants. A major use of yellow grease has been to add flavour and energy to livestock feeds. Brown grease is collected from traps and sewage plants and cannot be used for animal feed.

Yellow grease has the advantage of being cheaper than the original fats and oils and can be more easily converted to biodiesel than brown grease. Prices of fats and oils tend to fluctuate in parallel as shown in Fig. 7.1. Yellow grease, as with other sources, can be used both as the sole feedstock for biodiesel production or blended with soybean oil to reduce the input costs. Likely, as biodiesel production expands, prices on yellow grease will not only follow the major fats and oils prices but will be bid up closer to the prices of soybean oil and the other feedstock.

Methodology

The alternative projections for renewable fuel use were incorporated into AGMOD, an econometric-simulation model of US agriculture. The model is designed to generate annual forecasts of major variables for about a 10-year period (Ferris, 1991). The model, developed at Michigan State University, covers major commodities in the livestock, dairy, poultry and field crop sectors, including by-product feeds. The international sector focuses on coarse grain, wheat and oilseeds.

The variables in the model, of which there are over 700, can be classified as exogenous and endogenous. The endogenous variables are those generated within the model and measure

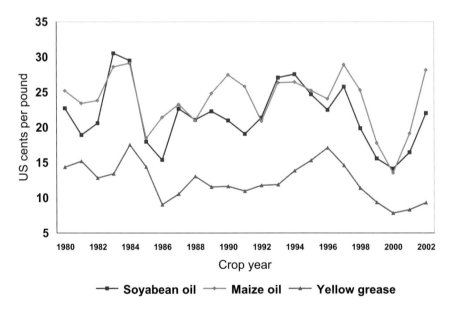

Source: Economic Research Service (2003) for soybean oil (crude, Decatur, Illinois) and maize oil (average, Chicago, Illinois); The Jacobsen Publishing Company (2004) for yellow grease (Illinois).

Figure 7.1. Prices of soybean oil, maize oil and yellow grease, 1980-2002 by crop years.

behavioural relationships such as how farmers respond to profits and how consumers respond to prices and the availability of products. The exogenous variables are those which are external to the model and enter the solution as assumptions, such as the renewable fuel scenarios. Other exogenous variables include: population, per capita incomes, inflation rates, energy prices, interest rates and exchange rates. Projections of these other exogenous variables were drawn from the USDA's 2004 baseline (US Department of Agriculture, 2004b). As in the USDA's baseline, the current 2002 farm bill is assumed to continue. Normal weather is also presumed.

The model was analysed considering three sets of baseline and scenario projections. The first set included the baseline and scenario for ethanol, keeping the biodiesel production at its baseline. The second set included the baseline and scenario projections for biodiesel, keeping ethanol production at its baseline. The third set compared the combination of the scenarios for both ethanol and biodiesel with their respective baselines. The third set was altered by forcing a reduction in soybean exports required to accommodate the increased crush for the expanded biodiesel market without a major increase in soybean oil imports. The results are presented as averages of key variables for 2006 to 2015 combined with the percentage changes between the baselines and the scenarios. A key for Tables 7.2 to 7.4 is in the Appendix at the end of this chapter.

Results

Ethanol

Table 7.2 contains the selected variables with annual averages for 2006 to 2015 under both the baseline and the scenario of expanded ethanol production. Biodiesel production was held at its low-level baseline in order to isolate the impact of expanding ethanol production separate from biodiesel. This is a view of how acceleration in ethanol production by nearly 37% over the 2006 to 2015 period would affect major agricultural variables.

Maize harvested acres would expand by a small amount (1.1 million acres) resulting in a 1% to 2% increase in production. In addition, there is a minor reduction in feed use and about a 6% reduction in exports, which is expected to accommodate the growing demand for maize as a feedstock in ethanol production. The price received by farmers for maize would be expected to average about 14 cents/bushel over the baseline, roughly a 6% increase. The gross margin over variable cash expenses per acre was estimated for the baseline at US$228. Under the scenario, the gross margin per acre would increase to US$246, nearly an 8% rise. Gross margins include both returns from sales of the product and direct government payments.

Most of the projected increase in maize acres would represent a shift out of soybeans. As indicated in Table 7.2, the net effect of ethanol expansion on soybeans would be small reductions in production and utilization, resulting in slightly higher prices on soybeans and its products of meal and oil.

To provide an aggregate view of the impact of expanded ethanol production, a measure of profitability for the combination of maize, other coarse grains, soybeans and wheat was constructed which encompasses both gross margins over cash variable costs per acre and the level of sales. This measure is shown in Table 7.2 as 'major crops' under 'aggregate gross

Table 7.2. Average annual projections of selected agricultural variables for 2006-2015 comparing the scenario of expanded ethanol production with the baseline.

Variable	Unit	Baseline	Scenario	% Change
Ethanol production	Million gallons	3,517	4,805	36.62
Maize				
Acres harvested	Million acres	73.5	74.6	1.50
Production	Million bushels	11,171	11,351	1.61
Feed use	Million bushels	6,154	6,103	-0.83
Ethanol use	Million bushels	1,268	1,733	36.67
Exports	Million bushels	2,398	2,257	-5.88
Farm price	US$/bushel	2.45	2.59	5.71
Gross margin	US$/acre	228	246	7.89
Soybeans				
Acres harvested	Million acres	73.1	72.4	-0.96
Production	Million bushels	3,100	3,070	-0.97
Crush	Million bushels	1,915	1,900	-0.78
Meal fed	Million tons	38.5	38.1	-1.04
Biodiesel use	Million bushels	23	23	0.00
Exports	Million bushels	1,001	982	-1.90
Farm price	US$/bushel	6.93	7.00	1.01
Gross margin	US$/acre	192	195	1.56
Soy meal price	US$/ton	215	217	0.93
Soy oil price	US cents/pound	25.0	25.3	1.20
Aggregate gross margin				
Major crops	Billion US$	36.1	38.2	5.82
Livestock	Billion US$	64.3	64.1	-0.31
Maize gluten feed				
Production	1000 metric tons	13,141	15,025	14.34
Amount fed	1000 metric tons	8,001	8,503	6.27
Price	US$/ton	77.59	80.04	3.16
Distillers' dried grains				
Production	1000 metric tons	5,724	7,416	29.56
Price	US$/ton	108.30	107.44	-0.79
Government cost	Million US$	5,589	5,252	-6.03
Meat supply per capita	Pounds	185.0	183.8	-0.65
Index of livestock prices	2000=100	96.23	97.86	1.69

margin'. Aggregate gross margin increases from US$36.1 billion in the baseline to US$38.2 billion in the scenario over the 2006 to 2015 period, about a 6% increase.

A similar measure was constructed for livestock. This represents the gross margin over feed costs for the following enterprises: (1) cow-calf, (2) cattle on feed, (3) farrow-to-finish on pigs, (4) broilers, (5) turkeys, (6) eggs and (7) dairy. The gross margins per unit were multiplied by production to derive aggregate figures. Note that expanded ethanol production had little impact on these measures.

Increased production of mid-protein feed by-products of ethanol production would partially offset the reduced availability of maize and soybean meal for livestock feed. By-products of the wet milling process are maize gluten feed which is about 21% protein and a much smaller quantity of meal which is about 60% protein. Another by-product of the wet milling industry is maize oil. The feed by-product of the dry milling industry is distillers' dried grains, which is about a 30% protein source. Expanded production of ethanol with the accompanied protein feeds and maize oil would tend to depress prices in the soybean complex. However, the impact of reduced soybean production and higher maize prices (which support the soybean meal price) counter the negative effect.

Historic data are available on the production and feeding of maize gluten feed and meal; however, similar information on distillers' dried grains has not been available in recent years. Data presented in this chapter represent the authors' estimates. Expansion in ethanol production in recent years has been primarily in dry milling and consequently, production of distillers' dried grain has also expanded. However, the increase in distillers' dried grains production is not proportional since some maize is processed into alcoholic beverages. This analysis assumed that the proportion of ethanol produced in dry mills increases from the current level around 40% to over 45% by 2015. This may not be a linear trend in that the wet mills are able to shift between ethanol and high fructose maize syrup as profit opportunities change.

Note that the production of maize gluten feed (including meal) in Table 7.2 is projected to increase by over 14% over the 2006 to 2015 period relative to the baseline, but the amount fed is expected to increase just over 6%. The reason is that nearly all of the US production of maize gluten feeds has been exported, mostly to the European Union (EU), and the overseas market is expected to continue as an outlet. However, beginning in the mid-1990s, exports of these maize by-product feeds to Europe have levelled off and trended lower. The reason relates to a modification in the Common Agriculture Policy (CAP) and exchange rates. With declining price supports on grain and weakening of the euro, the high grain prices in the EU have dipped closer to the world market. This lessened the price advantage of protein feeds which have been entering the EU market with little restrictions. When the CAP's grain supports were high relative to world prices for both grain and protein feeds, maize gluten feeds and other imported protein feeds were fed as a source of energy as well as protein (Hasha, 2002).

Exports of maize gluten feeds in the baseline are projected to be somewhat lower than in the past decade for the reasons provided above. Also relevant are the AGMOD model's projection for relatively strong grain prices during the analysis period.

As with soybean meal, higher maize prices are expected to offset the downward price pressure of the expansion in the maize gluten feed production and a nominal price increase is forecast. On distillers' dried grains, however, a 30% production increase is projected over the baseline. Prices are projected to drop about 6%, more than offsetting the positive effect of higher maize prices. While data are limited or unavailable on production of distillers' dried grains, the assumption in this study is that nearly all the increase in production will be fed domestically.

Direct government payments to US farmers under the *Farm Security and Rural Investment Act of 2002* (2002 farm bill) were tabulated for maize, soybeans, wheat and milk. For crops under both the baseline and the ethanol scenario, prices held above the respective marketing loan rates so no loan deficiency payments were triggered. The only way that expanded ethanol production could reduce governmental outlays to these crops is by reducing counter-cyclical payments which occur only if farm prices drop below the target price for each crop minus fixed payment rates. In the baseline, this occurred for maize and wheat but not for soybeans. Counter-cyclical payments declined in the ethanol scenario as prices rose. Also, somewhat higher milk prices in the ethanol scenario reduced the payments under the milk income loss contract (MILC) programme. In total, direct government payments for these programmes dropped by 6%, equivalent to US$337 million annually or US$3.37 billion over a 10-year period. Should crop prices be lower than projected in the baseline of this study, the impact of these renewable fuels scenarios will be much greater in reducing government programme costs.

Table 7.3. Average annual projections of selected agricultural variables for 2006-2015 comparing the scenario of expanded biodiesel production with the baseline.

Variable	Unit	Baseline	Scenario	% Change
Biodiesel production	Million gallons	33.3	304.7	815.02
Maize				
Acres harvested	Million acres	73.5	73.2	-0.41
Production	Million bushels	11,171	11,135	-0.32
Feed use	Million bushels	6,154	6,140	-0.23
Ethanol use	Million bushels	1,268	1,268	0.00
Exports	Million bushels	2,398	2,392	-0.25
Farm price	US$/bushel	2.45	2.47	0.82
Gross margin	US$/acres	228	230	0.88
Soybeans				
Acres harvested	Million acres	73.1	74.1	1.37
Production	Million bushels	3,100	3,141	1.32
Crush	Million bushels	1,915	1,939	1.25
Meal fed	Million tons	38.5	38.6	0.26
Biodiesel use	Million bushels	23	209	808.70
Exports	Million bushels	1,001	1,003	0.20
Farm price	US$/bushel	6.93	7.32	5.63
Gross margin	US$/acre	192	207	7.81
Soy meal price	US$/ton	215	200	-6.98
Soy oil price	US cents/pound	25.0	32.2	28.80
Aggregate gross margin				
Major crops	Billion US$	36.1	37.5	3.88
Livestock	Billion US$	64.3	64.8	0.78
Maize gluten feed				
Production	1000 metric tons	13,141	13,141	0.00
Amount fed	1000 metric tons	8,001	8,001	0.00
Price	US$/ton	77.59	74.85	-3.53
Distillers' dried grains				
Production	1000 metric tons	5,724	5,728	0.07
Price	US$/ton	108.30	101.91	-5.90
Government cost	Million US$	5,589	5,523	-1.18
Meat supply per capita	Pounds	185.0	185.2	0.11
Index of livestock prices 2000=100		96.23	96.05	-0.19

Expanding ethanol production accompanied with higher maize and soybean meal prices are projected to have nominal effects on the livestock industry, cutting production slightly and raising prices by 1% to 2%. A 1% to 2% increase in livestock prices translates into less than a 1% increase at retail stores considering that cattle prices have averaged about 46% of beef at retail in recent years, with pig prices about 27% and wholesale broiler prices about 56% of the retail price (US Department of Agriculture, 2004a). The impact on restaurant prices would even be less.

Biodiesel

Because of the relatively low baseline level for biodiesel production, the scenario represents over an 8-fold increase or 815% (Table 7.3). With ethanol production held at its baseline, the expansion in biodiesel involved a minor shift in acreage from maize to soybeans. This shift helped support maize prices.

Curiously, the AGMOD model did not generate enough of an increase in soybean crushing to accommodate the expansion in the demand for soybean oil. Normally, soybeans are crushed more for the meal rather than the oil. In the baseline, the value of meal/bushel of soybeans was 80% greater than the value of the oil. Since livestock numbers in the scenario were the same as in the baseline, the demand for meal was unchanged. Exports of meal are generally less than 20% of total production. The expansion in the availability of soybean oil was derived mainly by an increase in imports, a drop in exports and drawing down of stocks.

The tight domestic supplies raised soybean oil prices nearly 30% in the scenario. At the same time, some accumulation of soybean stocks put downward pressure on soybean meal prices, as prices fell 7%. However, the net effect was a 5% to 6% increase in the price of soybeans at the farm.

The crop sector benefited moderately (about a 4% increase in aggregate gross margins) from biodiesel expansion with the livestock industry grossing about the same in returns over feed costs. While the production of maize by-product feeds was unchanged, their prices followed the reduction in soybean meal prices.

Cost of the federal government programme was reduced slightly (about 1%) in the biodiesel scenario because soybean prices were above the marketing loan rate and the level that would have triggered counter-cyclical payments in both the baseline and in the scenario. Supplies and prices on livestock were unchanged in the scenario.

Ethanol and Biodiesel in Combination

The impact of the combined expansion in ethanol and biodiesel production is presented in Table 7.4. The scenario represents a 44% increase in the production of renewable fuels from maize and soybeans. The apparent strength of the gross margin over variable cash costs/acre on maize versus soybeans in the scenario (US$246 versus US$214) tended to pull acreage into maize and away from soybeans. This occurred, even though the gross margin on soybeans increased more than maize in the scenario. The changes in acreages, however, were relatively minor.

Increased biodiesel production from soybeans had little effect on the maize sector when comparing Table 7.4 with Table 7.2. The primary difference was that the lower price of soybean meal shown in Table 7.4 dropped prices on distillers' dried grains by 5%.

The impact on the soybean sector of the combined expansion can be observed by comparing Table 7.4 with Table 7.3. In addition to the reduction in soybean acreage and production, the higher price structure in the maize market supported soybean meal prices (the decline was less). This, in combination with slightly higher oil prices, helped strengthen the soybean market.

Questions might be raised about why soybean crush and exports remained about the same in the scenario versus the baseline. Econometric models such as AGMOD are designed to generate baseline projections with moderate variations. Major departures from the baselines provide challenges such as in this analysis. As in Table 7.3, about a third of the increase in the supply of soybean oil feedstock for biodiesel came from imports; the remainder from a reduction in exports, a draw down in stocks and a small reduction in food use.

Ethanol and Biodiesel in Combination with Adjustments

Adjustments were made in the AGMOD model to reduce soybean exports with a comparable increase in domestic crush in order to (1) keep soybean oil imports at the nominal level in the

Table 7.4. Average annual projections of selected agricultural variables for 2006-2015 comparing the scenario of ethanol and biodiesel production with the baseline.

Variable	Unit	Baseline	Scenario	% Change
Ethanol and biodiesel Production	Million gallons	3,550	5,110	43.94
Maize				
Acres harvested	Million acres	73.5	74.7	1.63
Production	Million bushels	11,171	11,351	1.61
Feed use	Million bushels	6,154	6,104	-0.81
Ethanol use	Million bushels	1,268	1,733	36.67
Exports	Million bushels	2,398	2,257	-5.88
Farm price	US$/bushel	2.45	2.59	5.71
Gross margin	US$/acre	228	246	7.89
Soybeans				
Acres harvested	Million acres	73.1	72.4	-0.96
Production	Million bushels	3,100	3,071	-0.94
Crush	Million bushels	1,915	1,900	-0.78
Meal fed	Million tons	38.5	38.1	-1.04
Biodiesel use	Million bushels	23	209	808.70
Exports	Million bushels	1,001	998	-0.30
Farm price	US$/bushel	6.93	7.48	7.94
Gross margin	US$/acre	192	214	11.46
Soy meal price	US$/ton	215	205	-4.65
Soy oil price	US cents/pound	25.0	32.9	31.60
Aggregate gross margin				
Major crops	Billion US$	36.1	40.1	11.08
Livestock	Billion US$	64.3	64.5	0.31
Maize gluten feed				
Production	1000 metric tons	13,141	15,025	14.34
Amount fed	1000 metric tons	8,001	8,508	6.34
Price	US$/ton	77.59	77.94	0.45
Distillers' dried grains				
Production	1000 metric tons	5,724	7,416	29.56
Price	US$/tons	108.30	102.80	-5.08
Government cost	Million US$	5,589	5,234	-6.35
Meat supply per capita	Pounds	185.0	183.9	-0.58
Index of livestock prices	2000=100	96.23	97.71	1.54

baseline and (2) to maintain stock levels. Exports of soybean oil were allowed to decline to relatively small amounts.

Exports of soybeans were reduced by 7% in order to increase the crush by about 5%. By retaining more soybeans domestically, the price of soybean oil dropped from 32.9 cents/pound as shown in Table 7.4 to 30.6 cents/pound. This was still about 22% over the baseline. With more soybean meal produced, there would be downward pressure on prices as the additional amounts would be headed for the export market.

Caveat on Price Projections

Of some concern in this analysis is that the world may face relatively low grain stocks in the not too distant future. In *World Agricultural Supply and Demand Estimates,* the USDA projected world stocks of coarse grains at the end of the 2004-05 crop year at 104 million metric tons (US Department of Agriculture, 2004c). This is 11% of annual use of these grains which include maize, sorghum, barley, oats and other. About 70% of world use of coarse grains both for livestock (65%) and for food (35%) is maize. To put this in perspective, a stock level of 11% at the end of the 2004-05 crop year is considerably lower than the 23% level experienced over the 10-year period 1993 to 2002. The situation for wheat, mainly a food grain and the other major world traded cereal, is much the same. The USDA's projection on ending stocks of world wheat at the end of the 2004-05 crop year is 21% of utilization. This is 11 percentage points lower than the 1993 to 2002 average of 32%.

The projections from the AGMOD model for 2006 to 2015 do not raise carryover levels above the near term projections for the 2004-05 crop year. Limited expansions in cropland harvested and projected linear extensions of crop yields do not overtake and exceed the demands implicit in the macro economic forecasts in the USDA baseline.

Grain and oilseed prices will be sensitive to abnormal weather and pest problems. Bidding for grain and oilseeds for food and feed rather than for fuel will be strong in situations with tight supplies. The prospect of relatively low grain carryover levels combined with uncertain weather leads to questions about what is happening to the demand for animal protein. Whether the rapid move to low-carbohydrate diets such as those promoted by Dr Robert Atkins or Dr Arthur Agatston are a fad or substantive, major adjustments were made in the AGMOD model to capture the shift in meat demand detected for 2004. Underlying the projections in this study is the assumption that this shift will continue, though at a lower level than evident in 2004.

If renewable fuels are to expand markedly in the future, the importance of increased funding for research, teaching and extension/outreach in the USA and abroad is paramount. The promise of biotechnology will be required.

Conclusions

Expansion in the use of maize for ethanol and soybeans for biodiesel can be accommodated without major increases in acreages or disruptions in the normal pattern of utilization of these crops. Higher maize and soybean prices will have minimal impacts on retail food prices. Most notable will be larger supplies of maize by-product feeds for domestic use and export. Renewable fuels programmes can help ensure an economically viable agriculture and minimize federal farm programme costs, benefits which will be even more pronounced if maize and soybean prices are lower than projected in the baseline.

References

Economic Research Service. (2003) *Oil Crops Situation and Outlook Yearbook*. OCS-2003. US Department of Agriculture, Economic Research Service, Market and Trade Economics Division, Washington, DC.

Economic Research Service. (2004) *Feed Situation and Outlook Yearbook*. FDS-2004. US Department of Agriculture, Economic Research Service, Washington, DC. [Accessed 2004.] Available from http://www.ers.usda.gov/publications/so/view.asp?f=field/fds-bby/

Ferris, J. (1991) *Understanding 'AGMOD' – An Econometric Model of U.S. and World Agriculture*. Staff paper #91-5. Michigan State University, Department of Agricultural Economics, East Lansing, Michigan.

Food and Agricultural Policy Research Institute. (2001) *Impacts of Increased Ethanol and Biodiesel Demand*. FAPRI-UMC Report #13-01. University of Missouri-Columbia, Food and Agricultural Policy Research Institute, Columbia, Missouri.

Foreign Agricultural Service. (2004) *PSDONLINE*. Production, Supply and Distribution Database. US Department of Agriculture, Foreign Agricultural Service, Washington, DC. [Accessed 2004.] Available from http://www.fas.usda.gov/psd/

Foreman, L. (2001) *Characteristics and Production Costs of U.S. Corn Farms*. Statistical Bulletin Number 974. US Department of Agriculture, Economic Research Service, Washington, DC.

Foreman, L. and Livezey, J. (2002) *Characteristics and Production Costs of U.S. Soybean Farms*. Statistical Bulletin Number 974-4. US Department of Agriculture, Economic Research Service, Washington, DC.

Hasha, G. (2002) *Livestock Feeding and Feed Imports in the European Union – A Decade of Change*. FDS-0602-01. US Department of Agriculture, Economic Research Service, Washington, DC.

The Jacobsen Publishing Company. (2004) Yellow Grease. *Market News: Animal* [Accessed 2004.] Available from http://www.thejacobsen.com/index.htm

Tyson, K. (2002) *Brown Grease Feedstocks for Biodiesel*. US Department of Energy, National Renewable Energy Laboratory, Golden, Colorado.

Urbanchuk, J. (2001) *An Economic Analysis of Legislation for a Renewable Fuels Requirement for Highway Motor Fuels*. AUS Consultants, Inc., Moorestown, New Jersey.

US Department of Agriculture. (2004a) *Agricultural Outlook: Statistical Indicators*. Table 8. (April) US Department of Agriculture, Economic Research Service, Washington, DC. [Accessed 2004.] Available from http:www.ers.usda.gov/publications/Agoutlook/AOTables

US Department of Agriculture. (2004b) *USDA Agricultural Baseline Projections to 2013*. Staff Report WAOB-2004-1. US Department of Agriculture, Office of the Chief Economist, Washington, DC. [Accessed 2004.] Available from http://www.ers.usda.gov/publications/waob041/

US Department of Agriculture. (2004c) *World Agricultural Supply and Demand Estimates*. WASDE-410 (May). US Department of Agriculture, Office of the Chief Economist, Washington, DC. [Accessed 2004.] Available from http://usda.mannlib.cornell.edu/reports/waobr/wasde-bb/2004/wasde410.pdf

US Department of Commerce. (2003a) *Fats and Oils: Production, Consumption and Stocks: 2001-Summary*. M311K(01)-13. US Department of Commerce, Economics and Statistics Administration, US Census Bureau, Washington, DC. [Accessed 2004.] Available from http://www.census.gov/industry/1/m311k0113.pdf

US Department of Commerce. (2003b) *Fats and Oils: Production, Consumption and Stocks: 2002-Summary*. M311K(02)-13. US Department of Commerce, Economics and Statistics Administration, US Census Bureau, Washington, DC. [Accessed 2004.] Available from http://www.census.gov/industry/1/m311k0213.pdf

US Department of Energy. (2004a) *Annual Energy Outlook 2004, With Projections to 2025*. DOE/EIA-0383 (2004). US Department of Energy, Energy Information Administration, Office of Integrated Analysis and Forecasting, Washington, DC.

US Department of Energy. (2004b) *Summary Impacts of Modeled Provisions of the 2003 Conference Energy Bill*. SR/OIAF/2004-02. US Department of Energy, Energy Information Administration, Office of Integrated Analysis and Forecasting. Washington, DC.

Appendix

Keys to Tables 7.2 – 7.4

Nearly all the data used in generating these projections originate with various agencies in USDA. A key publication is the monthly *World Agricultural Supply and Demand Estimates* (US Department of Agriculture, 2004c) in which the World Agricultural Outlook Board integrates information from the Agricultural Marketing Service (AMS), the Economic Research Service (ERS), the Farm Service Agency (FSA) and the Foreign Agricultural Service (FAS). Crop and livestock estimates and farm prices originate with the National Agricultural Statistics Service (NASS). Of particular value for international historical data has been the *PSDONLINE* (Foreign Agricultural Service, 2004).

The gross margin figures on maize and soybeans for US farmers per acre are constructed by using a procedure that generates the gross receipts/acre by adding the return from the market (US average farm price or the marketing loan rate, which ever is higher, times the US average crop yield) plus an estimate of direct government payments/acre. By subtracting the ERS estimates of variable cash costs/acre, the gross margins/acre are derived (Foreman, 2001; Foreman and Livezey, 2002).

The aggregated gross margins on crops are calculated by multiplying the gross margins/acre on maize, sorghum, barley, oats, wheat and soybeans by harvested acres with adjustment to derive sales. Sales rather than production were used to avoid double counting on livestock farms growing their own feed. Then on livestock, the aggregate gross margins represent the gross from sales less estimates of feed costs which include opportunity costs for home grown feeds. On dairy, direct payments under the national dairy market loss programme are included in gross receipts.

The soybean meal and oil prices are simple averages for the crushing seasons beginning October 1 at Decatur, Illinois. The oil price is for crude and the meal price is for 48% protein.

While prices on the maize processing by-products of maize gluten feed (Illinois Points) and distillers' dried grains (Lawrenceburg, Indiana) are collected and reported by USDA, the production statistics are not. On maize gluten feed and meal, the amounts fed by crop year are published in *Feed Situation and Outlook Yearbook* (Economic Research Service, 2004). Export data are available from FAS which are added to the amounts fed to derive total production. For the past couple of decades, production data on distillers' dried grains, not being available, have been derived from ERS's estimates of maize used for fuel alcohol with assumptions about the allocation between wet and dry milling.

For the estimates of government costs on feed grain, soybeans, wheat and dairy, the per unit payments (per acre on crops and per hundredweight on milk) were multiplied by harvested acres/production. Meat supply per capita represents production plus imports minus exports on beef, pork, broilers and turkeys per capita. The 'Index of livestock prices' is an average of market prices on cattle and pigs and wholesale prices on broilers and turkeys weighted by their respective per capita availabilities.

Chapter 8

Ethanol Policies, Programmes and Production in Canada

K.K. Klein, Robert Romain, Maria Olar and Nancy Bergeron

Introduction

In late 2002, the Canadian government ratified the Kyoto Protocol. This protocol calls for the reduction of greenhouse gas (GHG) emissions during the years 2008-2012 to 94% of their 1990 level. In Canada, GHG emissions are projected to increase to 841 million tons/year by 2010 under business as usual (BAU) conditions. The Kyoto target for Canada is 623 million tons/year, requiring a reduction of more than 35% below the BAU situation. While an implementation plan on how to achieve this reduction has not yet been agreed with the provinces and industries, increasing the use of ethanol and other biofuels will be integral to successfully achieving this target. BIOCAP Canada, a non-governmental organization that links Canadian university researchers looking for answers regarding GHG reduction, has estimated an ethanol production potential of 738 million gallons with the use of available grains and residues in Canada (BIOCAP Canada and Canadian Agri-Food Research Council, 2003). This production level would result in a reduction of GHGs of 4.67 million tons of carbon dioxide (CO_2) per year (Table 8.1).

An Ethanol Expansion Program[1] (EEP) was developed by the federal government with a national objective for increasing the consumption of fuel ethanol in Canada. The programme calls for at least 35% of Canadian consumption of petrol by 2010 (mid-point of the 2008-2012 period targeted by the Kyoto Protocol) to be E10 (10% ethanol and 90% petrol). If this target is met, ethanol production will have to increase to 370 million gallons/year by 2010 (from the existing 63 million gallons). It is estimated that the GHG reductions associated with this level of ethanol production will be about 1.9 million tons/year compared to the BAU scenario (Government of Canada, 2003).

The major agricultural provinces in Canada also have initiated programmes to increase the production and consumption of ethanol. While the programmes invariably are defended on the basis of environmental concerns, a major driver undoubtedly reflects the search for new markets for low-priced agricultural commodities in Canada.

---

[1]The Ethanol Expansion Program includes the Future Fuels Initiative program, a component of which is the National Biomass Ethanol Program.

© CAB International 2005. *Agriculture as a Producer and Consumer of Energy* (eds J.L. Outlaw, K.J. Collins and J.A. Duffield)

Table 8.1. Ethanol potential and GHG savings using Canadian cereal crops.

Raw material	Ethanol production potential (Million gallons)	GHG savings per year (Million tons CO_2)
Cereal grain	312.3	1.92
Maize grain	141.6	0.87
Cereal residues	158.5	1.49
Maize stover	46.2	0.40
Totals	658.6	4.67

Source: BIOCAP Canada and Canadian Agri-Food Research Council (2003), Table 2.

This chapter reviews the policies and programmes initiated in Canada to increase production of ethanol and their progress to date. Although biodiesel is likely to become an important alternative fuel in the future, to date, little activity to develop biodiesel has taken place in Canada.

Policies and Programmes for Ethanol in Canada

Federal Government

The federal government in Canada promotes the development of a fuel ethanol industry through two main instruments: (1) an excise petrol tax exemption and (2) the Ethanol Expansion Program (EEP). In addition to these two important initiatives, the federal government shows leadership by operating its E85 poly-fuels vehicle fleet, which can use all ethanol-petrol blends up to 85% ethanol (Mesly, 2004).

The Canadian federal excise petrol tax of US$0.28/gallon is not imposed on the portion of ethanol contained in gasohol (although there is no guarantee on how long this exemption will last). This is about half the level of support available in the USA. Imports of USA-produced fuel ethanol are eligible for the Canadian federal excise tax exemption on ethanol-blended fuels (Manitoba Energy, Science and Technology, 2002). All fuel ethanol use in Canada, whether from domestic sources, the USA, or other countries, enjoys the same tax exemption. North American Free Trade Agreement (NAFTA) countries also pay no import duty (D. Tupper, Agriculture and Agri-Food Canada, Ottawa, Ontario, June 2004, personal communication).

EEP provides support for ethanol in three ways: (1) contingent loan guarantees of US$102 million; (2) public awareness financing (to provide market information for consumers) of US$2.2 million; and (3) US$73 million devoted to the financing of fuel ethanol production facilities. The contingent loan guarantee programme was created to counter any reduction or elimination of the excise tax exemption that could affect the viability of new ethanol plants. The programme will come into effect if these changes are imposed prior to December 31, 2014 (Farm Credit Canada, 2004). In order to qualify for the loan guarantee, manufacturers would have to experience a reduction in cash flow due to a change in the excise tax treatment. Loans would be made directly to lenders in order for ethanol manufacturers to be able to restructure their long-term debts. The contingent loans would be repayable at commercial rates of interest (Government of Canada, 2001).

In addition to loan guarantees, the programme provides US$2.2 million over 5 years for a public outreach component. Its aim is to provide essential market information to consumers through such activities as public education on fuel ethanol, analysis of fuel ethanol markets and producer economics and liaison with provinces and industries interested in ethanol plant expansion (Natural Resources Canada, 2001). The contributions for fuel ethanol production capacities are offered in two rounds of funding, for a total of US$73 million over 3 years. The maximum amount payable to any applicant is US$36.5 million and cannot represent more than 25% of the total project costs minus other federal, provincial/territorial and municipal governmental contributions. The eligibility criteria include a minimum new or expanded production capacity of 2.63 million gallons/year and the condition to start production no more than 30 months after signing the contribution agreement. Contributions are repayable. Repayments must start 3 years after the date of the final contribution payment and must end 10 years after the date of the final contribution payment (Government of Canada, 2003).

Provinces

Provincial policies on fuel ethanol are driven mainly by the characteristics of the province's economies. The governments of Manitoba and Saskatchewan have a conciliatory ethanol expansion policy since they consider expansion of ethanol use as a potential boost for their rural economies. The government of Alberta offers lower subsidies than do other provinces, perhaps due to the importance of its oil industry. British Columbia is analysing the commercial feasibility of cellulose-based ethanol production technology because of its forest residues. Table 8.2 shows the provincial incentives for production of fuel ethanol.

On June 11, 2004, the Government of British Columbia announced that, effective July 1, 2004, ethanol blended with petrol will be exempt from the provincial fuel tax exemption of US$0.40/gallon, provided the blended product contains at least 5% and no more than 25% ethanol (Canadian Taxpayers Federation, 2004; British Columbia Ministry of Provincial Revenue, 2004).

The Alberta Ministry of Agriculture, Food and Rural Development has had an ethanol policy since 1993. The policy guarantees that the exemption of provincial fuel tax payable on vehicle fuel will continue for a period of 5 years after the start-up of an ethanol production plant. The exemption is currently US$0.25/gallon of ethanol sold in the province (Cheminfo Services Inc. *et al.*, 2000). The Government of Alberta maintains its ethanol fuel tax exemption for ethanol sold in the province regardless of where it is produced (D. Tupper, Agriculture and Agri-Food Canada, Ottawa, Ontario, June 2004, personal communication).

From 1991 to 1993, the government of Saskatchewan subsidized the Lanigan ethanol plant owned by Pound-Maker Agventures Ltd. at the level of US$1.11/gallon (Freeze and Peters, 1999). The Saskatchewan government's ethanol policy was changed in the March 2000 budget. The province reinstated an exemption of US$0.42/gallon for ethanol blended with petrol. Then, in March 2002, the Saskatchewan government announced a plan to develop an ethanol industry. The plan is called Greenprint for Ethanol Production in Saskatchewan (Government of Saskatchewan, 2002). One component of the plan is the *Ethanol Fuel Act* which was passed in 2002 and amended in 2004. This act sets a mandate that fuel volumes must contain 2% ethanol by May 1, 2005, and 7.5% ethanol by November 1, 2005 (Government of Saskatchewan, 2004). A second part of the plan is the obligation for distributors to buy at least 30% of their ethanol from plants that produce 6.6 million gallons/year or less (Briere, 2002). The regulation promotes producer-owned facilities.

The heterogeneity of the provincial tax exemptions (amounts, eligibility and duration) represents an important barrier to inter-provincial trade. For example, Alberta's single ethanol plant exports almost all its production to the USA because it does not have access to the

Table 8.2. Tax exemptions for fuel ethanol by province.

Province	Provincial fuel tax exemptions for ethanol (US$/gallon)	Eligibility for the subsidy	Duration
Alberta	0.25	No restriction on ethanol source	5 years after the start-up of an ethanol production plant
British Columbia	0.40	For E85 to E100 and E5 to E25 Ethanol must be produced in British Columbia	N/A
Ontario	0.40	No restriction on ethanol source	Until 2010
Saskatchewan	0.42	Ethanol must be produced and consumed in Saskatchewan	5 years
Quebec	0.44 to 0.55	Ethanol must be produced in Quebec	1999 - 2012
Manitoba	0.56 until August 2007 0.42, September 2010 – August 2010 0.28, September 2010 – August 2013 (in addition, 0.04 US$/gallon excise tax reduction for the gasoline blended with 10% Manitoba-made ethanol)	Ethanol must be produced and consumed in Manitoba	No duration specified
Federal	0.28	N/A	No duration specified

Source: Cheminfo Services Inc. *et al.* (2000); Government of Manitoba (2002).
Note: N/A = not available.

Saskatchewan market; Saskatchewan's tax exemption applies only to locally produced ethanol. On the other hand, Saskatchewan ethanol producers can sell their production in Alberta where the tax exemption does not impose any restriction on the source of the ethanol.

Manitoba has no oil refineries and imports all of its petrol (Government of Manitoba, 2002). In December 2003, the Government of Manitoba passed *The Biofuels and Gasoline Tax Amendment Act*. The act establishes a mandate for ethanol use in the province such that 85% of all petrol sold must contain 10% ethanol by September 2005. The act also outlines a petrol tax reduction for gasohol (E10) of US$0.056/gallon of gasohol until August 31, 2007, reduced to US$0.042/gallon of gasohol for the next 3 years and to US$0.028/gallon of gasohol for the following 3 years (Government of Manitoba, 2003b). Regulations that will accompany the act are currently being developed and will be shared with industry stakeholders prior to implementation (D. Tupper, Agriculture and Agri-Food Canada, Ottawa, Ontario, June 2004, personal communication). As in the case of Saskatchewan, the Manitoba subsidy is available only for ethanol that is produced and consumed in the province. As a result, an ethanol producer in Manitoba that is not engaged in the distribution or retail of gasohol does not qualify for the tax preference (Government of Manitoba, 2002). The Manitoba ethanol programme also provides a declining tax preference averaging approximately US$0.04/gallon of petrol that is blended with 10% Manitoba-made ethanol. This component of the programme is scheduled to end in 2013 (Government of Manitoba, 2003a).

Despite having the most generous incentives in Canada, the Manitoba ethanol industry has not changed in over two decades. However, the government of Manitoba states that since the announcement of an ethanol mandate, there has been renewed interest by the oil industry and ethanol producers from across North America in building ethanol plants in Manitoba (Government of Manitoba, 2002).

Since 1980, Ontario has provided an exemption from its road and usage tax on petrol for the ethanol portion of ethanol-blended fuels sold in the province. The current value of the exemption is US$0.40/gallon of ethanol (Seaway Valley Farmers Energy Co-operative Inc., 2004). Since October 1994, the province has entered into project-specific agreements, such as the Ontario Ethanol Manufacturers' Agreement, with ethanol producers that use renewable feedstock. This guarantees that the financial benefit of the US$0.41/gallon exemption to producers will remain until 2010, even if the tax structure is changed by administrative or legislative action in the interim (Seaway Valley Farmers Energy Co-operative Inc., 2004). Two other governmental initiatives for sustaining the domestic ethanol industry are a US$3.65 million grant to Commercial Alcohols Inc. for building its Chatham plant and the use of ethanol blends in the governmental vehicle fleets (Government of Ontario, 2002). In the spring of 2004, the Government of Ontario established a target to have ethanol represent 5% of the petrol pool by 2007 and 10% by 2010 (which would require around 370 million gallons of ethanol). Policy development is underway to determine how best to achieve this target (D. Tupper, Agriculture and Agri-Food Canada, Ottawa, Ontario, June 2004, personal communication).

There are no fuel ethanol plants in Quebec at this time; the one in Temiscaming produces industrial ethanol. However, it is expected that a fuel ethanol plant will be built in Varennes using the federal financial support offered through the EEP. The provincial tax policy to support the Varennes plant features a tax exemption amounting to 106% to 130% of the provincial road tax of US$0.42/gallon, or US$0.44 to 0.55/gallon of ethanol. The formula that will be applied to compute the exemption has yet to be finalized, so there is still some uncertainty with respect to the final amount of public support to this venture.

The governments of New Brunswick and Prince Edward Island are studying the feasibility of establishing ethanol production facilities in their provinces based on agricultural or forest resources. In June 2004, the Prince Edward Island government released its Energy Framework and Renewable Energy Strategy which mentions the pursuit of an ethanol and biofuels industry in the province. Through the Atlantic Energy Ministers' Forum, there may be an opportunity to develop a regional ethanol facility that improves economic feasibility (D. Tupper, Agriculture and Agri-Food Canada, Ottawa, Ontario, June 2004, personal communication).

Table 8.3. Canadian ethanol plant capacities (million gallons/year), 1976-2000.

Company	Location	1976	1980	1990	1995	2000
Ontario Paper	Thorold, Ontario	1.06	1.06	N/A	N/A	N/A
St. Lawrence Starch	Mississauga, Ontario	3.96	3.96	N/A	N/A	N/A
Commercial Alcohols	Varennes, Quebec	18.49	18.49	18.49	N/A	N/A
North West	Kerrobert, Saskatchewan	N/A	0.79	0.79	N/A	N/A
Mohawk Oil	Minnedosa, Manitoba	N/A	1.06	2.38	2.64	2.64
Commercial Alcohols	Tiverton, Ontario	N/A	N/A	3.17	5.81	5.81
Tembec Enterprises	Temiscaming, Quebec	N/A	N/A	4.76	4.76	4.76
Pound-Maker Agventures	Lanigan, Saskatchewan	N/A	N/A	N/A	2.64	3.17
API Grain Processing	Red Deer, Alberta	N/A	N/A	N/A	N/A	6.87
Commercial Alcohols	Chatham, Ontario	N/A	N/A	N/A	N/A	39.63
Total		23.51	25.36	29.59	15.85	62.87

Source: Cheminfo Services Inc. *et al.* (2000).
Note: N/A (not available).

Fuel Ethanol Production Capacity

Total ethanol production capacity in Canada currently is 63 million gallons/year. Fuel ethanol is the major product with 67% of total capacity while industrial ethanol constitutes the remaining 33% (Table 8.3). Production of fuel ethanol is concentrated in southeastern Ontario (72%) where Commercial Alcohols has two plants, a 40 million gallons/year production capacity at Chatham and a 5.8 million gallons/year at Tiverton (fuel ethanol represents 65% of total ethanol production of both plants).

The plant in Minnedosa, Manitoba, owned by Husky Energy Inc., was the first Canadian plant to produce fuel ethanol (Agriculture and Agri-Food Canada, 2001). This plant started its operations in 1980 with a capacity of 1.1 million gallons/year. It currently operates at a capacity of 2.64 million gallons/year. During 1976 to 1990, four other ethanol plants were operating with a total production capacity of 24.3 million gallons/year, but they only produced industrial ethanol.

Commercial Alcohols' Varennes plant in Quebec ceased production in 1991 after two decades of operation. EEP will help finance a US$77 million conversion of the plant that will bring it back into production with an annual capacity of 33.2 million gallons/year.

EEP is slated to finance six other plants. The total production capacity financed by the EEP will amount to 195 million gallons/year. Both total and average production capacities represent important increases relative to existing plants. Plans are in the works for 19 other ethanol plants that would provide an additional production of 317 million gallons/year. Parts of these plants are to be assisted financially by provincial governments.

Figure 8.1 shows the evolution of Canadian ethanol supply from 1976 to the present (continuous line) and projected supply for the next 10 to 20 years (based on EEP-financed plants and projected capacities of planned plants). Figure 8.1 shows that ethanol production capacities in Canada increased substantially from the mid-1990s to the present. This increase is expected to continue until 2020 due to the construction of plants financed partially by public funds.

Given the high costs of production and the availability of cheaper fuels, construction of ethanol plants is very sensitive to governmental regulations and public funding. Development of a commercial ethanol industry is likely to be highly dependent on technological improvements that reduce production costs and increase environmental benefits, as well as on the manifestation of environmental constraints such as climate change.

Feedstock

Ethanol can be produced from two main categories of feedstock – grains and cellulose. Ethanol is obtained from grains by fermentation of sugars (starch) and from cellulose by conversion of the cellulose into sugars and their fermentation afterwards. Grain-based ethanol is obtained mainly from sugarcane, maize, wheat or barley. Cellulose-based ethanol comes from waste biomass or dedicated energy crops like switchgrass, prairie grasses and fast-growing trees. Ethanol production in North America uses maize as the primary feedstock. The exception to this is western Canada where wheat has been the dominant feedstock. This is due to the lack of maize production in the prairies where wheat provides lower production costs than importing maize into the region. The area generally does not have enough heat degree-days and moisture for maize production. Exceptions are southern Manitoba and a very small part of the irrigated area of southern Alberta (Cheminfo Services Inc. *et al.*, 2000).

In Canada, the grain-based production process is used for 92% of actual production capacity. Ethanol plants in Canada presently use maize (73%), wheat (17%) and barley (3%) as grains (such as Commercial Alcohols, Husky Energy, API Grain processors, Pound-Maker Ethanol) and agricultural and forestry waste (7%) as cellulose (for example, Iogen Corporation and Tembec). Details on capacities, feedstock, co-products, type of ethanol produced, etc. are shown in Fig. 8.2 and Table 8.4.

Wheat has lower starch content but higher protein and fibre content than maize. Commercial ethanol yields from wheat typically vary from 98.6 gallons/ton to 145 gallons/ton of wheat (Cheminfo Services Inc. *et al.*, 2000). All classes of wheat are capable of being used for ethanol production, but the favoured feedstock for a dry-milling ethanol plant is medium hard Canadian Prairie Spring (CPS) wheat. A study published by Freeze and Peters

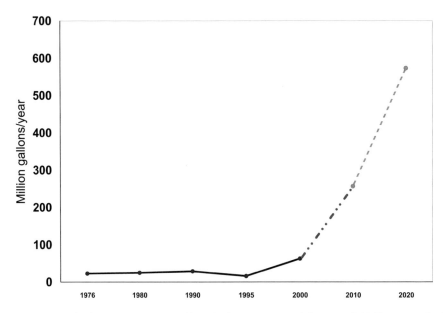

Source: Cheminfo Services Inc. *et al.* (2000); Government of Canada (2004), appendix 3.

Figure 8.1. Canadian ethanol production capacity.

(1999) found that CPS wheat cultivated on dark brown and black soils generated the highest revenues for an ethanol plant because of its high starch content and low price.

Barley can also be used for ethanol production and is the feedstock in the Saskatchewan plant at Lanigan. However, it has lower starch and higher fibre contents than wheat, making it less desirable. Some of the carbohydrates in barley are beta-glucans that are difficult to hydrolyse and ferment. Barley hulls can cause erosion of ethanol plant equipment. The lower cost of barley is insufficient to overcome the disadvantages of processing barley. This results in production costs for ethanol being higher for barley than for wheat (Cheminfo Services Inc. *et al.*, 2000).

Iogen Corporation, based in Ottawa, is a world leader in cellulose ethanol technology. Iogen has produced 0.79 to 1.06 million gallons of ethanol per year in a demonstration plant in Ottawa (using about 44 tons/day of feedstock). In April 2004, the company announced that it is searching for a location to site a commercial facility that will process about 1,653 tons/day of feedstock and produce around 44.9 million gallons of ethanol per year (Iogen Corporation, 2004). This will be the first commercial plant to process bio-wastes into ethanol in the world.

Co-products

The main co-product of cellulose-based ethanol is lignin, which is burned to produce steam for the ethanol production process, with the excess potentially being converted into electricity for sale to the power grid (Government of Manitoba, 2002).

Only the starch component of the grain is converted to ethanol. The fibre, protein, minerals, CO_2 and vitamins remain and are recovered as co-products. The co-products of grain-based ethanol may be described under two general categories of ethanol production: dry-milling and wet-milling. Dry-milling is the dominant process in Canada's relatively small

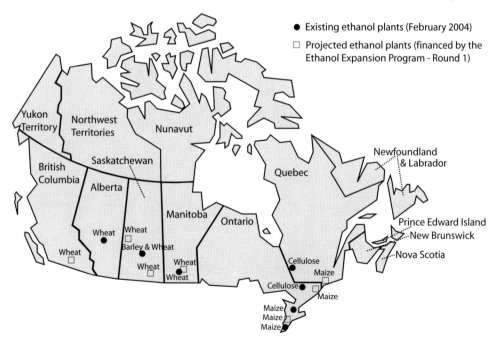

Figure 8.2. Feedstocks used in Canadian ethanol plants.

Table 8.4. Canadian ethanol plants, December 2003.

Producer	Location	Capacity (million gal/yr)	Feedstock (million gal/yr)	Production process	Co-products	Feedlot integration	Type of ethanol	Fuel ethanol retailer	Ownership
Iogen Corporation	Ottawa, Ontario	0.79-1.06 (demo plant)	Wheat straw (40 ton/day)	Not available	Lignin	Not available	Fuel ethanol	Petro-Canada, Shell	Not available
Husky Energy Inc.	Minnedosa, Manitoba	2.64	Wheat (2)	Dry-mill	Fibrotein (food-grade co-product)	No	Fuel ethanol	Husky Energy Inc.	An integrated Canadian energy company
Pound-Maker Ethanol Ltd.	Lanigan, Saskatchewan	3.17	Wheat (1.5), barley (3)	Dry-mill	Distillers grains	Yes	Fuel ethanol	Husky Energy, Inc.	Not available
Tembec	Temiscaming, Quebec	4.76	Pulping process waste	Not available	Lignin	Not available	Industrial ethanol	Not available	Forest products company
Commercial Alcohols Inc.	Tiverton, Ontario	5.81	Maize	Dry-mill	Wet distillers grains	No	65% fuel ethanol; 35% industrial ethanol	Suncor Energy, Inc.	The largest manufacturer of industrial/fuel ethanol in Canada
API Grain Processors /Permolex	Red Deer, Alberta	6.87	Canadian Prairie Spring Red Wheat	Wet-mill	Wheat gluten	Yes	85% fuel ethanol; 15% industrial ethanol	Not available	Agri Partners International, Inc. (API) and the Edmonton Pipe Industry Pension Trust Fund
Commercial Alcohols Inc.	Chatham, Ontario	39.63	Maize (15)	Dry-mill	Dried distillers grains (125,000 ton) wet distillers grains, syrup, CO_2 (100,000 ton)	No	65% fuel ethanol; 35% industrial ethanol	Suncor Energy, Inc.	The largest manufacturer of industrial/fuel ethanol in Canada

Source: Derived from various producer Internet sites.

ethanol plants. The only wet-milling ethanol plant is the one located in Red Deer, Alberta, owned by API Grain Processors/Permolex and integrated with a feedlot. The economics of production favour the dry-milling process, in contrast to many large US plants that are able to exploit the scale efficiencies of wet-milling technology (Agriculture and Agri-Food Canada, 2002).

There are three main co-products of the wet-milling production process: gluten meal, gluten feed and germ. In the case of wheat, the primary outlet for gluten is for bakery products. Gluten is added to white pan bread, rolls, diet breads and other products. The addition of gluten to baked goods improves dough-handling properties and quality of the finished product. Supplementing flours that have poor baking qualities and low protein content with gluten permits a reduction in the number of flour types required for baking and tends to increase production flexibility. Addition of gluten to buns and rolls improves hinge strength and produces the type of crust most desirable in commercial markets where buns are steamed. Gluten also finds some usage as a supplemental source of protein in breakfast cereals. Gluten can also be used as a texturizing protein and meat substitute in vegetable meat-like products (Cheminfo Services Inc. *et al.*, 2000).

While a vital gluten market for human consumption is expected to rise only in proportion to population, significant growth is anticipated in pet foods where it is used as a supplement and/or replacement for meat, due to its very high protein content (e.g., greater than 80%). Gluten is an attractive alternative to pet food processors because of the higher prices of meat (based on protein content) (Cheminfo Services Inc. *et al.*, 2000).

The market for wheat gluten in North America is dominated by less than half a dozen producers. The major Canadian producer of wheat gluten is ADM (Archer Daniels Midland) in Lachine, Quebec, but this facility does not make ethanol. Its capacity is estimated to be 22,046 tons of gluten per year, representing nearly twice the size of total Canadian demand. ADM exports gluten to the USA. The only ethanol facility in Canada producing wheat gluten as a co-product is API Grain Processing in Red Deer, Alberta. Its capacity is assumed to be smaller than ADM's, with a substantial portion of its production potentially used for its enriched flour products (Cheminfo Services Inc. *et al.*, 2000).

Ethanol Production Cost

Ethanol plants in Canada are very small relative to petroleum refineries; therefore, they are price takers and have no influence on ethanol prices (Shapouri, 2003). Ethanol prices are much higher than petrol prices in Canada, though the difference narrows with each increase in the price of oil. The main factors influencing ethanol prices are production costs and tax exemptions. The unit production cost for ethanol is the sum of feedstock and processing cost/volume of ethanol, after subtracting the value of the co-products (Baker *et al.*, 1990). Thomassin and Baker (2000) estimated that feedstock cost represented 57% of the total production cost for a 52.63 million gallons/year maize ethanol plant located in southern Ontario.

Heath (1989) estimated the cost of producing ethanol from maize feedstock (including processing costs) in the late 1980s at US$0.97/gallon. Using Heath's estimate of processing costs and Jerusalem artichokes (*Helianthus tuberosus*) as a feedstock, Thomassin *et al.* (1992), estimated the feedstock and processing costs of ethanol to be US$0.55/gallon for western Canada and US$1/gallon for Quebec. A study conducted for the Government of Alberta in 2000 estimated the breakeven ethanol price for a 26.4 million gallon/year facility to approximate US$0.78/gallon. The plant in the study used, advantageously, a raw material (wheat) price of US$80.45/ton and a co-product (Distiller's Dried Grain) price of US$129/ton (Cheminfo Services Inc. *et al.*, 2000). A study conducted for the Government of Manitoba (2002) underlines the economies of scale obtained by larger ethanol plants (see Fig. 8.3).

Concluding Note

Increased production of ethanol and other biofuels shows great promise for reducing GHG production in Canada. However, existing technologies do not appear to make the manufacture of ethanol a profitable venture for private sector companies without government assistance. Since Canada has committed to reduce GHGs through its ratification of the Kyoto protocol, a number of government regulations have been made that mandate minimal usages of ethanol for automobile fuels. In addition, both federal and provincial governments have put in place various types of subsidies to encourage development of a viable ethanol industry in Canada. A Canadian company, Iogen Corporation, has developed a promising technology to produce ethanol from cellulose. Following more than 25 years of research and several years of operating a demonstration plant, the company announced recently that it plans to develop a full-scale commercial plant as soon as it can determine the best location. The prospect that wastes and residues can be used in the commercial production of ethanol offers an exciting prospect for adding to and diversifying the feedstock for this important alternative fuel.

It is clear that the Canadian public, through their elected politicians, have been pushing for cleaner, more environmentally friendly fuels that can be produced from biological rather petrochemical processes. However, much research remains to be done to make the biologically produced fuels less costly to produce. Research is required on higher-yielding crops and

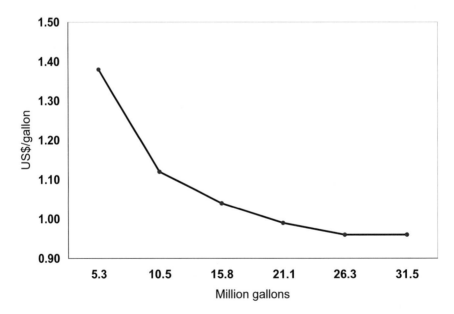

Source: Manitoba Energy, Science and Technology (2002).

Figure 8.3. Ethanol production costs by plant size.

biomass, development of new high-value co-products from the production process and improved methods to cheaply extract sugars from cellulosic raw materials.

References

Agriculture and Agri-Food Canada. (2001) Ethanol. *Bi-weekly Bulletin* (May 25) Vol. 14, No. 9. [Accessed 2004.] Available from http://dsp-psd.communication.gc.ca/Collection/A27-18-14-9E.pdf

Agriculture and Agri-Food Canada. (2002) Canada: Primary Processing of Grains, Oilseeds, Pulses and Special Crops. Bi-weekly Bulletin. (July 19) Vol. 15, No. 14. Available from http://www.agr.gc.ca/mad-dam/e/bulletine/v15e/v15n14e.htm

Baker, L., Thomassin, P. and Henning, J. (1990) The Economic Competitiveness of Jerusalem Artichoke (*Helianthus tuberosus*) as an Agricultural Feedstock for Ethanol Production for Transportation Fuels. *Canadian Journal of Agricultural Economics* 38: 981-990.

BIOCAP Canada and Canadian Agri-Food Research Council. (2003) *An Assessment of the Opportunities and Challenges of a Bio-Based Economy for Agriculture and Food Research in Canada*. BIOCAP Canada, Kingston, Ontario, and Canadian Agri-Food Research Council, Ottawa, Ontario. [Accessed 2004.] Available from http://www.biocap.ca/files/Position_Paper_OCBBE.pdf

Briere, K. (2002) Ethanol. *The Western Producer*. Special Report. November 21: 76-78.

British Columbia Ministry of Provincial Revenue. (2004) *Notice to Collectors: Changes to the Taxation of Alternative Motor Fuels*. British Columbia Revenue Programs Division, Victoria, British Columbia.

Canadian Taxpayers Federation. (2004) 2004 Tax Rates by Province. *Tax Facts*. Canadian Taxpayers Federation, Ottawa, Ontario. [Accessed 2004.] Available from http://www.taxpayer.com/Facts/index.html

Cheminfo Services Inc., (S&T)^2 Consultants Inc. and Cemcorp Ltd. (2000) *Ethanol Production in Alberta: Final Report*. Cheminfo Services, Toronto, Ontario. [Accessed 2004.] Available from http://www1.agric.gov.ab.ca/$department/deptdocs.nsf/all/agc6751/$FILE/ethanol%20production%20in%20alberta%20final.doc

Farm Credit Canada. (2004) National Biomass Ethanol Program. Farm Credit Canada, Regina, Saskatchewan. [Accessed 2004.] Available from http://www.fcc-sca.ca/en/Products/Partnerships/nbep_e.asp?main=2&sub1=partnerships&sub2=nbep

Freeze, B. and Peters, T. (1999) A Note on the Profitability of Wheat-ethanol-feedlot Production in Alberta. *Canadian Journal of Agricultural Economics* 47: 67-78.

Government of Canada. (2001) *Future Fuels Program National Biomass Ethanol Program*. Agriculture and Agri-Food Canada, Ottawa, Ontario. [Accessed 2004.] Available from http://www.climatechange.gc.ca/english/newsroom/2001/steps/steps_future.asp

Government of Canada. (2003) *Government of Canada Launches Ethanol Expansion Program, October 20, 2003*. Government of Canada, Natural Resources Canada, Ottawa, Ontario. [Accessed 2004.] Available from http://climatechange.gc.ca/english/newsroom/2003/20031020_ethanol.asp

Government of Canada. (2004) *Government of Canada Announces Successful Proposals in Ethanol Expansion Program*. Government of Canada, Natural Resources Canada, Ottawa, Ontario. [Accessed 2004.] Available from http://www.nrcan-rncan.gc.ca/media/newsreleases/2004/200402_e.htm

Government of Manitoba. (2002) *Ethanol: Made in Manitoba*. Government of Manitoba, Manitoba Ethanol Development Initiative, Ethanol Advisory Panel, Winnipeg, Manitoba. [Accessed 2004.] Available from http://www.gov.mb.ca/est/energy/ethanol/ethanol2-low.pdf

Government of Manitoba. (2003a) Province Introduces Proposed Legislation to Mandate 10 Per Cent Ethanol Blends by 2005. Government of Manitoba, News Media Services, Winnipeg, Manitoba. [Accessed 2004.] Available from http://www.gov.mb.ca/chc/press/top/2003/11/2003-11-21-03.html

Government of Manitoba. (2003b) *The Biofuels and Gasoline Tax Amendment Act*. Bill 2, 2nd Session, 38th Legislature. S.M. 2003, c. 5. Assented to December 4, 2003. [Accessed 2004.] Available from http://web2.gov.mb.ca/laws/statutes/2003/c00503e.php

Government of Ontario. (2002) *Ontario Supports Clean, Renewable Power Generation*. Government of Ontario, Ministry of the Environment, Toronto, Ontario. [Accessed 2004.] Available from http://www.ene.gov.on.ca/envision/news/2002/102401fs.htm

Government of Saskatchewan. (2002) *Greenprint for Ethanol Production in Saskatchewan*. Government of Saskatchewan, Saskatchewan Industry and Resources, Regina, Saskatchewan. [Accessed 2004.] Available from http://www.ir.gov.sk.ca/adx/asp/adxGetMedia.asp?DocID=3288,3286,2937,2936,Documents&MediaID=4734&Filename=Greenprint.pdf

Government of Saskatchewan. (2004) *Increased Ethanol in Fuel May 1st, 2005*. Government of Saskatchewan, Saskatchewan Industry and Resources, Regina, Saskatchewan. [Accessed 2004.] Available from http://www.gov.sk.ca/newsrel/releases/2004/06/10-354.html

Heath, M. (1989) *Towards a Commercial Future: Ethanol & Methanol as Alternative Transportation Fuels*. Canadian Energy Research Institute Study No. 29. University of Calgary Press, Calgary, Alberta.

Iogen Corporation. (2004) *A Fully Integrated Facility*. Iogen Corporation, Ottawa, Ontario. [Accessed 2004.] Available from http://www.iogen.ca/2200.html

Manitoba Energy, Science and Technology. (2002) Ethanol Backgrounder. *Homegrown Energy: Developing Manitoba's Ethanol Industry - Energy Development Initiative*. Manitoba Energy, Science and Technology, Energy Development Initiative, Winnipeg, Manitoba. [Accessed 2004.] Available from http://www.gov.mb.ca/est/energy/ethanol/ebckgrder2.html#Benefits

Mesly, P. (2004) Éthanol: les enjeux d'un carburant 'écolo'. *Coopérateur Agricole* (January) Vol. 33, No. 1. [Accessed 2004]. Available from http://www.coopfed.com/COOPERATEUR/contenu/janv_2004/page34.htm

Natural Resources Canada. (2001) *Future Fuels Initiative*. Natural Resources Canada, Office of Energy Efficiency, Transportation, Ottawa, Ontario. [Accessed 2004.] Available from http://oee.nrcan.gc.ca/vehiclefuels/ethanol/ethanol_futurefuels.cfm?PrintView=N&Text=N

Seaway Valley Farmers Energy Co-operative Inc. (2004) *Seaway Project Highlights*. Seaway Valley Farmers Energy Co-operative Inc., Cornwall, Ontario. [Accessed 2004.] Available from http://www.seaway-energy.on.ca/highlite.html

Shapouri H. (2003) *Ethanol Production Economics*. Presentation, World Summit on Ethanol for Transportation. Quebec City, Quebec. November 2-4.

Thomassin, P. and Baker, L. (2000) Macroeconomic Impact of Establishing a Large-scale Fuel Ethanol Plant on the Canadian Economy. *Canadian Journal of Agricultural Economics* 48: 67-85.

Thomassin, P., Henning, J. and Baker, L. (1992) Macroeconomic Impacts of an Agro-Ethanol Industry in Canada. *Canadian Journal of Agricultural Economics* 40: 295-310.

Chapter 9

Economic Analysis of Alternative Lignocellulosic Sources for Ethanol Production

Brian K. Herbst, David P. Anderson, Michael H. Lau, Joe L. Outlaw, Steven L. Klose and Mark T. Holtzapple

Introduction

With recent increases in oil prices leading to an increase in petrol prices, the demand for oxygenates as fuel extenders and octane boosters has grown tremendously. Methyl tertiary-butyl ether (MTBE) and ethanol have been the primary fuel oxygenates in petrol. MTBE has been recently banned by several states as it is linked to water contamination and will likely be banned in additional states, if not nationwide.

Production capacity for ethanol in the USA is expected to exceed 3.4 billion gallons (gal) during 2004 (Renewable Fuels Association, 2003). Ethanol production has increased rapidly over the past few years due to concern over the conflict in the Middle East, the need for reduction in air pollution, the ban on MTBE and suppressed commodity prices for maize (Herbst, 2003).

The increased demand for ethanol across the country and volatile feedstock prices have led people to look for different ways to economically produce ethanol. Processes for making ethanol from biomass other than grain have been in the existence for many years (Klass, 1998). But alcohol fuel production from any feedstock has not been an economical alternative to petrol production in the USA. Most US ethanol plants are located in the Corn Belt because of the large supply and relatively low price of maize. The abundant supply and lower cost of alternative ethanol feedstocks in other parts of the country has led to increased interest in lignocellulosic ethanol (ethanol from biomass other than grain). Of interest is how much potential exists for producing lignocellulosic ethanol at a lower cost than by feed grains.

The objective of this study is to determine the feasibility of a lignocellulosic ethanol plant in Texas. This research uses new engineering technology in biomass ethanol production that is currently being piloted. The second objective is to compare the financial feasibility of a lignocellulosic plant to a feed grain ethanol plant in Texas.

Background

There has been considerable research on lignocellulosic ethanol production (Hanson, 1985; Nienow *et al.*, 1999; Gallagher *et al.*, 2003; Mapemba and Epplin, 2004a,b). Past research has examined biomass resources, the national and global supply and demand for biomass ethanol, the uses of biofuels, as well as research on methods to process ethanol (Osburn and Osburn, 1993; Holtzapple et al., 1999; Gallagher et al., 2003; Holtzapple, 2003). In fact, producing alcohol from biomass is as old as the first brewery or distillery.

Biomass is any organic matter from plants that derives energy from photosynthesis conversion. Traditional sources of biomass include fuel wood, charcoal, animal manure and grains. Modern sources of biomass include energy crops, agricultural residue and municipal solid waste (Australian Cooperative Research Centre for Renewable Energy Ltd., 2004). Some of the benefits of biomass as an energy source are that it does not contribute to global warming, it has a neutral effect on carbon dioxide emissions, has low sulphur content and thus does not contribute to sulphur dioxide emissions and biomass is a domestic source not subject to world energy price fluctuations. Even today, about 14% of the world's primary energy is derived from biomass and about 0.3% of the total diesel and petrol consumption is fuel produced from biomass (Veringa, 2004).

Biomass can be produced at rates of up to 20 metric tons (MT)/acre of farm crops and trees and up to 50 tons/acre of algae and grass. Osburn and Osburn (1993) focused on biomass production from hemp, which can not be legally grown in Texas, or in most other states.

Many recent studies have examined the supply of biomass fuel production (Gallagher et al., 2003; Mapemba and Epplin 2004a,b). These studies have particularly focused on the amount of biomass produced by various crops, the costs of growing the crops, the potential production and the distribution system required. For example, biomass ethanol production ranges from 6 gal/ton from cheese whey to 89 gal/ton from raisins (Klass, 1998). The conversion rate for sugarcane to ethanol is 42 gal/ton to 49 gal/ton. Other crops that can be used for biomass to ethanol have conversion rates of 87 gal of ethanol/ton of barley and 25 gal/ton of potatoes.

There have also been studies that look at the feasibility of biomass used for electricity generation (Hanson, 1985; Nienow et al., 1999; English et al., 2004). Ethanol is not the only product from biomass. A whole host of chemicals and products have the potential to be produced (Holtzapple et al., 1999). El Bassam (2002) reported that potential energy products that can be made from biomass include biogas, biodiesel, ethanol, diesel, petrol, hydrogen and electricity. Biomass has the potential to replace fossil fuels in the production of these energy sources.

There have been many feasibility studies of the production of grain-based ethanol in the USA (Fruin and Halbach, 1986; Fruin et al., 1996; Evans, 1997; Food and Agricultural Policy Research Institute, 2001; Otto and Gallagher, 2001; Van Dyne, 2002; Herbst, 2003). The studies have been done throughout the USA, but focus mainly in the Midwest because of the quantity of maize that is readily available at a lower price. All of these feasibility studies have examined the effects of maize price, ethanol price and other input prices on the plant's success. The economic impact of the new plant on the local community has also been an important part of these studies.

This chapter makes several contributions to the literature. While there have been many feasibility studies of ethanol plants performed, this one examines production in Texas. This research examines two production processes by comparing the economics of grain based to that of lignocellulosic based ethanol production. The first production process is a standard dry-mill ethanol plant. The second production process is a new technology for lignocellulosic biomass conversion to ethanol. This research also highlights the importance of incorporating risk analysis into any economic feasibility study and so fills that gap in the literature.

The Lignocellulosic Process

This section discusses the lignocellulosic production process used in this analysis. Lignocellulosic processes can convert a wide variety of biomass materials such as sewer sludge, manure, agriculture residues and agriculture crops into acids and alcohol fuels using microorganisms, water, steam, lime and hydrogen through an anaerobic process (Holtzapple, 2003).

Figure 9.1 summarizes the lignocellulosic process utilized in this analysis. This process differs from the use of acid hydrolysis of biomass material to produce ethanol. The process calls for mixing biomass with a nutrient source such as manure or sewage sludge at a ratio of 80% to 20%. There are four phases to the process: pre-treatment and fermentation, dewatering, acid springing and hydrogenation.

During the pre-treatment phase, biomass, lime and calcium carbonate are blended and stored in a large pile. Air is blown up through the pile while water is trickled down through the pile. The combination of air and lime remove lignin from the biomass which reduces the pH rendering the bio-matter digestible. The pile is then inoculated with anaerobic microorganisms from saline environments. The microorganisms digest the biomass forming carboxylic acids commonly known as volatile fatty acids (VFAs) such as acetic, propionic and butyric acids. The VFAs combine with calcium carbonate to form carboxylate salts, which are extracted from the pile with water.

Four reactor piles are created of equal volume. Figure 9.2 and Fig. 9.3 show the schematic of the pre-treatment and fermentation facility. Each reactor is shaped like a cone to minimize material use. For a 44 ton/hour facility, each reactor has a base diameter of 397 feet and is 115 feet high. The fuel pile is covered with a geomembrane to resist the weather, wind and sun. The base consists of a 3 feet thick layer of gravel that is divided by bermed walls to collect the VFA solution.

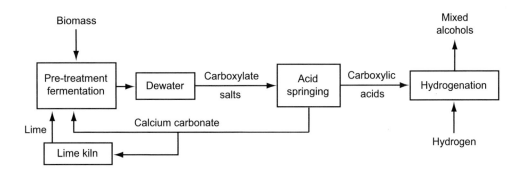

Source: Holtzapple (2003).

Figure 9.1. Schematic of the lignocellulosic process.

From fermentation, the VFA solution is concentrated using a vapour compression evaporator during the dewatering phase. The fermentation broth containing the VFAs is heated to 100°C and mixed with high-molecular-weight acid (e.g., heptanoic) to acidify the fermentation broth. Steam and lime are then used to remove non-condensable gases and

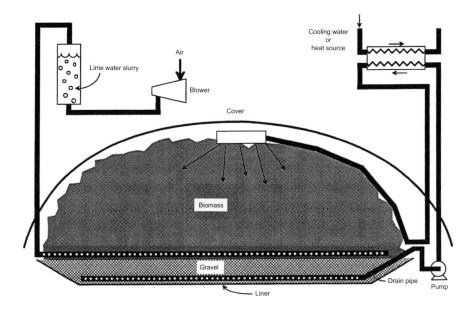

Source: Holtzapple (2003).

Figure 9.2. Schematic of fermentation facility.

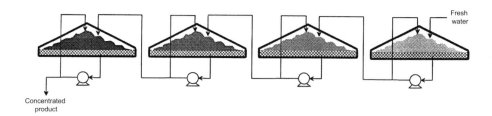

Source: Holtzapple (2003).

Figure 9.3. Schematic of lignocellulosic pre-treatment process.

calcium carbonate. The treated fermentation broth is heated to 212°C and water is evaporated from the solution concentrating the salts.

Acid springing converts the carboxylate salts into carboxylate acid and calcium carbonate. The concentrated broth is blended with carbon dioxide and a low-molecular-weight tertiary amine (triethyl) to form insoluble calcium carbonates and amine carboxylates. Approximately 75% of the calcium carbonate removed can be used in the pre-treatment and fermentation phase and the remaining 25% is converted to lime using a special lime kiln. Most of the water is then removed leaving a concentrated amine carboxylate.

The carboxylate acids are blended with high-molecular-weight alcohols to form esters and water. The water is evaporated and remaining esters are mixed with high-pressure hydrogen to form alcohols. The resulting ethanol fuel is cooled and stored for transportation to be mixed with petrol fuel. Large storage tanks are used to hold the ethanol fuel until shipping.

Co-products

The lignocellulosic process produces water, heat, carbon dioxide, calcium carbonate and residual biomass as co-products. The lignocellulosic facility can almost be self-sufficient in terms of producing part or all of the required inputs other than the feedstock, after the first year of operation, if the necessary equipment for lime production, water recycling, steam captures and boilers are in place. Water can be reused for the pre-treatment and fermentation phase. Calcium carbonate can be manufactured into lime and used in the pre-treatment and fermentation phase. The heat generated can be transferred to dryers to aid in the evaporation during the dewatering phase.

The lignocellulosic structure is completely sealed from the outside environment and all carbon dioxide gas produced can be collected. The carbon dioxide can be released once it is 'scrubbed' to remove odour or sold to oil refineries to be pumped into oil wells to aid in the collection of oil. However, the carbon dioxide market is very limited.

Residual biomass is the largest co-product produced. Lignocellulosic ethanol production differs from maize-based ethanol production in that maize produces distillers dried grains with solubles (DDGS) that be can be sold to livestock operations for feed. Approximately 20% of the biomass feedstock is residual biomass when the lignocellulosic process is complete. The residual biomass can be used internally to generate power and steam for the facility or it can be sold to coal-fired power plants as a fuel source to reduce sulphur emissions.

Net Energy Balance of Lignocellulosic Ethanol

The net energy balance of lignocellulosic ethanol is incomplete as it is dependent upon which feedstock is used as a fuel source. Dried biomass has an approximate heating value of 5,000-8,000 British thermal units (Btu)/pound (Osburn and Osburn, 1993). The efficiency of the lignocellulosic process highlighted in this research is also still under experiment. The lignocellulosic process has an alcohol yield per ton of biomass ranging from approximately 90-100 gal/ton to 130-140 gal/ton depending on the final output desired. The process can produce ethanol or higher energy alcohol products depending on which is desired.

Initial testing has shown ethanol produced from this lignocellulosic process has a slightly higher energy content than maize-based ethanol. A gallon of petrol contains approximately 125,000 Btu/gal and maize-based ethanol contains 84,000 Btu/gal (Holtzapple, 2003). The energy content of this lignocellulosic process-produced ethanol is approximately 95,000 Btu/gal because of the additional mixed alcohol fuels that are produced during the process. The energy content of the residual biomass co-product is similar to coal. It is substitutable for coal in co-firing energy production facilities and can reduce sulphur emissions.

Lignocellulosic Feedstock Requirements

In other parts of the world, lignocellulosic ethanol has been produced using sugarcane bagasse as a feedstock. However, the supply of sugarcane in Texas is uncertain and the industry is not large enough to support large-scale lignocellulosic ethanol production. The amount of feedstock required depends on the desired output amount for the facility. The feedstock is decomposed at the same rate for all crops and all plant sizes.

Lignocellulosic ethanol feedstock demand differs from grain-based ethanol feedstock demand as year-round supply is not necessary. This process only requires feedstock input once a year to build the fuel pile. This is advantageous when compared to other forms of biomass energy production.

Model

A simulation model of two similar-sized plants was developed to examine the feasibility of the lignocellulosic and feed grain ethanol production facilities. The economic feasibility was evaluated using capital budgeting analysis. Capital budgets were developed for construction and operating costs for each type of plant. Simulation analysis methods were used to incorporate risk into the model. This model simulates the economic activity for a 30 million gal/year lignocellulosic production facility and a 30 million gal/year feed grain production facility.

Stochastic simulation is a tool that allows for 'what if…' questions to be analysed in a probabilistic, or risky, framework. This method of financial and economic analysis recognizes that input and output prices are variable, affecting profitability. Stochastic simulation models are preferred over deterministic models because they not only provide an average financial outcome, but also a range of possible outcomes based on probabilities. Confidence intervals are developed around the average outcomes.

Simulation can be done both deterministically and stochastically. Deterministic results do not include risk on any variables and give one answer. Most ethanol studies, to date, have been deterministic. In a deterministic model, it is possible to look at best and worst case scenarios, as a sensitivity analysis, in addition to one answer. This best and worst case analysis has passed for risk analysis in most of the past studies and is inadequate at best. Deterministic modelling is also called perfect knowledge because the results come from only a single estimate for each variable, with no risk (Herbst, 2003).

Stochastic simulation incorporates risk in key variables that are uncertain. In the case of an ethanol plant, the significant uncertain variables are prices for biomass, feed grain (maize and grain sorghum), ethanol, DDGS, natural gas and electricity. Stochastic simulation allows the model to be evaluated over a wide range of possible prices. These variables were simulated using the multivariate empirical (MVE) distribution to insure that the future values are correlated the same way they were correlated in the past and the relative risk of the simulated variables is equal to their historical relative risk. The model developed for this study simulates 500 iterations of the production and financial activity of an ethanol plant.

Historical data is used to create MVE distributions for the stochastic variables (Richardson et al., 2000). The MVE distribution was chosen because it can be used with limited data where a normal distribution would not work. The MVE also provides for full correlation of a system of stochastic variables.

Instead of assuming a certain type of business structure (e.g., corporation, cooperative, limited liability company, partnership) this analysis assumes a generic entity. Profits are taxed at 30%, which is consistent with shareholders and/or partners paying taxes on their earnings. Dividends are calculated as 30% of after-tax net income paid any time net income is greater than zero. If the plant experiences losses, the analysis assumes that there is unlimited financing available. While this is not realistic, it is important for long-run evaluation purposes that the plant is allowed to operate without having to shut down because of a cash shortage. This assumption also allows each scenario to run for the full analysis period eliminating the need to evaluate present values for different length projects.

This analysis used the SIMETAR© simulation package, developed by Richardson et al. (2004). SIMETAR© is an add-in to Microsoft® Excel and was developed in Visual Basic for Applications®. An advantage of this software is that capital budgets can be developed and then manipulated for each type of ethanol plant in one Excel spreadsheet. Risk can then be added to selected stochastic variables within the capital budgeting framework.

Three alternative scenarios for the lignocellulosic ethanol plant are evaluated. Alternative scenarios for initial investment levels (US$20 million, US$40 million and US$60 million) are analysed for lignocellulosic ethanol production. At this time, there are no commercial size and scale lignocellulosic ethanol plants of this type in operation. However, one pilot plant is in operation and extensive work has been done developing operating and capital costs (Holtzapple, 2003). The use of three initial investment levels allows for the analysis to use a range of possible capital costs depending on final plant construction costs.

Three key indicator variables are reported for each scenario. The variables are:

1. Net Cash Income – Net cash income (NCI) is defined as revenues minus operating expenses minus depreciation expense.
2. Ending Cash Before Borrowing – Ending cash (EC) before borrowing is the ending cash flow (total cash inflows less outflows). This value does not reflect short-term borrowing to cover cash flow deficits.
3. Net Present Value – Net present value is calculated through 15 years of operation. An 8% discount rate is used in the analysis.

Net present value (NPV) is the present value of the annual returns to the investment plus the salvage value as defined by the present value of ending net worth minus the initial investment cost. NPV (Equation 1) is specifically defined as:

$$\text{NPV} = -\text{Initial Equity Investment} + \sum_{t=1}^{15}\left(\frac{\text{Dividends}}{(1+i)^t}\right) + \left(\frac{\text{Ending Net Worth}}{(1+i)^{15}}\right)$$

The discounting of future returns and ending net worth allows for the comparison of initial capital investment to returns that occur in different time periods based on the purchasing

Table 9.1. Comparison of variable cost (US$/gal) for lignocellulosic and feed grain ethanol.

	Lignocellulosic	Feed grain
Lime/denaturant	0.048	0.037
Inhibitor/enzymes	0.011	0.055
Hydrogen/chemicals	0.053	0.028
Natural gas	0.027	0.152
Electricity	0.026	0.042
Steam/maintenance materials	0.049	0.037
Cooling water/miscellaneous costs	0.060	0.028
Labour	0.042	0.092
Administrative cost	0.015	0.046
Total variable cost/gallon (excluding feedstock)	0.331	0.517

power of dollars in 2004. Included in the discount rate of 8% are the combined assumptions of future inflation and the investors' required real rate of return. A positive NPV indicates returns over and above 8%.

Assumptions

This study is designed to assess the feasibility of a 30 million gal/year lignocellulosic ethanol production facility in Texas at three different initial investment cost scenarios for the same size plant. No location-specific costs or alternative plant sizes were considered.

The plant is assumed to operate 8,000 hours/year and convert 45 tons of biomass per hour. The plant converts the biomass at 93 gal of fuel per ton of feedstock. Initial investment cost scenarios are of US$20 million, US$40 million and US$60 million, with 50% borrowed capital at 8% for 15 years, for a 30 million gal/year production facility at full capacity. The plant would operate at half capacity (15 million gal/year) in year 1 due to construction time. The plant uses sorghum silage as a feedstock. The feedstock conversion rate is held constant for the 15-year planning horizon. All production and process-related reference data were obtained from Holtzapple (2003).

The assumptions for feed grain ethanol came from industry standards and the Dumas Texas Area Ethanol Feasibility Study (Bryan and Bryan International, 2001). The feed grain ethanol plant's feedstock is grain sorghum. The plant has the capacity of 30 million gal/year and is estimated to cost US$42 million to build with the same assumptions of 50% borrowed capital and 15-year planning horizon at 8% interest. The feed grain ethanol is produced at a conversion rate of 2.75 gal of ethanol/bushel of feed grain. The feed grain plant has US$250,000 in capital replacement costs per year that the lignocellulosic plant does not.

The lignocellulosic production facility used for comparison to the grain ethanol plant is the US$40 million initial investment level. Results are presented for each initial investment level, but only the US$40 million plant is used for comparison. Table 9.1 contains variable costs, by category, excluding feedstock costs, for lignocellulosic and feed grain-based ethanol. Variable costs for lignocellulosic-based ethanol are US$0.186/gal less than feed grain-based ethanol. Feedstock cost per gallon of ethanol produced is US$0.53 for lignocellulosic ethanol versus US$0.79 for feed grain ethanol. The price used for ethanol was stochastic and averaged US$1.117/gal. The lignocellulosic-based ethanol plant receives US$2.3 million, annually, in revenues for selling the residue from the feedstock to be used at a power plant for electricity production. The feed grain-based plant sells 96,111 tons of DDGS at an average price of

Table 9.2. Net present value sample statistics generated by the economic analysis of the lignocellulosic- and grain-based ethanol plants (million US$).

Initial investment	Net present value			
	Mean	Standard deviation	Minimum	Maximum
Lignocellulosic				
US$20 million	19.4	4.1	5.0	30.7
US$40 million	5.8	4.2	-9.1	17.2
US$60 million	-8.8	5.2	-30.6	3.6
Grain				
US$40 million	1.6	12.3	-59.5	30.7

US$97.85/ton, totalling US$9.4 million/year. The projected DDGS price is modelled as a function of maize and soybean meal prices and is stochastic.

Results

This section discusses the simulation results for each of the three different initial capital investment amounts and then compares the US$40 million initial investment plant to the feed grain ethanol plant.

Net Present Value

Table 9.2 contains the mean, standard deviation and the range of NPV for the lignocellulosic-based ethanol plant under each investment scenario analysed. As expected, the NPV for the US$20 million investment is the highest and the US$60 million investment is the lowest. The standard deviation, or financial risk, increases as the investment level increases. This is caused by an increase in the initial debt level due to the higher cost facilities.

Net Cash Income

Figure 9.4 contains the NCI and the probability that NCI is negative for each initial investment level over the planning horizon. Net cash income for the US$20, US$40 and US$60 million plants averages US$7.6 million, US$5.6 million and US$3.5 million in 2005, respectively, during the first full year of production. The average annual NCI is projected to decrease each year up to 2018 to US$4.9 million, US$2.9 million and US$610,000. Income levels decrease over time as the input costs are inflated while the average ethanol price and ethanol yield per unit of biomass are not inflated.

The probability of a negative NCI increases over time for each initial investment level as average NCI declines. For the US$20 million initial investment plant, the probability of a negative NCI remains at about 9% over the entire period. The probability of a negative NCI for the US$60 million plant approaches 50% by the end of the period.

Ending Cash Balance

EC ranges from US$5.1 million to US$500,000, in 2005, for the US$20 and US$60 million plants, respectively (Fig. 9.5). EC continues to build at the lower investment levels through 2018. However, it peaks in 2013 for the largest investment level and begins to decline as debt service overwhelms net revenues. It is important to remember that EC is what remains after all expenses are paid. This includes 8% on equity capital put into the business. So, the business is pulling dividends and profits out of the operation.

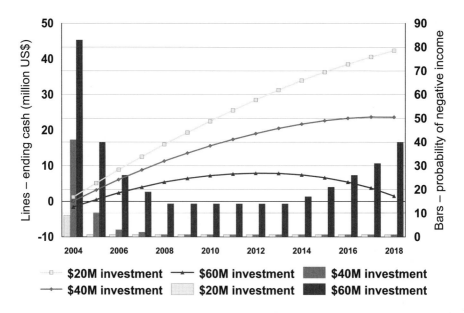

Figure 9.4. Projected average annual net cash income and probability of negative net cash income for US$20, US$40 and US$60 million initial investment levels for the lignocellulosic ethanol plant.

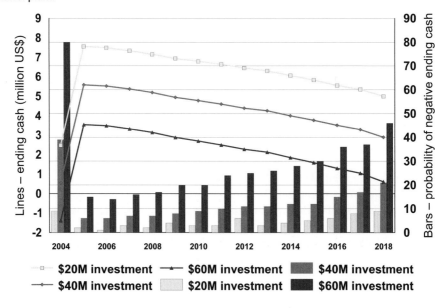

Figure 9.5. Projected average annual ending cash and probability of negative ending cash for US$20, US$40 and US$60 million initial investment levels for the lignocellulosic ethanol plant.

The probability of having a negative EC balance decreases to only 1% over time for the lowest investment level. The US$60 million plant has a 40% probability of a negative EC balance by the end of the planning horizon.

Comparison to Ethanol Production

Net cash income for both types of plants declines over the planning period (Fig. 9.6). However, the lignocellulosic plant has a higher NCI in each year, except 2018, when compared to the feed grain plant. Although the lignocellulosic plant has a higher NCI than the grain-based plant in all years but 2018, the NCI decreases at a faster rate than for the grain-based plant. Net cash incomes converge due to assumptions on the long-term feedstock prices. Projected grain prices obtained from the Food and Agricultural Policy Research Institute are used for the 2004-2011 period and then are held constant at their 2011 level for the remainder of the planning horizon (Food and Agricultural Policy Research Institute, 2003). Biomass feedstock prices are held constant throughout the period. Over the entire period the lignocellulosic plant averages about US$1.2 million more in net cash income, annually.

The probability of negative net income increases over time for both the lignocellulosic and feed grain plants. The feed grain plant has a higher probability of negative net income increasing from 23% in 2005 to 26% in 2018. The lignocellulosic plant increases from 6% in 2005 to 21% in 2018.

Comparison of Ending Cash Balance

Both lignocellulosic and feed grain ethanol are projected to have positive average EC balances over the planning period (Fig. 9.7). EC balance for lignocellulosic ethanol increases significantly while feed grain ethanol EC balance remains relatively constant over most of

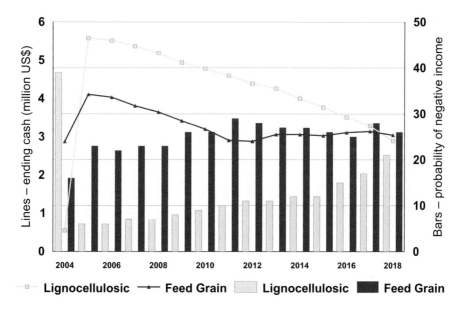

Figure 9.6. Comparison of net cash income for a 30 million gallon/year lignocellulosic ethanol plant and a 30 million gallon/year feed grain ethanol plant.

the time period, after initial growth. EC balance for lignocellulosic and feed grain ethanol plants are projected to be US$3.2 million and US$4.4 million in 2005, respectively. The lignocellulosic plant's EC increases to US$23.6 million in 2018 while the feed grain plant's EC increases to US$10.2 million. The probability of a negative EC balance over the planning period increases for the feed grain plant and decreases for the lignocellulosic plant. By 2018 the grain fuel plant has a 27% chance of negative EC versus only a 1% chance for the lignocellulosic plant.

Conclusions

The results of the study support the notion that lignocellulosic-based ethanol production may be a profitable enterprise in Texas, given the assumptions made. This study compared lignocellulosic ethanol production with initial investment levels of US$20 million, US$40 million and US$60 million. The analysis indicates that the profitability of the lignocellulosic plant will depend greatly on the initial investment costs.

The lignocellulosic ethanol plant returned positive NPV and net cash income at each investment level. EC balance increases at the US$20 million and US$40 million initial investment level, but decreases at the US$60 million initial investment level with a 40% probability of being negative at the end of the planning period.

Comparison of a 30 million gal/year feed grain ethanol plant and a 30 million gal/year lignocellulosic ethanol plant shows the lignocellulosic plant to have a higher projected net cash income, real net worth and net present value. The lignocellulosic-based ethanol plant has lower probabilities of negative net income and negative EC balances (1% compared to 27%).

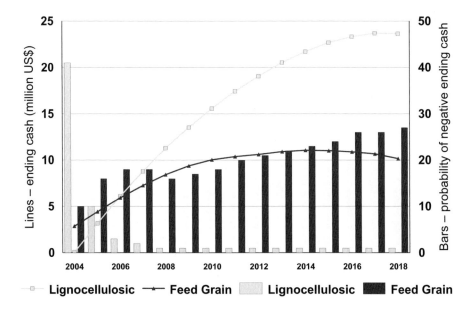

Figure 9.7. Comparison of ending cash balance for a 30 million gallon/year lignocellulosic ethanol plant and a 30 million gallon/year feed grain ethanol plant.

The results show that the lignocellulosic-based ethanol has a higher probability of financial success than the feed grain-based ethanol in Texas. There is ongoing research in the field of lignocellulosic-based ethanol and new processes are being developed. It would appear that there is potential in this type of ethanol production as an alternative to today's commercial ethanol production.

References

Australian Cooperative Research Centre for Renewable Energy Ltd. (2004) *What is Biomass?* Murdoch University, Murdoch, Australia. [Accessed 2004.] Available from http://web.archive.org/web/20040207052648/http://www.acre.murdoch.edu.au/refiles/biomass/text.html

Bryan and Bryan International. (2001) *Dumas Texas Area Ethanol Feasibility Study*. Bryan and Bryan International, Cotopaxi, Colorado.

El Bassam, N. (2002) *Global Potential of Biomass for Transport Fuel*. Presented paper, First World Renewable Energy Policy and Strategy Forum, Towards Johannesburg: Renewable Energies: 'Agenda 1 of the Agenda 21', June 13-15, Berlin, Germany.

English, B., Jensen, K., Menard, J., Walsh, M., De La Torre Ugarte, D., Brandt, C., Van Dyke, J. and Hadley, S. (2004) Economic Impacts Resulting from Co-Firing Biomass Feedstocks in Southeastern Untied States Coal-Fired Plants. Presented paper, American Agricultural Economics Association Meeting, August 1-4, Denver, Colorado.

Evans, M. (1997) *The Economic Impact of the Demand for Ethanol*. Presented paper, The Midwestern Governors' Conference, February, Lombard, Illinois.

Food and Agricultural Policy Research Institute. (2001) *Impacts of Increased Ethanol and Biodiesel Demand*. FAPRI-UMC Report #13-01. University of Missouri-Columbia, Food and Agricultural Policy Research Institute, Columbia, Missouri.

Food and Agricultural Policy Research Institute. (2003) *Baseline Projections, July 2003*. University of Missouri-Columbia, Food and Agricultural Policy Research Institute, Columbia, Missouri.

Fruin, J. and Halbach, D. (1986) *The Economics of Ethanol Production and Its Impact on the Minnesota Farm Economy*. Staff Paper (P86-12). University of Minnesota, Department of Applied Economics, St. Paul, Minnesota.

Fruin, J., Rotsios, K. and Halbach, D. (1996) *Minnesota Ethanol Production and Its Transportation Requirements*. Staff Paper (P96-7). University of Minnesota, Department of Applied Economics, St. Paul, Minnesota.

Gallagher, P., Dikeman, M., Fritz, J., Wailes, E., Gauther, W. and Shapouri, H. (2003) *Biomass from Crop Residues: Cost and Supply Estimates*. Agricultural Economics Report No. 819. US Department of Agriculture, Economic Research Service, Washington, DC.

Hanson, G. (1985) Financial Analysis of a Proposed Large-Scale Ethanol Cogeneration Project. *Southern Journal of Agricultural Economics* 17(02): 67-76.

Herbst, B. (2003) *The Feasibility of Ethanol Production in Texas*. MS thesis, Texas A&M University, Department of Agricultural Economics, College Station, Texas.

Holtzapple, M. (2003) MixAlco Process. Texas A&M University, Department of Chemical Engineering, College Station, Texas.

Holtzapple, M., Davidson, R., Ross, M., Aldrett-Lee, S., Nagwani, N., Lee, C., Lee, C., Adelson, S., Kaar, W., Gaskin, D., Shirage, H., Chang, N., Chang, V. and Loescher, M. (1999) Biomass Conversion to Mixed Alcohol Fuels Using the MixAlco Process. *Applied Biochemistry and Biotechnology* 77-79: 609-631.

Klass, D. (1998) *Biomass for Renewable Energy, Fuels and Chemicals*. Academic Press, San Diego, California.

Mapemba, L. and Epplin, F. (2004a) *Lignocellulosic Biomass Harvest and Delivery*. Presented paper, Southern Agricultural Economics Association Meeting, February 14-18, Tulsa, Oklahoma.

Mapemba, L. and Epplin, F. (2004b) *Use of Conservation Reserve Program Land for Biorefinery Feedstock Production*. AEP-0403. Oklahoma Agricultural Experiment Station, Stillwater, Oklahoma.

Nienow, S., McNamara, K., Gillespie, A. and Preckel, P. (1999) A Model for the Economic Evaluation of Plantation Biomass Production for Co-Firing with Coal in Electricity Production. *Agricultural and Resource Economic Review* 28(1): 106-118.

Osburn, L. and Osburn, J. (1993) *Biomass Resources for Energy and Industry*. Access Unlimited, Frazier Park, California. [Accessed 2004.] Available from http://www.ratical.org/renewables/biomass.html

Otto, D. and Gallagher, P. (2001) *The Effects of Expanding Ethanol Markets on Ethanol Production, Feed Markets and the Iowa Economy*. Iowa State University, Department of Economics, Ames, Iowa.

Renewable Fuels Association. (2003) *Ethanol Industry Outlook 2003: Building a Secure Energy Future*. Renewable Fuels Association, Washington, DC. [Accessed 2004.] Available from http://www.ethanolrfa.org/outlook2003.html

Richardson, J., Klose, S. and Gray, A. (2000) *An Applied Procedure for Estimating and Simulating Multivariate Empirical (MVE) Probability Distributions in Farm Level Risk Assessment and Policy Analysis*. Presented paper, Southern Agriculture Economics Association Annual Meeting, January 28-February 2, Lexington, Kentucky.

Richardson, J., Schumann, K. and Feldman, P. (2004) *SIMETAR©: Simulation for Econometrics to Analyze Risk*. Simetar, Inc., College Station, Texas. [Accessed 2005.] Available from http://www.simetar.com

Van Dyne, D. (2002) *Employment and Economic Benefits of Ethanol Production in Missouri*. University of Missouri-Columbia, Department of Agricultural Economics, Columbia, Missouri.

Veringa, H. (2004) *Advanced Techniques for Generation of Energy from Biomass and Waste*. Energy Research Centre of the Netherlands, ECN Biomass, Petten, Netherlands. [Accessed 2004.] Available from http://www.ecn.nl/_files/bio/Advanced_techniques_for_generation_of_energy_from_biomass_and_waste__October_2004.pdf

Chapter 10

The Supply of Maize Stover in the Midwestern United States

Richard G. Nelson, Marie E. Walsh and John J. Sheehan

Introduction

Interest in alternative energy sources has increased in recent years due to an increasing recognition of the economic, environmental and energy security impacts that result from continued reliance on fossil fuels such as oil, coal and natural gas. Production of electricity, transportation fuels and organic chemicals (e.g., lactic acid, plastic precursors, etc.) from renewable energy resources such as biomass offer an opportunity to reduce fossil fuel use. Biomass resources include, but are not limited to: forestry and mill residues; urban wood wastes (municipal solid wastes, construction and demolition wastes); dedicated energy crops such as switchgrass (*Panicum virgatum*), hybrid poplar (*Populus* x) and willow (*Salix* sp.); and agricultural residues such as maize stover (the non-grain components of a maize plant such as the stalks, leaves, cobs, roots, etc.) and the straw from small grains (wheat, oats, rice, barley, rye).

Life cycle analyses indicate that bioenergy and bioproducts have lower environmental impacts, a more favourable energy balance (ratio of energy output to total energy inputs) and can provide for economic development and enhanced energy security compared with many fossil fuel technologies (Delucchi, 1994, 1997; Riley and Tyson, 1994; Sheehan *et al.*, 1996; Wang *et al.*, 1999; Mann and Spath, 2001; Sheehan, 2002).

Agricultural residues and maize stover in particular, offer a potentially large, existing and untapped biomass resource that can be used to produce bioenergy and bioproducts that can simultaneously address several economic, environmental and energy security issues.

Removal of maize stover from the field, however, raises several issues with respect to sustaining soil tilth (organic matter), protecting the soil surface from water and wind erosion, controlling nutrient (fertilizer) and chemical runoff, maintaining soil moisture and providing for long-term sustainable crop production. Maize stover use for bioenergy and bioproducts requires leaving enough residue on the ground to protect the soil.

The cost of collecting and delivering maize stover to a user facility at a competitive price is crucial to developing a bioenergy and bioproducts-based industry using this resource. These costs depend on the quantities of stover that can be removed at any given site, a profit margin to farmers and transportation and storage costs as a function of the user-facility location and maize stover demand level.

This chapter describes a methodology to estimate agricultural crop residue supply curves (quantity at a given price) taking into consideration residue quantities needed to keep soil erosion at or below the tolerable limit (T). T is defined as the maximum rate of soil erosion that will *not* lead to prolonged soil deterioration and/or loss of productivity by the US Department of Agriculture (USDA) Natural Resources Conservation Service (NRCS). This study limits the application to, and discusses the results of, the methodology for a continuous maize rotation in the top ten maize-producing states (Illinois, Indiana, Iowa, Kansas, Minnesota, Missouri, Nebraska, Ohio, South Dakota and Wisconsin). Maize is the largest agricultural crop produced, both in acres and quantities, in the USA. Nearly 85% of US maize acres and production occurs in those ten states where it is produced most commonly in a continuous maize or maize-soybean rotation.

For each county in each state, all cropland soil types were identified. Acres, field topology characteristics (percentage low and high slopes), erodibility and the tolerable soil-loss limit for each soil type were obtained from NRCS (Natural Resources Conservation Service, 1995). These data were used in the rain and wind erosion equations described later in this chapter. Three tillage scenarios (conventional, reduced/mulch and no-till) were analysed for each soil type. Conventional tillage scenarios consist mainly of mouldboard ploughing and/or heavy discing and leave less than 30% of the field covered; reduced/mulch tillage scenarios include light discing and chisel ploughing and leave a minimum of 30% of the field surface covered with residue; and the no-till scenarios use field operations that provide little or no disturbance to the field surface. Harvest, planting, tillage and chemical application dates for each field operation reflect the most likely time of year and month they are expected to occur.

This chapter extends and improves the methods used in previous studies by providing a more detailed and rigorous approach to estimating the quantities of residues that can be removed and the costs of collecting residues as a function of the removable residues (Larson, 1979b; Lindstrom *et al.*, 1979; Christensen *et al.*, 1983; Rooney, 1998; Kerstetter and Lyons, 2001; Gallagher *et al.*, 2003).

Methodology

Removal of maize stover for bioenergy and bioproduct use is directly influenced by a number of factors including grain yield, crop rotation, field management practices (e.g., type and timing of tillage and other management practices), climate and physical characteristics of the soil (e.g., soil type, erodibility index, topology, etc.).

The next several sections describe the methodology used to estimate the county level quantities of maize stover that are produced and the amount that must remain to maintain soil erosion at or below T and can be removed for bioenergy and bioproduct use; the cost of collecting removable quantities; and develop a regional maize stover supply curve.

Estimate Maize Stover Quantities Produced

The quantity of maize stover produced is directly related to the maize grain yield. County-level harvested maize acres and grain yields for the years 1999-2003 were obtained from USDA National Agricultural Statistical Service (NASS) (National Agricultural Statistics Service, 2003). A simple average over the time period was used to represent the county acres and yields used in this analysis. Table 10.1 presents the weighted average maize acres and yields for the ten states in the analysis. These average yields were converted to stover quantities

Table 10.1. Average state maize acres, yields and percentage acres by tillage practice.

State	Harvested maize acres (million acres)	Average grain yield (bushels/acre)	Percentage acres in conventional tillage	Percentage acres in mulch till	Percentage acres in no-till
Illinois	10.92	148.4	73	11	16
Indiana	5.49	140.3	69	10	20
Iowa	11.82	152.4	58	26	16
Kansas	2.85	127.2	44	38	18
Minnesota	6.55	145.6	81	17	2
Missouri	2.68	117.2	54	18	27
Nebraska	7.83	137.4	35	44	21
Ohio	3.12	131.0	67	8	25
South Dakota	3.51	107.6	55	27	18
Wisconsin	2.79	133.2	62	27	11

Source: Maize acres and grain yields from National Agricultural Statistics Service (2003) and are weighted averages for the years 1999-2003. Percentage acres by tillage practice from Conservation Technology Information Center (2000) and are weighted averages for the years 1995-2000.

using fresh grain weight to bushel factors and dry weight residues to fresh grain weight ratios. For maize, these factors were 56 pounds/bushel (lb/bu) of grain and a 1:1 ratio of dry stover to fresh grain weight (Heid, 1984; Larson, 1979a,b; Brown, 2003). Thus for a maize grain yield of 150 bu/acre, the stover produced was 4.2 dry tons (150 bu/acre multiplied by 56 lb/bu multiplied by 1/2000 lb/ton).

Estimate Maize Stover Quantities that must Remain to Maintain Soil Erosion At or Below the Tolerable Soil Loss Limit (T)

In order to quantify the amount of residue that can be sustainably removed, quantities of residues that must be left on the field to maintain rain and/or wind erosion at or below tolerable soil-loss levels (T) were estimated. The amount of soil erosion that agricultural cropland experiences is a function of field management (tillage practices in particular), rainfall, wind, temperature, solar radiation, soil type, field topology and the amount of residue (cover) left on the field from harvest until the next planting. Soil erosion due to water (rainfall) and/or wind directly affects most agricultural cropland in the USA. Water erosion dominates in the eastern two-thirds of the USA with wind erosion being more prominent in the western USA.

The revised universal soil loss equation (RUSLE) and the wind erosion equation (WEQ) were used to estimate the maize stover quantities that must be left on the field (Skidmore *et al.*, 1970, 1979; Skidmore, 1994; Renard *et al.*, 1997). They are designed primarily to estimate long-term, average annual soil erosion on a site-specific field characterized by a particular soil type, slope and runoff length, field length, cropping and management practices used and localized climate conditions. Stover quantities that must be left on the field under a continuous maize rotation were estimated for each soil type and tillage combination for rain and wind erosion, with the higher of the two remaining quantity estimates being those which were used as the quantities that must remain on the field.

Estimate Maize Stover Quantities that must Remain to Maintain Rain Erosion At or Below the Tolerable Soil Loss Limit (T)

Maize stover quantities needed to control rain erosion were estimated using RUSLE (equation 1).

$$A = R * K * S * L * C * P \qquad (1)$$

where A is the average annual soil loss (metric tons/hectare/year), R the rainfall-runoff erodibility factor (location/county specific), K the soil erodibility factor, S the slope steepness factor, L the slope-length factor, C the cover-management factor and P the support-practices factor. The A in RUSLE can be replaced by a tolerable soil loss limit value (T) specific to each soil type examined. P was assumed to be 1.0, which provides the most conservative estimate for stover removal. All factors except C were independent of crop grain yield or crop management.

C is a function of the harvest grain yield and is directly influenced by field operations that affect field surface cover throughout the year (i.e., tillage). In order to estimate the annual erosion and quantities of removable maize stover attributable to specific field operations and harvest yields, the C-factor must be determined in relation to these conditions. RUSLE can be re-written as equation 2.

$$C = (R * K * L * S * P)/T \qquad (2)$$

where C is the only unknown factor. To solve for C, the RUSLE C-Batch Program, developed by the USDA NRCS National Soil Survey Center, Lincoln, Nebraska, was used. C-Batch estimates C for various crop rotations, crop grain yield variations and tillage operations. For this analysis, maize grain yields of 50 bu/acre, 80 bu/acre, 100 bu/acre, 125 bu/acre and 150 bu/acre were assumed. The estimated maize stover quantities that must remain for each soil type were weighted by the percentage of acres of each soil type in a county and aggregated to the county level to create representative county level residue remaining functions for each tillage practice.

Estimate Maize Stover Quantities that must Remain to Maintain Wind Erosion At or Below the Tolerable Soil Loss Limit (T)

WEQ (equation 3) was used to estimate the quantities of maize stover that must remain on the field to maintain wind erosion at or below T.

$$E = f (WI, WK, WC, WL, WV) \qquad (3)$$

where E is the average annual soil loss (metric tons/hectare/year). WI is the wind erodibility index (a measure of soil susceptibility to detach and be transported by wind) and varies by individual soil type. WK is the soil ridge-roughness factor and describes the condition of the field surface at a particular time. WC is the climate factor and represents the amount of erosive wind energy present at a particular (county-level) location. WL is a function of wind direction, field length and width and is the unsheltered median travel distance of wind across a field. WV is the vegetative factor. The relationship between E and the other variables is highly non-linear.

The amount of stover that must remain was estimated by analysing total soil loss attributable to wind forces in each field-management period (time between each field operation) for all individual soil types and then summing across all field-management periods including crop growth. A detailed discussion of the application of WEQ to agricultural crop residue removal is provided in Nelson (2002).

Estimate Maize Stover Quantities that can be Removed while Maintaining Soil Erosion At or Below T

The quantity of stover that can be removed was calculated as the difference in county-level stover production in dry tons/acre (dt/ac) minus the higher of the maize stover quantity that must be left to maintain water or wind erosion at or below T for each of the three tillage practices. Multiplying the total harvested maize acres by the stover removal quantities (dt/ac) for each tillage practice provided an estimate of the amount of maize stover that could be available for bioenergy and bioproducts if all maize acres used the associated tillage practice. Multiplying the total stover quantity for each tillage practice by the percentage of maize acres using each tillage practice (Conservation Technology Information Center, 2000) provided a weighted average quantity of available maize stover given the current mix of tillage practices.

Estimate Cost of Collecting Maize Stover as a Function of the Removable Yield

Once the quantities (dt/ac) of maize stover that can be collected by tillage practice were estimated, the farmgate or edge of field cost of collecting the stover was calculated. Maize stover was assumed to be collected in the form of large round bales. Collection costs included the cost of baling the stover, picking up the bales and moving them to the edge of the field (staging) and stacking them at field's edge for storage. Equipment cost estimates used methodology recommended by American Agricultural Economics Association (2000) and agricultural equipment parameters and engineering performance equations provided by American Society of Agricultural Engineers (2001).

Equipment costs included repairs, fuel/lubrication/oil, capital costs (depreciation, interest), labour, taxes, housing and insurance costs. Different collection practices (combinations of windrowing, mowing, raking and baling operations) and equipment configurations (use of a crop processor or not) are assumed depending on removable quantity level. Figure 10.1 shows the collection costs as a function of removable (available) maize stover quantities (dt/ac). Costs are in 2002 US$.

Delivered costs to a user facility include not only the cost of collecting the maize stover, but also a premium paid to farmers to encourage collection and cover any additional nutrients that may be needed for the next crop year and the costs of transporting and storing the stover. These costs are not included in this analysis.

Construct a Regional Maize Stover Supply Curve for a Continuous Maize Rotation

The maize stover supply curve was constructed by ordering the tillage weighted county quantities from lowest to highest value and applying the quantities to each collection cost level (US$12.50/dt, US$17.50/dt, US$20/dt, US$25/dt, US$30/dt, US$35/dt, US$40/dt, US$45/dt and US$50/dt). Quantities reported in each price range were the sum of the cumulative quantities in the preceding range, plus the quantities in the selected price range (i.e., at US$35/dt, quantities are those available between US$30/dt and US$35/dt plus those available at less than US$30/dt).

Results

Table 10.2 presents the state weighted average quantities of maize stover that were produced, the quantities of stover required to remain on the field to keep rain and/or wind erosion at or below T by tillage practice and the removable stover by tillage practice. The quantities that must be left on the fields decrease as one goes from conventional to no-till practices. US states characterized by flat lands, such as Minnesota, require fewer maize residues to be left than do states with more steeply sloping lands, such as Missouri. Counties in western states where wind erosion is a major consideration (Kansas, Nebraska and South Dakota) require that all of the stover produced be left on the ground under a conventional tillage system. In South Dakota, all stover must be left on the field under a mulch till system as well. The stover quantities that can be removed by tillage practice in each state vary substantially by county. Table 10.3 shows the range of removable maize stover quantities and the number of counties in which no stover can be removed by tillage practice.

The estimated quantities (dt) of maize stover available for bioenergy and bioproduct use for nine collection cost values are shown in Table 10.4. No maize stover quantities are available at collection costs of less than US$17.50/dt. At a collection cost of between US$17.50/dt and US$20.00/dt, 35 million dt of maize stover could be collected. Available stover quantities increased from 35 million dt at a cost of less than or equal to US$20/dt to 98.9 million dt at a collection cost of less than or equal to US$35/dt. Nearly all of the available maize stover could be collected at a cost of less than US$35/dt. Above this level, quantities increased only slightly (an additional 2.5 million dt are available at costs of US$50/dt). It

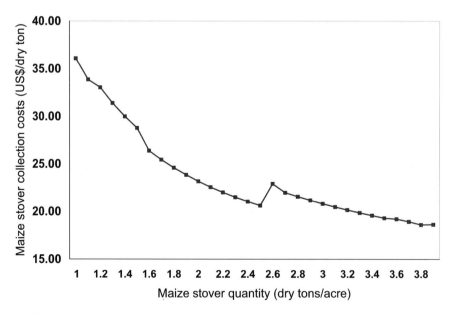

Source: Walsh (2004).

Figure 10.1. Maize stover collection costs.

Table 10.2. State average stover produced, quantities that must remain to control erosion by tillage practice and removable stover quantities by tillage practice (dry tons/acre).

	Average residue produced	Average residues that must remain to control erosion by tillage practice			Average residues that can be removed by tillage practice		
		Conventional tillage	Mulch till	No-till	Conventional tillage	Mulch till	No-till
Illinois	4.16	2.60	1.53	0.70	1.78	2.64	3.45
Indiana	3.93	2.54	1.46	0.70	1.68	2.50	3.23
Iowa	4.27	2.66	1.66	0.78	1.74	2.61	3.48
Kansas	3.56	82.96	3.50	2.37	0.00	0.41	1.24
Minnesota	4.08	1.27	0.74	0.27	2.81	3.33	3.81
Missouri	3.28	4.95	3.02	1.63	0.26	0.60	1.65
Nebraska	3.85	62.19	2.40	1.32	0.00	1.49	2.53
Ohio	3.67	3.01	1.80	0.79	1.16	1.94	2.88
South Dakota	3.01	67.08	5.37	1.26	0.00	0.00	1.77
Wisconsin	3.73	3.15	2.04	1.00	0.82	1.69	2.73

Source: Nelson (2004).

Table 10.3. Range of available maize stover (dry tons/acre) within each state.

	Conventional tillage		Mulch till		No-till	
	Range[a]	Number of counties[b]	Range[a]	Number of counties[b]	Range[a]	Number of counties[b]
Illinois	0 to 3.59	35	0 to 3.94	13	0.0008 to 4.44	0
Indiana	0 to 3.88	36	0 to 4.12	16	0 to 4.28	5
Iowa	0 to 3.55	33	0 to 4.00	3	1.38 to 4.39	0
Kansas	0	105	0 to 1.71	63	0 to 2.41	13
Minnesota	0 to 3.85	10	0 to 3.99	7	0 to 4.34	7
Missouri	0 to 3.64	109	0 to 4.00	90	0 to 4.22	30
Nebraska	0	93	0 to 3.59	26	0 to 3.80	5
Ohio	0 to 3.83	50	0 to 3.99	29	0 to 4.08	15
South Dakota	0	66	0	66	0 to 2.94	16
Wisconsin	0 to 2.93	27	0 to 3.19	11	0 to 3.50	9

[a] Range of removable residues.
[b] Number of counties with zero available residues.
Source: Nelson (2004).

should be noted that these estimates were based on collection costs only and assumed that all of the available stover was collected. Thus, the estimated quantities represented the maximum available given the current tillage mix and a continuous maize rotation. Lower quantities due to yield variations from weather and less than full participation by farmers would decrease the total quantities available at any given cost level. Additionally, the estimated costs do not include a premium to farmers for transportation and storage costs. Inclusion of these costs could add another US$10/dt to US$40/dt.

The analysis only evaluated the quantities of maize stover than can be removed and used for bioenergy and bioproducts controlling for erosion impacts. Impacts of agricultural residue removal on soil carbon levels are also an important consideration. A recent life cycle analysis in Iowa indicated that generally, if sufficient quantities of maize stover were left on the field to keep erosion at or below T, soil carbon levels would continue to increase (Sheehan, 2002). However, there are exceptions. Leaving sufficient quantities of stover on the field to control erosion on lands that are flat and consist primarily of soils with low erodibility factors can lead to a decrease in soil carbon levels. Thus, under some circumstances, soil carbon

Table 10.4. State quantities of maize stover available for bioenergy and bioproducts by collection cost (US$/dry ton).

	Quantity (million dry tons)								
	12.50	17.50	20.00	25.00	30.00	35.00	40.00	45.00	50.00
Illinois	0.0	0.0	7.6	19.9	21.5	22.5	22.6	22.8	22.8
Indiana	0.0	0.0	3.6	10.1	10.7	11.1	11.3	11.4	11.4
Iowa	0.0	0.0	11.7	24.2	25.4	26.1	26.4	26.5	26.7
Kansas	0.0	0.0	0.0	0.1	0.4	0.6	0.8	0.8	0.9
Minnesota	0.0	0.0	9.1	18.8	19.0	19.1	19.2	19.2	19.2
Missouri	0.0	0.0	0.5	1.0	1.4	1.7	1.7	1.7	1.8
Nebraska	0.0	0.0	0.8	7.0	8.7	9.1	9.3	9.3	9.3
Ohio	0.0	0.0	1.4	4.1	4.5	4.7	4.9	4.9	4.9
South Dakota	0.0	0.0	0.0	0.7	0.8	0.8	0.9	0.9	0.9
Wisconsin	0.0	0.0	0.2	2.3	2.9	3.2	3.3	3.3	3.4
Total	0.0	0.0	35.0	88.1	95.4	98.9	100.3	100.8	101.4

Source: Walsh (2004).

needs will dominate erosion needs. To have a fuller understanding of the quantities of maize stover that can be removed, both erosion and soil carbon issues must be examined.

The analysis which has been presented is static and does not allow for changes in maize acres and location that might result from increased farm income from the sale of maize stover. As of this writing, the authors are conducting an analysis to include agricultural residues into the POLYSYS dynamic agricultural sector model that has already been modified to include dedicated energy crops such as switchgrass, hybrid poplar and willow (Walsh et al., 2003). An evaluation of soil carbon needs is also planned.

Summary and Discussion

Interest in the use of non-fossil fuel energy sources has been increasing in recent years. An option to decrease the use of fossil fuels for the production of transportation fuels, electricity and organic chemicals includes producing these products from biomass resources such as maize stover. However, removal of maize stover from the field raises many questions with respect to soil quality.

This chapter presented a methodology to estimate the supply (quantity at a given price) of maize stover taking into consideration the need to limit soil erosion impacts that result from stover removal. Also discussed were the results of using a continuous maize rotation and three tillage practices by soil type for the ten largest maize producing states. This analysis indicates that up to 98 million dt of stover (equivalent to 1.37 quads of primary energy or 1.5% of US current use) could be collected at a cost of less than $35/dt and 35 million dt (0.49 quads) could be collected at costs of less than $20/dt. Farmer premiums, transportation and storage costs are not included and could add an additional $10/dt to $40/dt to the cost of the stover. These costs will be a function of the quantities needed by the user facility, given the maize stover distribution and collection costs.

The study also presented a rigorous and detailed methodology to estimate the quantities and collection costs of maize stover that can be removed while still maintaining erosion at or below T. The analysis is readily amenable to being incorporated into other modelling systems (soil carbon models, dynamic agricultural sector models, life cycle analyses models, regional economic impact models, etc.) to examine a range of constraints and benefits that could occur from the use of maize stover for bioenergy and bioproducts. The analysis indicates that in several locations, sufficient quantities of maize stover were available at reasonable costs to produce bioenergy and bioproducts.

It should be noted that the analysis presented has been static and included an evaluation of erosion constraints only. Additionally, the removal of maize stover on soil carbon levels was not considered. Furthermore, this chapter presents results that were aggregated to the state level, but substantial variation in removable stover quantities and collection costs exist within each state.

Acknowledgement

Funding for this research was provided by the US Department of Energy under National Renewable Energy Laboratory Contract Number ACO-1-31103-01.

References

American Agricultural Economics Association. (2000) *Commodity Costs and Returns Estimation Handbook.* American Agricultural Economics Association, Ames, Iowa.

American Society of Agricultural Engineers. (2001) *ASAE Standards 2001,* ASAE S495. American Society of Agricultural Engineers, St. Joseph, Michigan.

Brown, R. (2003) *Biorenewable Resources: Engineering New Products from Agriculture.* Iowa State Press, Ames, Iowa.

Christensen, D., Turhollow, A., Heady, E. and English, B. (1983) *Soil Loss Associated with Alcohol Production from Corn Grain and Corn Residue.* CARD Report 115. Iowa State University, Center for Agricultural and Rural Development, Ames, Iowa.

Conservation Technology Information Center. (2000) *National Crop Residue Management Survey.* Conservation Technology Information Center, West Lafayette, Indiana.

Delucchi, M. (1994) *Emissions of Greenhouse Gases from the Use of Transportation Fuels and Electricity.* US Department of Energy, Argonne National Laboratory, Argonne, Illinois.

Delucchi, M. (1997) *A Revised Model of Emissions of Greenhouse Gases from the Use of Transportation Fuels and Electricity.* UCD-ITS-RR-97-22. University of California-Davis, Institute of Transportation Studies, Davis, California.

Gallagher, P., Dikeman, M., Fritz, J., Wails, E., Gauthier, W. and Shapouri, H. (2003) Supply and Social Cost Estimates for Biomass from Crop Residues in the United States. *Environmental and Resource Economics* 24(4) (April): 335-358.

Heid, W. (1984) *Turning Great Plains Crop Residues and Other Products into Energy.* Agricultural Economic Report No. 523. US Department of Agriculture, Economic Research Service, Washington, DC.

Kerstetter, J. and Lyons, J. (2001) *Logging and Agricultural Residue Supply Curves for the Pacific Northwest.* Washington State University Energy Program, Olympia, Washington, DC.

Larson, W. (1979a) Crop Residues: Energy Production or Erosion Control? *Journal of Soil and Water Conservation* 34(2): 74-76.

Larson, W. (1979b) *Effects of Tillage and Crop Residue Removal on Erosion, Runoff and Plant Nutrients.* Soil Conservation Society of America, Journal of Soil and Water Conservation Special Publication No. 25, Ankeny, Iowa.

Lindstrom, M., Gupta, S., Onstad, C., Larson, W. and Holt, R. (1979) Tillage and Crop Residue Effects on Soil Erosion in the Corn Belt. *Journal of Soil and Water Conservation* 34(2): 80-82.

Mann, M. and Spath, P. (2001) *Life Cycle Assessment of Biomass Co-firing in a Coal-Fired Power Plant.* NICH Report No. 29457. US Department of Energy, National Renewable Energy Laboratory, Golden, Colorado.

National Agricultural Statistics Service. (2003) Crops. *Agricultural Statistics Data Base.* US Department of Agriculture, National Agricultural Statistics Service, Washington, DC. [Accessed 2004.] Available from http://www.nass.usda.gov:81/ipedb/

Natural Resources Conservation Service. (1995) *Soil Survey Geographic (SSURGO) Database.* Miscellaneous Publication No. 1527. US Department of Agriculture, Natural Resources Conservation Service, Washington, DC.

Nelson, R. (2002) Resource Assessment and Removal Analysis for Corn Stover and Wheat Straw in the Eastern and Midwestern United States – Rainfall and Wind-Induced Soil Erosion Methodology. *Biomass and Bioenergy* 22: 349-363.

Nelson, R. (2004) *Rainfall and Wind Erosion-Based Removal Analysis and Resource Assessment for Corn Stover and Wheat Straw for Selected Cropping Rotations in the United States,* Draft Final Report. Enersol Resources, Manhattan, Kansas.

Rooney, T. (1998) *Lignocellulose Feedstock Resource Assessment.* NEOS Corporation. Lakewood, Colorado.

Renard, K., Foster, G., Weesies, G., McCool, D. and Yoder, D. (1997) *Predicting Soil Erosion by Water: A Guide to Conservation Planning with the Revised Universal Soil Loss Equation.* Agriculture Handbook 703. US Department of Agriculture, Agricultural Research Service, Washington, DC.

Riley, C. and Tyson, K. (1994) *Alternative Fuels and the Environment.* Lewis Publishers, Ann Arbor, Michigan.

Sheehan, J. (2002) *Life-Cycle Analysis of Ethanol from Corn Stover.* NICH Report No. PO-510-31792. US Department of Energy, National Renewable Energy Laboratory, Golden, Colorado.

Sheehan, J., Duffield, J., Coulon, R. and Camobreco, V. (1996) *Life-Cycle Assessment of Biodiesel versus Petroleum Diesel Fuel.* Presented paper, 1996 Intersociety Energy Conversion Engineering Conference, Washington, DC.

Skidmore, E. (1994) *Soil Erosion Research Methods.* Soil and Water Conservation Society of America, Ankeny, Iowa.

Skidmore, E., Fisher, P. and Woodruff., N. (1970) Wind Erosion Equation: Computer Solution and Application. *Soil Science Society of America Journal* 34: 931-935.

Skidmore, E., Kumar, M. and Larson,W. (1979) Crop Residue Management for Wind Erosion Control in the Great Plains. *Journal of Soil and Water Conservation* 34 (2): 90-94.

Walsh, M. (2004) Unpublished data. Department of Agricultural Economics, University of Tennessee, Oak Ridge, Tennessee.

Walsh, M., De La Torre Ugarte, D., Shapouri, H. and Slinsky, S. (2003) The Economic Impacts of Bioenergy Crop Production on U.S. Agriculture. *Environmental and Resource Economics* 24(4): 313-333.

Wang, M., Saricks, C. and Santini, D. (1999) *Effects of Fuel Ethanol Use on Fuel-Cycle Energy and Greenhouse Gas Emissions.* ANL/ESD-38. US Department of Energy, Argonne National Laboratory, Argonne, Illinois.

Chapter 11

Economic Modelling of a Lignocellulosic Biomass Biorefining Industry

Francis M. Epplin, Lawrence D. Mapemba and Gelson Tembo

Introduction

A biorefinery is a facility that converts (refines) biological material (biomass) into products. Breweries and wineries are examples of firms that convert biological material (i.e., grain, grapes) into relatively high value products including beer and wine. A processing firm that produces ethanol from maize grain is another example of a biorefinery. In some respects, a biorefinery is similar to a petroleum refinery that uses crude oil as a feedstock and produces fuels and other products. A crude oil refinery may produce a wide array of petroleum products including petrol, diesel fuel, jet fuels, kerosene, heating oils, asphalt and propane gas. Petrochemicals processed from crude oil are used to make many products including cosmetics, clothes and plastics. Plastics made from crude oil are used in a wide variety of products.

Technology designed to enable conversion of lignocellulosic biomass (LCB) feedstock into useful products is under development (Lynd *et al.*, 1991; Phillips *et al.*, 1994; Wyman, 1994; Gallagher and Johnson, 1999; National Research Council, 1999; McAloon *et al.*, 2000; Rajagopalan *et al.*, 2002). Plants contain lignin, cellulose and hemicellulose. Lignin is largely responsible for the strength and rigidity of plants. Cellulose is the chief constituent of the cell wall in all green plants. The term lignocellulosic is formed from the words lignin and cellulose and is used to refer to plant material such as that contained in crop residues and mature perennial grasses. The economic success of an unsubsidized LCB biorefinery will depend upon its ability to either produce unique valuable products or to produce products that are cheaper than fossil-based substitutes.

Experience from conventional crude oil refineries, electric power generating plants and maize grain-based ethanol biorefineries, suggest that the cost of delivered feedstock is a major component of the cost to produce products. To achieve economies of size often evident in industrial processing, a rather substantial quantity of LCB may be required. The term economies of size is used to describe a situation in which the average cost of production per unit declines as the size of the operation increases. To achieve economies of size, some electric power generating plants use 10,000 tons of coal/day. Similarly, a medium-sized LCB biorefinery may require 2,000 dry tons of feedstock/day. Assuming that trucks could haul 20 tons of LCB with an average dry matter content of 85% (17 tons of dry matter/truck), a 2,000 tons/day biorefinery would require approximately 118 truckloads of LCB/day (approximately five trucks/hour), 7 days per week.

In the USA, the infrastructure for production, harvest, storage, transportation and price risk management of feed grains such as maize is well developed. Unlike maize grain, a well-developed harvest, storage and transportation system does not exist for LCB. While some farmers have harvest machines and equipment that might be used to harvest LCB, it is unlikely that most regions would have a sufficient investment in harvesting machinery that could provide massive quantities of LCB in a consistent form and provide an orderly flow of LCB to a biorefinery throughout the year.

If an LCB-based biorefinery that plans to operate throughout the entire year relied strictly on a single feedstock with a narrow harvest window, a massive quantity of material must be harvested in a relatively short period of time. Storage space must be sufficient to store a year's supply of material. For example, in the US Corn Belt (US Midwest), maize stover, the residue remaining after harvest of maize grain, is plentiful in the autumn of the year after maize grain harvest. However, maize stover harvest and storage might be complicated by mud, snow, fire and a relatively narrow harvest window (Glassner et al., 1998; Schechinger, 2000). Schechinger, who was involved with the management of a pilot maize stover collection project conducted near Harlan, Iowa, has written that the collection, storage and transportation of a continuous flow of maize stover was a '...logistical nightmare...' (2000, p. 6). Some of these potential problems could be mitigated if a biorefinery could use a variety of feedstocks. Harvest windows differ across species enabling the use of harvest and collection machinery throughout many months and reducing the fixed costs of harvest machinery per unit of feedstock.

The economic viability of an LCB biorefinery will depend critically upon the total cost to deliver a continuous flow of feedstock throughout the year. The cost will depend upon a number of factors including: type of feedstock, yield, moisture content, the number of harvest days during the harvest window, harvest method, storage, storage losses, transportation, feedstock inventory management, biorefinery size and biorefinery location.

The objective of the chapter is to describe a modelling system that may be used to determine for a specific region, the most economical sources of LCB, timing of harvest and storage, inventory management, biorefinery size and biorefinery location. Optimal combinations of LCB feedstocks will be determined from among grasses produced on Conservation Reserve Program (CRP) land, crop residue (maize stover and wheat straw), perennial grasses (indigenous grasses from native range and introduced grasses from pastures) and dedicated feedstock such as switchgrass (*Panicum virgatum*). The model accounts for LCB acquisition cost, harvest cost, harvest work days by month, in-field storage, storage losses, on-site storage, transportation costs, monthly requirements, biorefinery size and location. The model could be used to develop an optimal plan for providing a steady flow of LCB to an optimally located biorefinery.

Procedure

A modelling system may be designed and used to answer a number of very specific questions essential to determine the economics of an LCB biorefinery. Examples of these questions include: Where would LCB be produced? What feedstock or combination of feedstocks is economically optimal? How much LCB harvest machinery would be required? How much of what species should be harvested in each period (month)? What quantity of LCB should be placed in field storage in each period (month)? What quantity of LCB should be placed in storage at the biorefinery in each period (month)? What quantity of LCB should be removed from each storage location in each month? What is the optimal transportation flow of LCB from the field and from field storage to the biorefinery? What is the optimal size of the biorefinery? Where should the biorefinery be located?

A number of researchers have used modelling to address some of these issues (English et al., 1981; Nienow et al., 1999; Nilsson, 1999; Graham et al., 2000; Kadam et al., 2000; Kaylen et al., 2000). However, the model developed for the analysis conducted herein addresses all of these issues. The economic optimization model used to conduct the study is an enhancement of a model developed by Tembo (2000) and described by Tembo et al. (2003). For a given case study area, the model was designed to determine LCB-based biorefinery processing capacity that maximizes industry worth over expected life and the optimum quantities of LCB stocks and flows. It is a multi-region, multi-period, mixed integer mathematical programming model, designed to identify key cost components, potential bottlenecks and reveal opportunities for reducing costs and prioritizing research. Mathematical programming is a powerful analytical technique consistent with the optimization principles of marginal economics (Hillier and Lieberman, 1980; McCarl and Spreen, 1996). It can be used to determine economically optimal combinations of activities, such as feedstock sources, subject to specified physical limitations, such as acreage constraints.

The model and case study considered a variety of feedstocks and recognized that:

- A LCB biorefinery would require a steady flow of feedstock and broke the year into 12 discrete periods (months);
- Different feedstocks have different harvest windows and that the dry matter yield of species depends upon the time (month) of harvest;
- Storage losses will occur and depend upon location and time of storage; and
- It was important to include multiple biorefinery sizes and locations that enable investigation of the trade-off between economies of biorefinery size and feedstock transportation costs (Tembo et al., 2003).

The model was used to determine, for specific regions in Oklahoma, the most economical source of LCB, inventory management, biorefinery size and biorefinery location. It was innovative, but contained several limitations. The analysis used conventional agricultural machinery cost estimation software to compute LCB harvest costs on an acre rather than ton harvested basis. These charges were assessed independent of yield. It did not place any restrictions on the number of acres that could be harvested during a time period. The model's methodology resulted in two potential problems. First, harvest costs varied by ton since they were fixed per acre for each species independent of expected yield. Second, since harvest capacity was not constrained, the base model determined that it was optimal to harvest more than 80% of total LCB tonnage required for an entire year in the month of September. The analysis assumed that the market would provide harvest machines in a timely manner. However, the assumed capacity does not currently exist and a large investment in harvest machines would be required to achieve the capacity necessary to harvest the annual quantity of required LCB in a short time period. In effect, the analysis did not appropriately account for harvest costs.

Thorsell et al. (2004) designed a coordinated harvest unit that provides a capacity to harvest a given number of tons of LCB per time period. The harvest unit includes a coordinated set of harvest machines consisting of mowers, rakes, balers, tractors and bale transporters. It is assumed that field speeds of machines may be adjusted with crop yield to achieve the throughput capacity. The coordinated harvest unit may result in substantial size economies associated with harvest machines. The cost estimates were developed under the assumption of a coordinated set of harvest machines operated by specialized harvest crews with extended harvest windows.

For this study, the harvest unit as designed by Thorsell *et al.* (2004) is incorporated into the model as an integer activity for an annual cost (depreciation, insurance, interest, taxes, repairs, fuel, oil, lubricants and labour) that provides capacity to harvest a given tonnage per month. Monthly capacity depends upon the number of harvest days per month and the number of endogenously determined harvest units. The model breaks the year into 12 discrete periods (months) enabling a flow of feedstock to a biorefinery and recognizes that the expected dry matter yield depends upon the time (month) of harvest and that storage losses will occur and depend upon location of storage and time of storage.

The model contains what McCarl and Spreen (1996) denote as sequencing activities in that harvest, storage and transportation are sequenced to provide a flow of material to the biorefinery. The model contains storage and inventory, in that LCB from a particular species may be harvested and placed in storage during a harvest month and LCB may be removed from storage for use in each of the 12 months. Alternatively, LCB may be transported and processed in the harvest month. Decisions regarding LCB production, harvest, storage and transportation are assumed to be made repeatedly in all years of biorefinery life. Each year is referred to as a representative single period. This type of model is appropriate when resource, technology and price data are assumed to be constant and a long-run steady state solution is acceptable. Biorefinery location is endogenously determined (Tembo *et al.*, 2003).

Data

The state of Oklahoma was considered for the case study. The state has a variety of potential LCB feedstocks, including plant residue, indigenous native prairies and improved pastures. In addition, cropland could be used to produce dedicated feedstock crops such as switchgrass. Oklahoma has 14.9 million acres that are in native prairie grass, 4.7 million acres in improved pasture, one million acres in CRP and 8.5 million acres of harvested cropland. The soils and climate of the state are such that harvest of potential feedstock could be conducted across many months. The CRP has demonstrated that landowners are willing to engage in long-term land leases. Each of the state's 77 counties was considered to be a potential LCB production source. Eleven potential biorefinery locations within the state were identified. Data from the agricultural census and Oklahoma Agricultural Statistics were used to determine existing acres of wheat, maize, native prairies, improved pastures and cropland. CRP acres were based upon 2003 enrolment (Farm Service Agency, 2003). A survey of professional forage specialists was conducted to disaggregate native prairie acreage into acres of tall grass, mixed and short grass prairies by county. Acres of improved pastures were disaggregated into acres of tall fescue (*Festuca arundinacea*), old world bluestem (*Bothriochloa* sp.) and bermudagrass (*Cynodon dactylon*).

Removable quantities of wheat straw and maize stover were based upon estimated relationships between residue yield and grain yield with consideration given to fulfil conservation compliance requirements. Yield estimates for native prairies and for grasses produced on CRP acres were obtained from a survey of professional agronomists in the respective production regions (counties). Yield estimates for switchgrass and improved pastures under different fertility regimes were obtained from consultations with agronomists. Yield adjustment factors that account for relative differences in expected yields depending upon harvest month were also obtained from professional agronomists.

Harvest costs were based upon the harvest unit as described by Thorsell et al. (2004). A harvest unit is defined as a coordinated set of harvest machinery, which includes ten labourers, nine tractors, three mowers, three rakes, three balers and a field transporter. The annual ownership and operating cost of one harvest unit is estimated to be US$580,000. A single harvest unit provides a throughput capacity of 341 tons per harvest day. Potential harvest months vary by species. The number of harvest days per month depends upon the weather. Harvest days per month were based upon monthly mean field-workday estimates for Oklahoma (Reinschmiedt, 1973).

The biorefinery is expected to operate 350 days per year and expected to have a feedstock requirement of either 1,000, 2,000, or 4,000 dry LCB tons/day. Storage at the biorefinery is limited to the amount that could be used in a 3-week period (21,000 tons, 42,000 tons and 84,000 tons for the 1,000 tons/day, 2,000 tons/day and 4,000 tons/day biorefineries, respectively). Field storage is not restricted. Field storage cost was estimated at US$2/ton/month and field storage losses were estimated at 0.5%/month. Minimum LCB inventory at the biorefinery was assumed to be equal to zero and storage losses at the biorefinery were assumed to be equal to 0.1% per month.

Estimates of field to biorefinery transportation distances were based upon map miles from cities located near the centre of the two counties. Bhat et al. (1992) estimated the cost of transporting a 17 dry ton truckload of LCB as $TC_{ij} = 34.08 + 1.00\ d_{ij}$ where TC_{ij} is the estimated cost of transporting LCB from production region i to biorefinery j and d_{ij} is the round-trip distance in miles. The average per dry ton transportation cost can then be determined by dividing by the assumed truck capacity of 17 dry tons. A feedstock dry matter content of 85% is assumed.

The provisions of the *Farm Security and Rural Investment Act of 2002* (FSRIA) allow managed harvest of CRP grassland acres a maximum of once every 3 years (Office of the Federal Register, 2003). Amendments included in the FSRIA provide for harvesting of biomass produced on CRP acres for biorefinery feedstock. With current regulations, it is likely that removal of biomass from CRP grasslands in Oklahoma could be conducted over a 120-day period beginning July 2. The legislation requires that acres used for grazing, hay or biomass harvest shall be assessed with a 25% annual rental payment reduction. The average rental rate for Oklahoma CRP land is US$32/acre (Farm Service Agency, 2003). An access and acquisition fee of US$10/acre was assessed in the model to compensate landowners for the reduction in CRP payment and removal of LCB.

A charge of US$10/ton was assessed to compensate landowners for removal of all feedstocks other than feedstock produced on CRP land. Constraints were included in the model to limit harvest to 25% of CRP acres per county. Use of other feedstocks was limited to no more than 10% of the available acres by county.

Results

A total of five models were solved. These models were differentiated by biorefinery feedstock requirements (either 1,000 tons, 2,000 tons, or 4,000 tons of LCB/day) and by feedstock source (multiple sources meaning LCB could be procured from the most economical source from among wheat straw, maize stover, native prairies, improved pastures, or from cropland converted to switchgrass; CRP meaning LCB could be procured only from CRP land). In Oklahoma, harvest of CRP land is currently restricted to a 120-day harvest season beginning July 2. In the absence of policy restrictions, for the region of the study, LCB could be harvested from multiple sources from June to February. Given the restriction that CRP acres can only be harvested in one of three years and given the model assumption that restricted CRP harvest to 25% of enrolled acres, there is insufficient CRP production to support a 4,000 tons/day biorefinery if production is limited to Oklahoma and CRP is the only source of feedstocks.

Table 11.1 includes a summary of results from the five models. As expected, as the size of the biorefinery is increased from 1,000 tons/day to 4,000 tons/day, the average one-way distance to transport LCB from the fields to the biorefinery increases from 49 miles to 78 miles for the multiple feedstock scenarios. This increases the transportation cost from US$7.73/ton to US$11.13/ton. This increase in transportation cost increases the cost to deliver a steady flow of feedstock from US$32/ton to US$35/ton. The results are similar for the CRP-limited scenarios. Average transportation distance increases from 98 miles to 147 miles and, transportation cost increases from US$13.53/ton to US$19.35/ton as biorefinery size increases from 1,000 tons/day to 2,000 tons/day. The average feedstock transport distance for all five scenarios is graphed in Fig. 11.1.

A coordinated set of harvest machines was defined as a harvest unit and included as an integer investment activity in the model. For a 4,000 tons/day multiple feedstock biorefinery, the model selects 26 harvest units as optimal (Table 11.1). Since a harvest unit includes three mowers, three rakes, three balers, nine tractors and one transport stacker, the 4,000 tons/day multiple feedstock biorefinery would require 234 tractors, 78 mowers, rakes and balers and 26 transport stackers. The estimated average investment in these harvest machines exceeds US$15 million. If feedstock production were limited to that grown on CRP land, a 2,000 tons/day biorefinery would require an investment in harvest machines of more than US$17 million. The use of multiple feedstocks that extends the harvest window to 9 months substantially reduces the required investment in harvest machines.

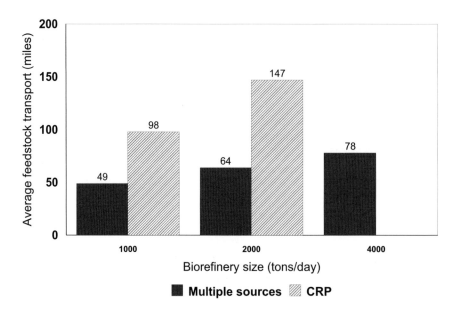

Figure 11.1. Estimated average distance to transport lignocellulosic biomass to a biorefinery located in Oklahoma from multiple feedstock sources and from CRP acres.

Table 11.1. Results of models solved to determine the cost to deliver a steady flow of lignocellulosic biomass to a biorefinery in Oklahoma.

Item	Multiple feedstocks			Feedstock restricted to CRP land	
Biorefinery size (tons/day)	1,000	2,000	4,000	1,000	2,000
Acquisition and field cost (US$/ton)	11.53	11.48	11.90	18.57	18.52
Harvest cost (US$/ton)	11.56	10.73	10.73	24.94	25.22
Field storage cost (US$/ton)	1.19	1.38	1.38	6.54	6.03
Transportation cost (US$/ton)	7.73	9.50	11.13	13.53	19.35
Total cost of feedstock (US$/ton)	32.01	33.10	35.15	63.59	69.12
Harvested acres	192,738	340,656	665,688	167,932	259,064
Harvest units (number)[a]	7	13	26	15	29
Average investment in harvest machines (US$,000)	4,130	7,670	15,340	8,850	17,110
Harvest months (number)[b]	9	9	9	4	4
Feedstock source					
Switchgrass (tons)	187,629	365,435	854,369	N/A[c]	N/A
Native grasses (tons)	106,529	262,963	406,131	N/A	N/A
Wheat straw (tons)	47,047	60,397	108,976	N/A	N/A
CRP land (tons)	9,541	11,594	17,736	348,821	666,823
Bermudagrass (tons)	0	0	5,941	N/A	N/A
Fescue (tons)	0	0	7,691	N/A	N/A
Maize stover (tons)	483	2,431	2,352	N/A	N/A
Total biomass harvested (tons)	351,229	702,819	1,405,638	348,821	666,823
Average distance hauled (miles)	49	64	78	98	147

[a] A harvest unit includes ten labourers, three mowers, three rakes, three balers, nine tractors and one transport stacker.
[b] Harvest of CRP land is restricted to July to October.
Note: Given the restriction that CRP acres can be harvested only once in three years, and given the model assumption of restricted CRP harvest to 25% of enrolled acres, there is insufficient CRP production to support a 4,000 tons/day biorefinery if production is limited to Oklahoma.
[c] Not applicable.

As described in Table 11.1 and shown in Fig. 11.2, for the multiple feedstock scenarios the cost to deliver a ton of LCB is estimated to be from US$32/ton to US$35/ton depending upon biorefinery capacity. Limiting feedstock to that produced on CRP land and limiting harvest to 120 days increases both harvest and storage costs. For a 2,000 tons/day biorefinery, the estimated cost to deliver a ton of LCB increases from US$33 for the multiple feedstock scenario to US$69 for the CRP-only scenario.

Based upon the assumptions of the model, every cost component is greater for the CRP-only scenario. The greater acquisition and field cost results from the assumption that multiple feedstocks could be procured at a cost of US$10/ton of harvested material whereas CRP feedstock was assessed a charge of US$10/acre. The CRP charge was assessed to offset the 25% reduction in payment to CRP contract holders. Harvest costs are estimated to be more than twice as much per ton for the CRP-only feedstock. This results from the 120-day CRP harvest window policy restriction. The multiple feedstock scenarios enable harvest from June to February, requiring less than half as many harvest units. The harvest window restriction increases the storage requirements and substantially increases the estimated average storage cost per ton for the CRP-only scenarios. Since CRP acres are widely dispersed across the state, transportation distances and costs are also substantially greater for the CRP-only scenarios.

Figure 11.3 includes a chart of the estimated quantity of feedstock harvested per month for a 2,000 tons/day biorefinery from both the multiple feedstock and the CRP-only scenarios.

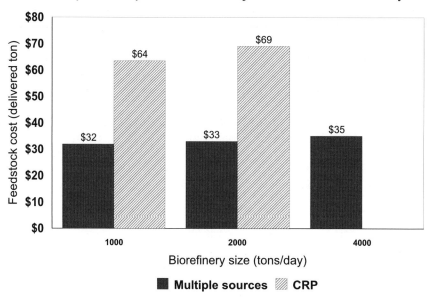

Figure 11.2. Estimated cost to deliver a ton of lignocellulosic biomass to a biorefinery located in Oklahoma from multiple feedstock sources and from only CRP acres located in Oklahoma.

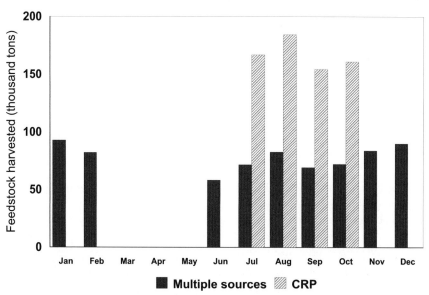

Note: CRP Harvest is limited by policy to 120 days beginning July 2. If multiple feedstocks are used, harvest may extend over 9 months.

Figure 11.3. Estimated quantity of feedstock harvested per month for a 2,000 tons per day biorefinery located in Oklahoma from multiple feedstock sources and from CRP acres.

Monthly harvest is restricted by both the number of expected harvest days and by the endogenously determined number of harvest units.

The model contains storage and inventory activities. LCB may be harvested and placed in storage for 9 months for the multiple feedstock scenarios and 4 months for the CRP-only scenarios. LCB may be removed from storage for use in each of the 12 months. Alternatively, LCB may be transported and processed in the harvest month. The model enables feedstock storage at the biorefinery and storage in fields at remote sites. Figure 11.4 includes a chart of the estimated quantity of feedstock stored per month at field sites for a 2,000 tons/day biorefinery from both multiple feedstock and CRP-only scenarios. For the CRP-only situation, replenishment of storage reserves begins with the first permissible harvest month of July. Harvest and increase of field storage inventory continues throughout August, September and October. At the end of October, when regulations require the harvest season must be completed, the combined field and biorefinery storage inventory must be sufficient to provide feedstock until harvest may be resumed in the following July. For the multiple feedstocks model, field inventory storage increases more gradually from August to February. The maximum quantity of required field storage for the multiple feedstock model is less than half of that required for the CRP-only model.

Figure 11.5 includes a chart of the estimated quantity of feedstock stored per month at the biorefinery site for a 2,000 tons/day biorefinery from both the multiple feedstock and CRP-only situations. As shown in Fig. 11.5, inventory is reduced to zero at the end of June in anticipation of a resumption of harvest in July. Feedstock is removed from field storage until the end of June when inventory of both field storage and storage at the biorefinery are reduced to zero. Minimum inventory constraints at the biorefinery were set to zero.

Figure 11.6 includes a chart of the specific feedstocks selected by the multiple feedstocks model for use by the 2,000 tons/day biorefinery. The optimal combination of feedstocks includes switchgrass produced on converted cropland (52%), indigenous grasses produced

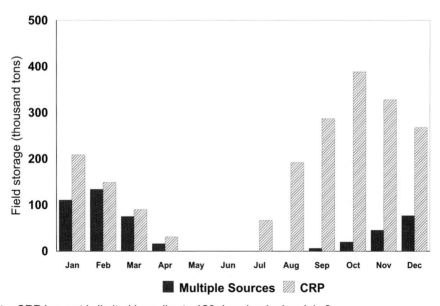

Note: CRP harvest is limited by policy to 120 days beginning July 2.

Figure 11.4. Estimated quantity of feedstock stored per month at remote sites for a 2,000 tons per day biorefinery located in Oklahoma from multiple feedstock sources and from CRP acres.

on native prairies (37.4%), wheat straw (8.6%), perennial grasses produced on CRP land (1.6%) and maize stover (0.3%). The most economical system would process a variety of feedstocks. Harvest would extend over as many months as permitted by weather and species.

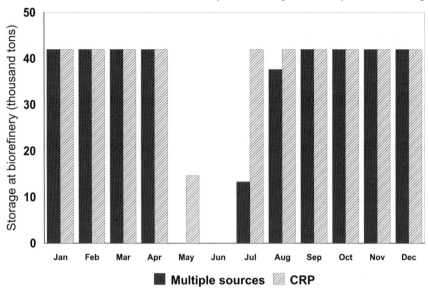

Note: CRP harvest is limited by policy to 120 days beginning July 2. Storage at the 2,000 tons per day biorefinery is limited to 42,000 tons.

Figure 11.5. Estimated quantity of feedstock stored per month at the biorefinery sites for a 2,000 tons per day biorefinery located in Oklahoma from multiple feedstock sources and from CRP acres.

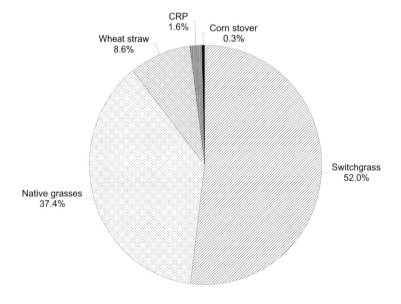

Figure 11.6. Feedstocks selected by the model for use by the 2,000 tons per day biorefinery with multiple sources.

Conclusions

CRP acres are widely dispersed in the study region and have relatively low expected yields. Harvest on CRP acres is limited to one in three years, the harvest window is limited to 120 days and harvesting requires the forfeiture of 25% of the annual payment in the year of harvest. In the study region, significant barriers exist for a biorefinery to rely on CRP lands to harvest a majority of its feedstock.

The research suggests that prior to investing in a biorefinery, it would be prudent to contract production to ensure a reliable flow of feedstock. This could be achieved by contracting with individual growers, or with a group of growers through a cooperative arrangement, or through long-term land leases similar to CRP leases. Given the quantity of LCB required and the lack of an existing infrastructure to harvest and transport a continuous flow of massive quantities of LCB, it is likely that an integrated and centrally controlled harvest and transportation system would develop. Public policy that restricts business ties between feedstock production and feedstock processing is likely to hinder the development of an LCB biorefinery industry.

It was determined that the estimated cost to deliver a flow of feedstock to a biorefinery ranged from US$32/ton to US$69/ton depending upon the size of the biorefinery and whether feedstock could be procured from multiple sources or from only CRP land. Increasing biorefinery feedstock requirements from 1,000 tons/day to 4,000 tons/day increases required transportation distances and increases the expected cost by US$3.14/ton for the multiple feedstocks model.

Given the underlying assumptions of the model, for the case study region, restricting harvest to CRP acres imposes a rather substantial cost on the industry. For a 2,000 tons/day biorefinery, limiting feedstock production to CRP land would increase the expected cost to deliver a ton of LCB to US$69 compared with a cost of US$33 for the multiple feedstock model. The economic model suggests that the cost to deliver a continuous flow of LCB to a biorefinery will depend critically upon the logistics of procuring, harvesting, storing and transporting the LCB.

Acknowledgements

The authors acknowledge the assistance of Raymond L. Huhnke, Charles M. Taliaferro and other personnel of the Biobased Products and Energy Center at Oklahoma State University, Stillwater, Oklahoma. This material is based upon work supported in part by Aventine Renewable Energy, Inc., USDA-CSREES IFAFS Competitive Grants Program award 00-52104-9662, USDA-CSREES Special Research Grant award 2002-34447-11908 and the Oklahoma Agricultural Experiment Station, project H-2403. Support does not constitute an endorsement of the views expressed in the chapter by Aventine Renewable Energy or by the US Department of Agriculture.

References

Bhat, M., English, B. and Ojo, M. (1992) Regional Costs of Transporting Biomass Feedstocks. In: Cundiff, J. (ed.) *Liquid Fuels from Renewable Resources: Proceedings of an Alternative Energy Conference, December 14-15, 1992*. American Society of Agricultural Engineers, St. Joseph, Michigan.

English, B., Short, C and Heady, E. (1981) The Economic Feasibility of Crop Residues as Auxiliary Fuel in Coal-Fired Power Plants. *American Journal of Agricultural Economics* 63(4): 636-644.

Farm Service Agency. (2003) *Conservation Reserve Program — Monthly Summary: September, 2003.* US Department of Agriculture, Farm Service Agency, Washington, DC. [Accessed 2004.] Available from http://www.fsa.usda.gov/dafp/cepd/stats/Sep2003

Gallagher, P. and Johnson, D. (1999) Some New Ethanol Technology: Cost Competition and Adoption Effects in the Petroleum Market. *The Energy Journal* (20)2: 89-120.

Glassner, D., Hettenhaus, J. and Schechinger, T. (1998) *Corn Stover Collection Project.* Presented paper, Great Lakes Regional Biomass Conference, BioEnergy 98, October 4-8, Madison, Wisconsin. [Accessed 2004.] Available from http://www.afdc.doe.gov/pdfs/5149.pdf

Graham, R., English, B. and Noon, C. (2000) A Geographic Information System-based Modeling System for Evaluating the Cost of Delivered Energy Crop Feedstock. *Biomass and Bioenergy* 18: 309-329.

Hillier, F. and Lieberman, G. (1980) *Introduction to Operations Research.* 3rd ed. Holden-Day, Inc., Oakland, California.

Kadam, K., Forrest, L. and Jacobson, W. (2000) Rice Straw as a Lignocellulosic Resources: Collection, Processing, Transportation, and Environmental Aspects. *Biomass and Bioenergy*, 18: 369-389.

Kaylen, M., Van Dyne, D., Choi, Y. and Blase, M. (2000) Economic Feasibility of Producing Ethanol from Lignocellulosic Feedstocks. *Bioresource Technology* 72: 19-32.

Lynd, L., Cushman, J., Nichols, R. and Wyman, C. (1991) Ethanol from Cellulosic Biomass. *Science* 251(March): 1318-1323.

McAloon, A., Taylor, F., Yee, W., Ibsen, K. and Wooley, R. (2000) *Determining the Cost of Producing Ethanol from Corn Starch and Lignocellulosic Feedstocks.* Technical Report NREL/TP-580-28893. US Department of Agriculture, Eastern Regional Research Center, Wyndmoor, Pennsylvania, and US Department of Energy, National Renewable Energy Laboratory, Golden, Colorado. [Accessed 2004.] Available from http://www.afdc.nrel.gov/pdfs/4898.pdf

McCarl, B. and Spreen, T. (1996) *Applied Mathematical Programming Using Algebraic Systems.* Texas A&M University, Department of Agricultural Economics, College Station, Texas. [Accessed 2004.] Available from http://agecon.tamu.edu/faculty/mccarl/books.htm

National Research Council. (1999) *Review of the Research Strategy for Biomass-Derived Transportation Fuels.* National Academy Press, Washington, DC. [Accessed 2004.] Available from http://books.nap.edu/html/biomass_fuels

Nienow, S., McNamara, K., Gillespie, A. and Preckel, P. (1999) A Model for the Economic Evaluation of Plantation Biomass Production for Co-Firing with Coal in Electricity Production. *Agricultural and Resource Economics Review* 28: 106-118.

Nilsson, D. (1999) SHAM—A Simulation Model for Designing Straw Fuel Delivery Systems Part 1: Model Description. *Biomass and Bioenergy* 16: 25-38.

Office of the Federal Register. (2003) 2002 Farm Bill—Conservation Reserve Program—Long-Term Policy; Interim Rule. *Federal Register* Vol. 68, No. 89, (Thursday, May 8) p. 24830.

Phillips, J., Clausen, E. and Gaddy, J. (1994) Synthesis Gas as Substrate for the Biological Production of Fuels and Chemicals. *Applied Biochemistry and Biotechnology* 45/46: 145-157.

Rajagopalan, S., Datar, R. and Lewis, R. (2002) Formation of Ethanol from Carbon Monoxide via a New Microbial Catalyst. *Biomass and Bioenergy* 23: 487-493.

Reinschmiedt, L. (1973) *Study of the Relationship Between Rainfall and Fieldwork Time Available and its Effect on Optimal Machinery Selection.* MS thesis, Oklahoma State University, Department of Agricultural Economics, Stillwater, Oklahoma.

Schechinger, T. (2000) *Current Corn Stover Collection Methods and the Future.* BioMass Agri Products, Harlan, Iowa. [Accessed 2004.] Available from http://www.afdc.doe.gov/pdfs/4922.pdf

Tembo, G. (2000) Integrative Investment Appraisal and Discrete Capacity Optimization Over Time and Space: The Case of an Emerging Renewable Energy Industry. PhD dissertation, Oklahoma State University, Department of Agricultural Economics, Stillwater, Oklahoma.

Tembo, G., Epplin, F. and Huhnke, R. (2003) Integrative Investment Appraisal of a Lignocellulosic Biomass-to-Ethanol Industry. *Journal of Agriculture and Resource Economics*, 28(3): 611-633.

Thorsell S., Epplin, F., Huhnke, R. and Taliaferro, C. (2004) Economics of a Coordinated Biorefinery Feedstock Harvest System: Lignocellulosic Biomass Harvest Cost. *Biomass and Bioenergy* 27, 327-337.

Wyman, C. (1994) Ethanol from Lignocellulosic Biomass: Technology, Economics and Opportunities. *Bioresource and Biotechnology* 50: 3-16.

Chapter 12

Economic Impacts of Ethanol Production from Maize Stover in Selected Midwestern States

Burton C. English, R. Jamey Menard, Daniel G. De La Torre Ugarte and Marie E. Walsh

Introduction

Recently, concern has been expressed in the rural USA about fluctuating net farm income, agriculture resource rigidity, higher energy prices and a need for increased economic development. Although net farm income has improved recently, income volatility remains a concern. According to the US Department of Agriculture (USDA), the 10-year average (1994-2003) for net farm income was US$48.2 billion. Income variation for this 10-year average was US$6.7 billion. From 1994 to 2003, net farm income ranged from a low of US$35.3 billion in 2002 to a high of US$57.8 billion in 1996 (Economic Research Service, 2003; Economic Research Service, 2004).

Unlike other non-farm economic sectors, resources in agriculture are not very mobile. Once the resources are employed by the agricultural sector they tend to remain there. US farmers use all of their productive capacity regardless of expected commodity prices. Even though a farmer may go out of business, the land usually remains in agricultural production. Historically, agriculture has been plagued by surpluses and low commodity prices (Ray *et al.*, 2003). 'Other industries would throttle back production and/or decrease productive capacity' (Ray, 2004, p. 39).

World oil prices have increased from US$22.68/barrel to US$33.40/barrel, a 47.3% increase, from January 2000 to May 2004 (Energy Information Administration, 2004d). According to the US Department of Energy (US DOE), the average retail price (May 2004) for regular unleaded petrol was US$2.01/gallon (gal). Adjusted for inflation, petrol prices have not been in that range since the autumn of 1985. This lack of stability in oil prices contributes greatly to the difficulty for consumers and businesses to plan and budget (Energy Information Administration, 2004c).

Public pressure has increased toward establishing value-added operations in rural areas. Interest in economic development of rural areas has traditionally focused on manufacturing opportunities and has neglected agricultural value-added prospects. Thus, rural communities either shipped raw commodities out or fed the raw agricultural commodities and shipped livestock from the region. Recent contributions to incomes and employment in rural areas have occurred through the development of an ethanol industry relying on agricultural feedstocks.

In a study conducted for US DOE, several reports were reviewed that analysed the economic impacts of fuel ethanol. In this analysis, it was found that 'These assessments all predicted substantial economic benefits from increased production of fuel ethanol' (Energetics, Inc. and NEOS Corporation, 1994, p. 1). For instance, Energetics, Inc. and NEOS Corporation (1994) report that:

- A 1993 USDA study estimated that increasing ethanol production to 2 billion gallons would create 28,000 new jobs.
- The National Corn Growers Association estimated that currently projected expansion of the ethanol industry through 2000 would create over 273,000 jobs throughout the USA.
- The US General Accounting Office (US GAO) estimated that an increase in ethanol production to the 2.0 billion gal to 5.0 billion gal level would increase net farm income by 1.3% per year or an average of US$415 million over the 8-year period of US GAO's analysis.

Biomass feedstocks – such as maize fibre (hull from a kernel of maize), maize stover (residue left from grain harvest), bagasse (residue left from the crushed stalks of sugarcane) and rice straw – contain cellulose, which can be converted to sugars that are then fermented to ethanol. New technologies are in development that will convert maize stover to ethanol more efficiently. In this process, the agricultural producer harvests the maize and windrows the residues. Following the harvest, the residues are baled, wrapped in a plastic mesh and transported to the edge of the field. Once at the field's edge, the stover is transported to the ethanol production facility in such a manner that there is 10 days of inventory kept at the ethanol plant. This process creates a by-product for the farmers to market. Producers will have to receive the total costs of harvesting and transporting the crop, plus an incentive payment, to entice their participation. The cost of harvesting and transporting the residue depends on the per acre residue yield.

Ethanol is a fuel substitute produced from renewable feedstocks. The number of alternative-fuelled vehicles or primarily fleet-operated vehicles using E85 (85% ethanol and 15% petrol) has increased from 1,527 in 1995 to an estimated 133,776 in 2003. In terms of fuel combustion, one and a half litres of ethanol has the same combustion energy as one litre of petrol. Ethanol has an octane rating of 110 and is used directly as fuel or as an octane-enhancing petrol additive. Blends of 10% ethanol with petrol can be used in all petrol-powered cars without engine or carburettor modification (Energy Information Administration, 2004b).

The use of ethanol itself, or blended with other automotive fuels, would result in less polluting carbon monoxide and less use of toxic compounds currently used to enhance automotive fuel octane levels. The potential for global warming would decrease because of lower hydrocarbon emissions in the air. Less dependency on foreign oil to satisfy US consumption needs would increase energy and economic security. In addition, as a renewable feedstock, potential benefits from an expanded market and additional economic opportunities for farmers and the rural areas of the USA could be realized.

To be competitive with existing subsidized fuels, ethanol facilities need to be efficient and economical. Once the maize stover enters the plant it must be handled and pre-treated. More specifically, the stover must go through a hydrolysis, fermentation and distillation process. Next, the stover must pass through an ethanol fermentation and enzyme production process. The product is then purified and the wastewater is treated. One solid by-product of the conversion process is lignin, which can be burned to produce steam or electricity. To reduce costs, further research to improve existing separation and conversion technologies and process systems is required. In addition, genetic engineering of agricultural crops, such as maize, to increase carbohydrate levels, is a potential option (Wooley *et al.*, 1999).

The majority of ethanol plants in the USA are located in the Midwest. Currently, the USA has 75 ethanol plants, with an additional 12 planned for construction. Current ethanol production capacity is estimated at 3.211 billion gal/year. Additional expansion and new plant capacity is estimated at 447 million gal. The top five states with the largest ethanol production capacity are: Iowa (714 million gal/year), Illinois (734 million gal/year), Nebraska (405 million gal/year), South Dakota (377 million gal/year) and Minnesota (418 million gal/year). These five states contain 82% of the production capacity for the USA (Hart, 2004). Ethanol demand is expected to increase. In 2002, US ethanol production, with maize as the primary feedstock, was 139,000 barrels/day. US DOE projects production to double by 2025. About 27% of the growth is projected to occur from conversion of cellulosic biomass (i.e., stover). In the high renewables case, all the projected growth is from cellulose, a result of more rapid improvement in the technology (Energy Information Administration, 2004a).

The objectives of this research are to provide estimates of economic impacts if stover-to-ethanol plants are established in the current maize producing states of the USA. The economic impact indicators used in the analysis include total industry output, employment and value-added. Analysis includes both the impacts that occur when the first most likely plant is constructed and in operation and when all feasible plants are in operation.

Methodology

There are numerous levels of impacts to a state's economy as a result of developing a stover-to-ethanol industry. This section reports the methodologies used to estimate these impacts at two levels. Economic impact analyses for ten single plants (one in each state) are initially conducted. Using this information and information from a geographic information system (GIS), statewide estimates are provided identifying the extent that the industry can expand under pre-determined ethanol prices. This information is then used to provide estimates of the likely economic impacts that would occur given a mature industry. The estimated economic impacts include those that occur as a result of changes in agriculture, transportation and plant operation. In addition, impacts occur as a result of the construction of the plants. This chapter only reports the major results of this study.

Models Employed in the Analysis

To evaluate the economic impacts of establishing a stover-to-ethanol industry for this study, four major models were incorporated into the analysis. The models include: 1) computer spreadsheets of alternative plant sizes; 2) Policy Analysis System (POLYSYS); 3) Oak Ridge Bioenergy Analysis System (ORIBAS); and 4) IMPLAN®. Both the POLYSYS and ORIBAS models were used to develop estimates of stover costs both at the farm and plant. Specifically, ORIBAS supplied, for each state and each stover conversion plant, data on harvesting costs, transportation costs, plant location and quantity of stover supplied from each county. A more detailed summary of each of these models follows.

Computer Spreadsheets of Alternative Plant Sizes

Ethanol plant needs for two different production alternatives were provided by the National Renewable Energy Laboratory (NREL). These alternatives were constructed for a 2,000 metric ton (MT)/day and a 1,000 MT/day plant for the year 2010. NREL provided engineering

Table 12.1. Quantities of maize stover available as feedstocks for ethanol production (dry tons).

State	Year 2005
Illinois	13,482,715
Indiana	6,615,635
Iowa	15,195,004
Kansas	1,675,661
Minnesota	9,723,571
Missouri	2,248,977
Nebraska	10,511,742
Ohio	3,295,542
South Dakota	2,216,419
Wisconsin	3,777,238

Source: English et al. (2001).

economics and ethanol production capacity summary spreadsheets for plants capable of consuming 1,000 MT and 2,000 MT of maize stover per day. These spreadsheets were used in several ways. They provided the necessary information required to model the investment and operating impacts within IMPLAN®. In addition, the spreadsheets determined the ethanol output price for pre-specified prices of the feedstock maize stover. Thus, the spreadsheets were used to estimate the amount that the plant could afford to pay for stover under assumed ethanol price scenarios.

POLYSYS

The Policy Analysis System (POLYSYS) modelling framework was developed to simulate changes in policy, economic, or resource conditions and estimate the resulting impacts for the US agricultural sector. The POLYSYS modelling framework is capable of endogenously (originating internally) considering a wide variety of region-specific crop rotations and management practices. Crops endogenously considered in POLYSYS include maize, grain sorghum, oats, barley, wheat, soybeans, cotton, rice and hay. Endogenous livestock commodities include beef, pork, lamb and mutton, broilers, turkeys, eggs and milk (Ray et al., 1998; Tiller et al., 1999).

Using the maize yields and acres for 2005 as estimated by POLYSYS, quantities of maize stover available as feedstocks for ethanol production are estimated for each county in the ten states (Table 12.1). Maize acres classified as highly erodible (e.g., an erosion index of 8 or higher) are excluded from consideration (Natural Resources Conservation Service, 2004). Assumed quantities required to remain to maintain soil quality are subtracted from the total quantities of stover produced – a maximum of 45% of the residues generated are allowed to be collected (D. Lightle, USDA/Natural Resources Conservation Service, Washington, DC, 1997, personal communication).

ORIBAS

The Oak Ridge Integrated Bioenergy Analysis System (ORIBAS), a GIS-based transportation model, is used to estimate the delivered costs of biomass to hypothetical ethanol facilities. ORIBAS includes a complete road network for each state. Maize stover quantities are evenly distributed across each county. ORIBAS locates facilities based on delivered feedstock costs with the first plant having the lowest delivered costs for quantities sufficient to meet its feedstock demands. Subsequent facilities have increasing costs as they must either purchase feedstocks from areas that are more expensive and/or transport feedstocks farther to satisfy their feedstock needs. The cost of delivering residues is estimated along with the location of the stover.

IMPLAN®

The study relied heavily on input-output analysis to derive economic impacts for constructing and operating various ethanol production facilities, including farm level stover production and transportation costs, for selected Midwestern states. Input-output analysis creates a picture of a regional economy to describe flows to and from industries and institutions. Input-output analysis is a useful tool to predict changes in overall economic activity as a result of some marginal change. Impacts estimated include *total industry output* (the annual value of the production by the industry examined*), employment* (total wage and salary employees, both full and part-time, as well as self-employed) and *total value-added* (employee compensation, proprietary income, other property type income and indirect business taxes). For each impact category, direct (changes in the demand for a sector's product), indirect (inter-industry purchases resulting from changes in demand for a sector's product) and induced (changes in household spending power resulting from changes in demand for a sector's product) changes are estimated.

The analyses for the study were based on the IMPLAN Professional® version 2.0 model using 2001 data (Minnesota IMPLAN Group, Inc., 1999; Olson and Lindall, 1999). IMPLAN® provides economic activity information for approximately 509 industry sectors based on the North American Industry Classification System (NAICS) code system. The ten states chosen for the study were Illinois, Indiana, Iowa, Kansas, Minnesota, Missouri, Nebraska, Ohio, South Dakota and Wisconsin. The breakeven macroeconomic impacts of the first ethanol facility in each state are estimated. Breakeven costs assume no premium is paid to the farmers to ensure participation (only the cost of collection and transport are included in the feedstock costs) and that the value of the ethanol is equal to the production cost.

Two ethanol production process technology plants were analysed with stover feed rates of 1,000 MT/day for one plant and 2,000 MT/day for the other. Technology for the plants is assumed at levels expected to exist by the year 2010. The process engineering design and economic information to provide input data for the input-output model were obtained from NREL (Wooley *et al.*, 1999). There are economies of scale to the plant (Fig. 12.1). If one assumes a feedstock price of US$30/ton of stover, the 1,000 MT/day facility requires a breakeven ethanol price of US$1.38/gal and the 2,000 MT/day facility needs a US$1.17/gal price to breakeven.

Total Project Investment (Plant Construction)

Based on Aden *et al.* (2002), project investment is comprised of total equipment cost, total installed cost (equipment cost plus warehouse and site development), indirect costs (field expenses, home office and construction fees and project contingencies) and total capital investment. Assignment of these cost categories and their corresponding values to the appropriate IMPLAN® industry sectors were required (Table 12.2).

Project investment costs for the two ethanol production facilities and the appropriate IMPLAN® sectors assigned are presented in Table 12.2. Project investment costs are considered a one-time start-up expense for constructing the ethanol production facility, not an annual reoccurring economic impact. Each state's baseline economy obtained from the IMPLAN® model was used to generate the economic impacts for this cost category.

Table 12.2. Project investment costs (year 2001, million US$).

		IMPLAN®	Year 2010	
			Metric tons/day	
Code	Sector		1,000	2,000
37	Manufacturing & industrial buildings		42.97	68.35
150	Other basic inorganic chemical manufacturing		0.04	0.06
171	Other miscellaneous chemical product manufacturing		0.01	0.02
238	Power boiler & heal exchanger manufacturing		6.56	10.59
239	Metal tank, heavy gauge, manufacturing		5.62	9.01
240	Metal can, box & other container manufacturing		0.37	0.58
255	Miscellaneous fabricated metal product manufacturing		0.33	0.51
257	Farm machinery & equipment manufacturing		1.12	1.70
269	All other industrial machinery manufacturing		5.79	8.88
273	Other commercial & service industry machinery manufacturing		0.84	1.49
275	Air purification equipment manufacturing		9.36	14.12
276	Industrial & commercial fan & blowing manufacturing		0.17	0.25
277	Heating equipment, except warm air furnaces		0.63	1.02
278	AC, refrigeration & forced air heating		1.88	3.05
285	Turbine & turbine generator sets unit manufacturing		7.04	11.50
288	Pump & pumping equipment manufacturing		3.96	6.79
289	Air & gas compressor manufacturing		0.88	1.11
292	Conveyor & conveyor equipment manufacturing		13.51	20.54
294	Industrial truck, trailer & stacker manufacturing		0.08	0.17
301	Scales, balances & misc. general purpose machinery		0.22	0.30
316	Industrial process variable instrument		0.44	0.59
425	Nondepository credit intermediation & related activities		11.86	18.83
460	Waste management & remediation services		13.72	22.85
	Total		127.43	202.32

Source: Aden et al. (2002).

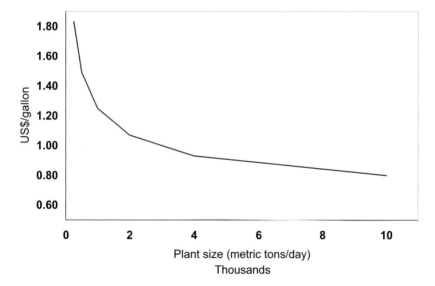

Source: Aden et al. (2002).

Figure 12.1. Economies of scale for ethanol production using stover as a feedstock (US$30/ton).

Table 12.3. Ethanol raw material input requirements and IMPLAN® sector assigned.

Raw material	IMPLAN® sector
Electricity (credit)	30 Power generation & supply
Makeup water	32 Water, sewage & other systems
Clarifier polymer, sulphuric acid, boiler chemicals, cooling tower chemicals, wastewater chemicals & polymers	150 Other basic inorganic chemical manufacturing
Maize steep liquor, cellulase enzyme	151 Other basic organic chemical manufacturing
Diammonium phosphate	157 Phosphatic fertilizer manufacturing
Hydrated lime	196 Lime manufacturing
Waste stream disposal	460 Waste management & remediation services
Feedstock	507 Rest of the world industry*

*Feedstock was assigned to Sector 507 to alleviate potential double counting of inputs from other industry sectors.
Source: Aden *et al.* (2002).

Table 12.4. Annual variable operating costs, (year 2001, million US$).

	IMPLAN®	Year 2010	
		Metric tons/day	
Code	Sector	1,000	2,000
30	Power generation & supply	-2.61	-6.54
32	Water, sewage & other systems	0.21	0.41
150	Other basic inorganic chemical manufacturing	1.20	1.72
151	Other basic organic chemical manufacturing	4.58	9.18
157	Phosphatic fertilizer manufacturing	0.11	0.21
196	Lime manufacturing	0.79	1.57
460	Waste management & remediation services	1.04	2.08
507	Rest of the world industry*	11.58	23.17
	Total	16.90	31.80
151	Other basic organic chemical manufacturing	44.29	74.15

*Feedstock was assigned to Sector 507 to alleviate potential double counting of inputs from other industry sectors.
Source: Aden *et al.* (2002).

Annual Operating Costs (Variable and Fixed)

The ethanol operating costs were added to IMPLAN®'s *Other Basic Organic Chemical Manufacturing* industrial sector. Gross absorption coefficients for this sector were modified for each state included in the analysis. This was required to account for the new input requirements for the ethanol plant.

The NREL supplied a detailed listing of raw material input requirements and expenses for each of the production facilities. Table 12.3 contains the raw material input requirements and the appropriate IMPLAN® sectors each were assigned.

The annual variable operating costs for the ethanol facilities are presented in Table 12.4. In addition, values to conduct the impact analysis for each production scenario were derived by multiplying ethanol production costs (US$/gal) and quantity of ethanol produced (million gal/year). This information was available from the ethanol production process engineering analysis sheets supplied by the NREL.

Fixed annual operating expenses required adjustments for personnel salaries, plus insurance and tax expenses. Modifications for these costs were adjusted in IMPLAN®'s *Other Basic Organic Chemical Manufacturing* total value-added categories for employee

Table 12.5. Annual fixed operating costs, (year 2001, million US$).

	IMPLAN®	Year 2010	
		Metric tons/day	
Code	Sector	1,000	2,000
427	Insurance carriers	0.58	0.67
438	Accounting & bookkeeping services	0.58	0.67
485	Commercial machinery repair & maintenance	1.50	2.38
VA adjustment	Employee compensation	1.86	2.15
VA adjustment	Indirect business taxes	1.15	1.82
	Total	5.67	7.69

VA=Value-added.
Source: Aden *et al.* (2002).

compensation and indirect business taxes, respectively. For each ethanol producing facility in the study, an additional 77 employees were also assigned to the model to reflect plant personnel requirements. IMPLAN®'s total industrial output was adjusted for the additional revenue generated from ethanol production (ethanol production costs multiplied by ethanol production in million gal/year) for each of the prototype plants. Annual fixed operating costs for each of the production facilities are presented in Table 12.5.

Agriculture and Transportation Costs

To properly allocate the agricultural costs for the production of maize stover, the quantity demanded for each of the prototype plants had to be determined. The US DOE's Oak Ridge National Laboratory (ORNL), using ORIBAS, provided this information, including farm-gate prices and transportation costs from the farm-gate to the plant-gate for each state. The term, *total farm value* (TFV) was calculated by multiplying the farm-gate price and the quantity of tonnes demanded. The agricultural or collection cost percentage for maize stover was a percentage of the TFV. The collection costs and their corresponding distribution percentages are capital costs (depreciation 24%), nonland capital costs (opportunity cost of ownership 7%), repair costs (34%), fuel/lube costs (7%), labour (13%) and mesh wrap (15%).

Similar to the ethanol production facility analysis, proper allocation of agricultural costs to the appropriate IMPLAN® industry sector for maize production was required. Maize production in IMPLAN® is represented in the sector – *Feed Grains*. The gross absorption coefficients for the feed grain sector's production function required adjusting. Labour and nonland capital cost values were assigned to IMPLAN®'s total value-added categories employee compensation and other property income categories, respectively.

One final adjustment to the input-output model was required. Transportation costs for each plant were estimated by ORIBAS and were assigned to the sector – *Motor Freight Transport and Warehousing*.

Initial Plant Analysis

The ORIBAS model provided two data sets for both ethanol production process scenarios analysed (year 2010 and 1,000 MT/day and 2,000 MT/day feed rates). The first data set contained pixel information including cost of harvesting biomass, cost of transporting the biomass to the plant and the quantity of biomass harvested. The second set of data identified the county location of the plant. This data set contains the average cost of the biomass delivered

Table 12.6. Ethanol breakeven maize stover 2005 prices (US$/metric ton dry matter).

Plant Size	Ethanol Price		
	US$1.15/gallon	US$1.25/gallon	S$1.35/gallon
1,000 MT/day	18.75	27.60	36.60
2,000 MT/day	37.35	46.20	55.14

to the plant and the state-county Federal Information Processing Standards (FIPS) location code.

Average and marginal costs of supplying biomass to the plant from both the farm and the transportation firm were estimated for each plant and scenario combination. Total farm-gate costs (*TFGC*) for plant (*j*) were estimated using a weighted sum of the farm-gate breakeven price (*FGP*) for each pixel (*i*) provided by ORIBAS multiplied by the quantity (*QCS*) supplied at each pixel (*i*) to plant (*j*) (equation 1).

$$TFGC_j = \sum_i FGP_i * QCS_{i,j} \quad (1)$$

Total transportation costs (TTC) were established in a similar manner. Transportation costs (*TC*) from pixel (*i*) to plant (*j*) were multiplied by the amount transported (*QCS*) to plant (*j*) from pixel (*i*) (equation 2).

$$TC_j = \sum_i TC_{i,j} * QCS_{i,j} \quad (2)$$

State Analysis

Using the pixel data for each of the plants located in the ten states provided by ORIBAS, total impact estimates are developed for each plant size and three ethanol price scenarios (US$1.15/gal, US$1.25/gal and US$1.35/gal). Information developed from Wooley et al. (2002) was used to determine the amount the two different 2010 plant sizes could afford to pay for maize stover (Table 12.6). Prices were then compared to GIS-generated and the adjusted average stover price delivered to the plant. If the plant-average stover price was less than the breakeven maize stover price, the plant was included. Agriculture and operating impacts for a single plant were multiplied by the number of plants. These impacts were added to transportation sector impacts adjusted to reflect the increase in transportation costs and economic activity as more plants are incorporated.

Once the number and location of ethanol production plants was determined for each plant size, price and profit scenario, the amount of maize stover required and the amount of ethanol produced on a yearly basis was determined for a mature ethanol-from-stover industry for each state. Depending on the size of facility, 1,000 MT/day or 2,000 MT/day, either 363,000 MT or 726,000 MT of stover are required for operation each year, respectively. This will result in 34.6 million gal of ethanol production/year for the 1,000 MT/day plant or 69.3 million gal of ethanol production/year for the 2,000 MT/day plant.

Results

Estimated total annual impacts for the 1,000 MT/day plant to each state ranged from US$94.1 million (South Dakota) to US$120.3 million (Kansas) (Table 12.7). Of this value, approximately

Table 12.7. Economic impact of the initial ethanol plant by state (year 2001, million US$).

State	Agriculture		Transportation		Operating		Total	
	Direct	Total	Direct	Total	Direct	Total	Direct	Total
2010 – 1,000 MT/day ethanol facility								
Illinois	6.53	13.54	3.43	8.76	44.30	95.00	54.26	117.30
Indiana	6.99	13.19	3.66	8.48	44.30	81.40	54.95	103.07
Iowa	6.41	11.89	3.51	7.34	44.30	75.30	54.22	94.54
Kansas	5.99	12.21	6.29	15.55	44.30	92.60	56.58	120.36
Minnesota	6.48	13.15	3.50	8.64	44.30	93.20	54.28	114.99
Missouri	6.27	13.05	4.30	10.81	44.30	90.20	54.86	114.06
Nebraska	6.19	11.98	3.36	7.73	44.30	78.60	53.85	98.31
Ohio	7.08	13.38	4.46	10.09	44.30	80.90	55.84	104.37
South Dakota	7.30	12.79	4.28	9.34	44.30	72.00	55.88	94.13
Wisconsin	7.01	13.59	4.21	9.93	44.30	82.70	55.53	106.23
2010 – 2,000 MT/day ethanol facility								
Illinois	13.13	27.24	8.57	21.88	74.10	158.40	95.80	207.52
Indiana	14.04	26.49	9.01	20.90	74.10	135.10	97.15	182.50
Iowa	12.85	23.84	8.42	17.63	74.10	125.00	95.37	166.46
Kansas	12.70	25.89	23.03	56.91	74.10	153.80	109.83	236.60
Minnesota	13.00	26.39	8.74	21.60	74.10	154.60	95.84	202.58
Missouri	14.62	30.46	19.56	49.22	74.10	150.60	108.27	230.28
Nebraska	12.95	25.08	8.10	18.62	74.10	130.00	95.15	173.70
Ohio	14.16	26.77	11.13	25.19	74.10	135.30	99.39	187.26
South Dakota	15.44	27.04	10.79	23.51	74.10	118.60	100.33	169.15
Wisconsin	14.33	27.78	10.20	24.07	74.10	137.90	98.63	189.75

US$55 million is directly spent to pay for operating inputs. Increasing the plant size results in an estimated total impact of between US$166.4 (Iowa) to nearly US$236.6 million (Kansas) depending on the state where the facility is located. The density of available maize stover has the largest impacts for transportation since less dense areas (Kansas and Missouri) will have to increase transportation distances and, hence, costs to supply the 2,000 MT/day required by this facility.

For the 1,000 MT/day plant, assuming an ethanol price of US$1.35/gal, a total of 101 plants are feasible, with Iowa constructing 27, followed by Illinois (23) and Nebraska (16) (Table 12.8). When the ethanol price decreases to US$1.25/gal, only four 1,000 MT/day plants are projected as being feasible. A further decline in ethanol price to US$1.15/gal results in no projected 1,000 MT/day plants

As a result of the economies of size, under the 2,000 MT/day facility, more ethanol is produced than with the small facility. While only 72 2,000 MT/day plants (compared with 101 for the 1,000 MT/day plants) are constructed under the $1.35/gal ethanol price scenario, residue use increases from 101,000 MT/day to 142,000 MT/day. Even when the ethanol price is US$1.15/gal, 2,000 MT/day plants are feasible with the construction of 47 plants projected.

Since it is only the US$1.35/gal ethanol price scenario that supports the 1,000 MT/day plant, the remainder of the chapter will focus on that price level. With full adoption of this technology – and assuming that no market competition evolves for maize stover – the stover-to-ethanol industry would result in estimated total economic impacts for the 101 1,000 MT/day plants of US$11 billion (Table 12.9). Nearly 37 million tons of residues are converted into 3.1 billion gal of ethanol. This level of ethanol production equals the estimated 2004 production level of 3.3 million gal (Hart, 2004). If the 2,000 MT/day technology were adopted, 52.3 million tons of stover would be converted to 4.4 billion gal of ethanol creating an economic impact to the ten-state economy of US$14 billion. Please note that these estimates assume an

Table 12.8. Number of projected ethanol plants by state by alternative ethanol price.

State	Metric tons/day					
	1,000	2,000	1,000	2,000	1,000	2,000
Ethanol price	US$1.15		US$1.25		US$1.35	
	Number of feasible plants					
Iowa	0	13	1	16	27	18
Illinois	0	11	0	13	23	15
Indiana	0	4	0	6	10	7
Kansas	0	0	0	0	2	1
Minnesota	0	7	0	10	17	10
Missouri	0	0	0	0	2	2
Nebraska	0	9	3	10	16	11
Ohio	0	1	0	3	1	3
South Dakota	0	1	0	1	1	1
Wisconsin	0	1	0	3	2	4
Total	0	47	4	62	101	72

Table 12.9. Economic impacts of a mature ethanol industry with 1,000 or 2,000 metric tons/day plants assuming an ethanol price of US$1.35/gallon (year 2001, million US$).

State	1,000 metric ton/day plant				2,000 metric ton/day plant			
	Direct	Indirect	Induced	Total	Direct	Indirect	Induced	Total
Illinois	1,274	827	664	2,765	1,485	954	796	3,234
Indiana	564	288	213	1,065	711	359	279	1,349
Iowa	1,501	617	512	2,630	1,779	725	623	3,127
Kansas	118	80	55	253	119	78	62	259
Minnesota	942	585	476	2,003	988	601	510	2,099
Missouri	115	70	56	241	236	143	131	510
Nebraska	891	407	343	1,641	1,097	492	438	2,027
Ohio	57	30	21	108	306	159	115	579
South Dakota	57	23	17	97	106	41	34	182
Wisconsin	115	61	45	221	418	220	176	814
Total	5,635	2,987	2,403	11,025	7,245	3,772	3,164	14,181

ethanol price of US$1.35/gal. If the price falls to US$1.15/gal, the number of feasible 2,000 MT/day plants decreases from 72 to 47.

This increased economic activity increases jobs throughout the economy. As expected, as the economic impacts increase, the numbers of jobs also increase. In the US$1.35/gal ethanol price scenario, a mature industry created with 1,000 MT/day plants creates 15,600 jobs directly and another estimated 53,000 jobs through the economic activity generated in the state's economies (Table 12.10).

The 2,000 MT/day plant industry results in an estimated total of 94,000 jobs with nearly 24,000 jobs created directly. For each plant, the number of jobs generated directly is a result of additional work required at all levels of production including the production of the feedstock, the transportation of the feedstock to the plant and the processing of the feedstock into ethanol. The activity that would be generated after the production of the ethanol through distribution and retail sales is not incorporated into the study.

Table 12.10. Number of jobs created by state and plant size, US$1.35/gallon ethanol price scenario.

State	Direct impact		Total impact	
	Metric tons/day			
	1,000	2,000	1,000	2,000
Illinois	3,170	4,309	15,657	19,372
Indiana	1,704	2,487	6,647	8,975
Iowa	4,031	5,381	16,616	20,597
Kansas	378	509	1,714	1,991
Minnesota	2,757	3,298	12,581	13,826
Missouri	498	1,354	1,819	4,391
Nebraska	2,351	3,282	10,810	13,962
Ohio	188	1,133	671	3,792
South Dakota	160	348	629	1,270
Wisconsin	416	1,782	1,527	6,067
Total	15,653	23,884	68,671	94,243

Conclusions

This study responds to two major current concerns in rural USA: adding value to basic agricultural commodities and increasing farm income. Increasing ethanol production provides a means to address the economic needs of rural areas. Economic analyses of increased ethanol production have indicated the potential for substantial benefits (Energetics, Inc. and NEOS Corporation, 1994). Moreover, new technologies that will convert maize stover to ethanol more efficiently are on the horizon.

The objectives of this research were to estimate the economic impacts of ethanol plants utilizing maize stover that would be established in the current maize producing states of the USA. These impacts are based on the construction and operation phase of the plants and include estimates of the total industry output, number of new jobs and value-added generated in each of the states.

The results of the study indicate that an ethanol plant provides substantial estimated economic impacts for total industry output and employment. For example, the number of new jobs in a state ranges from 160 to 4,000 for the 1,000 MT/day plants. In the case of an ethanol plant processing 2,000 MT/day, the number of new jobs created ranges from slightly over 600 to 16,600.

The study also showed that the number of feasible ethanol plants in each state could vary substantially based on the prices of ethanol and maize stover and plant size. The smaller plant size, 1,000 MT/day of maize stover, is much more sensitive to the prices of ethanol than to the price of the maize stover. While no plants are feasible if the ethanol price is at US$1.15/gal and the maize stover is at the breakeven price, 101 plants are feasible if the price of ethanol is US$1.35/gal at a breakeven stover price. The economies of size present in the larger plant, 2,000 MT/day, make this plant less sensitive to the changes in prices as the number of plants ranges from 47 to 72 in the corresponding two price scenarios outlined above for the 1,000 MT/day plant.

Finally, the economic potential of increased ethanol production can be better summarized when considering the case where producers are guaranteed US$1.35/gal at a breakeven price scenario for a 2,000 MT/day plant. For this scenario, an estimated 72 plants would be constructed, 8.8 billion gal of ethanol would be produced, US$1.0 billion in gross income to agricultural producers would occur and an estimated economic impact of US$14 billion in rural economies of the ten-state region would be realized.

While the one-time impacts of construction were also estimated, they were not incorporated into this chapter.

Acknowledgement

This research was partially funded by US Department of Energy, Oak Ridge National Laboratory Contract Number 4500010956.

References

Aden, A., Ruth, M., Ibsen, K., Jechura, J., Neeves, K., Sheehan, J., Wallace, B., Montague, L., Slayton, A. and Lukas, J. (2002) *Lignocellulosic Biomass to Ethanol Process Design and Economics Utilizing Co-Current Dilute Acid Prehydrolysis and Enzymatic Hydrolysis for Corn Stover.* NREL/TP-510-32438. US Department of Energy, National Renewable Energy Laboratory, Golden, Colorado. [Accessed 2004.] Available from http://www.nrel.gov/docs/fy02osti/32438.pdf

Economic Research Service. (2003) Farm Income Statements, 2000-2004F. *U.S. and State Farm Income Data.* US Department of Agriculture, Economic Research Service, Washington, DC. [Accessed 2004.] Available from http://www.ers.usda.gov/data/FarmIncome/finfidmu.htm

Economic Research Service. (2004) 2004 Forecast: Income for the Farm Sector and Farm Households. *Income for the Farm Sector and Farm Households.* US Department of Agriculture, Economic Research Service, Washington, DC. [Accessed 2004.] Available from http://www.ers.usda.gov/features/farmincome/2004/

Energetics, Inc. and NEOS Corporation. (1994) *Fuel Ethanol: A Review of Recent Economic Impact Analysis.* Energetics, Inc., Columbia, Maryland. Prepared for the US Department of Energy, Western Regional Biomass Energy Program and Great Lakes Regional Biomass Energy Program. [Accessed 2004.] Available from http://www.ethanol-gec.org/fueleth.pdf

Energy Information Administration. (2004a) Annual Energy Outlook 2004 with Projections to 2025. *Market Trends – Oil and Natural Gas.* US Department of Energy, Energy Information Administration, Washington, DC. [Accessed 2004.] Available from http://www.eia.doe.gov/oiaf/aeo/gas.html

Energy Information Administration. (2004b) Estimated Number of Alternative-Fueled Vehicles in Use in the United States, by Fuel, 1995-2004. *Alternative Fuels.* US Department of Energy, Energy Information Administration, Washington, DC. [Accessed 2004.] Available from http://www.eia.doe.gov/fuelalternate.html

Energy Information Administration. (2004c) This Week in Petroleum. *U.S. Gasoline.* US Department of Energy, Energy Information Administration, Washington, DC. [Accessed 2004.] Available from http://www.eia.doe.gov/oil_gas/petroleum/ info_glance/gasoline.html

Energy Information Administration. (2004d) World Crude Oil Prices 5/7/2004. *U.S. Petroleum Prices.* US Department of Energy, Energy Information Administration, Washington, DC. [Accessed 2004.] Available from http://www.eia.doe.gov/oil_gas/petroleum/info_glance/prices.html

English, B., Menard, J. and De La Torre Ugarte, D. (2001) *Using Corn Stover for Ethanol Production: A Look at the Regional Economic Impacts for Selected Midwestern States.* University of Tennessee, Agriculture Policy Analysis Center, Knoxville Tennessee. [Accessed 2004.] Available from http://web.utk.edu/~aimag/pubs/cornstover.pdf

Hart, C. (2004) Ethanol: Policies, Production, and Profitability. *Iowa AG Review* 10(2) (Spring): 6-7.

Minnesota IMPLAN Group, Inc. (1999) *IMPLAN System* (data and software), Minnesota IMPLAN Group, Inc., Stillwater, Minnesota.

Natural Resources Conservation Service. (2004) *Excessive Erosion on Cropland, 1997*. US Department of Agriculture, Natural Resources Conservation Service, Washington, DC. [Accessed 2004.] Available from http://www.nrcs.usda.gov/technical/land/meta/m5083.html

Olson, D. and Lindall, S. (1999) *IMPLAN Professional Software, Analysis, and Data Guide*. Minnesota IMPLAN Group, Inc., Stillwater, Minnesota.

Ray, D. (2004) Agricultural Policy for the Twenty-First Century and the Legacy of the Wallaces. *APAC Presentations – 2004*. [Accessed 2004.] Available from http://agpolicy.org/ppap/pp04/Pesek%20ISU%20Mar%203%202004.pdf

Ray, D., De La Torre Ugarte, D., Dicks, M. and Tiller, K. (1998) The POLYSYS Modeling Framework: A Documentation. *POLYSYS-Policy Analysis System*. University of Tennessee, Agriculture Policy Analysis Center, Knoxville, Tennessee. [Accessed 2004.] Available from http://agpolicy.org/poly/doccom.pdf

Ray, D., De La Torre Ugarte, D. and Tiller, K. (2003) *Rethinking US Agricultural Policy: Changing Course to Secure Farmer Livelihoods Worldwide*. University of Tennessee, Agriculture Policy Analysis Center, Knoxville, Tennessee. [Accessed 2004.] Available from http://agpolicy.org/blueprint/APAC%20Report%208-20-03%20WITH%20COVER.pdf

Tiller, K., Ray, D. and De La Torre Ugarte, D. (1999) The POLYSYS Modeling Framework: An Overview. *POLYSYS-Policy Analysis System*. University of Tennessee, Agriculture Policy Analysis Center, Knoxville, Tennessee. [Accessed 2004.] Available from http://agpolicy.org/poly/format.pdf

Wooley, R., Ruth, M., Sheehan, J., Ibsen, K., Majdeski, H. and Galvez, A. (1999) *Lignocellulosic Biomass to Ethanol Process Design and Economics Utilizing Co-Current Dilute Acid Prehydrolysis and Enzymatic Hydrolysis Current and Futuristic Scenarios*. NREL/TP-580-26157. National Renewable Energy Laboratory, Golden, Colorado. [Accessed 2004.] Available from http://www.nrel.gov/docs/fy99osti/26157.pdf

Chapter 13

Livestock Watering with Renewable Energy Systems

R. Nolan Clark and Brian D. Vick

Background

Mechanical windmills have been used to pump water for people and livestock in the US Great Plains for the past 120 years. While many mechanical windmills are still being used, most are close to 50 years old and are becoming too costly to maintain. Some farmers and ranchers are turning to solar photovoltaic (solar-PV) and/or wind-electric water pumping systems to supply their needs. Also, some utilities and electric cooperatives are beginning to investigate stand-alone wind-electric and solar-PV water pumping systems because of the high cost of maintaining rural transmission lines. Providing water in remote locations is also a major problem throughout the world (Postel and Vickers, 2004). It has been reported that over 30% of the world's population do not have a safe drinking water supply and many locations do not have access to electrical power from a utility (May, 2004). An attractive alternative is a stand-alone wind or solar water pumping system.

Stand-alone wind-electric water pumping systems and solar-PV pumping systems have been tested over the past 15 years by the US Department of Agriculture, Agricultural Research Service, Conservation and Production Research Laboratory at Bushland, Texas. Studies have been conducted to properly match electric motors to various size power units. Clark and Mulh (1992) showed that a wind-electric pumping system will pump twice as much water as a mechanical pumping system at a comparable cost. The development of a controller to maintain a proper electrical match between the generating device and the pump motor has been the greatest challenge. These wind pumping systems are described in papers by Clark and Vick (1994) and Clark and Vick (1995). Similarly, solar-PV pumping experiments have been conducted and are described in papers by Clark (1994) and Vick and Clark (1996) and Vick *et al.* (2003). This study includes data from two wind-electric water pumping systems and two solar-PV pumping systems.

Objectives

The chapter reports on research that had the following objectives: (1) To develop a wind-powered water pumping system that is reliable, low maintenance and cost competitive with current water pumping systems; (2) to develop a solar-PV powered water pumping system that is reliable, low maintenance and cost competitive with current water pumping systems;

Table 13.1. Specifications of wind-powered water pumps.

Wind turbine	Aermotor	Bergey 1500	Bergey XL
Rated power	> 1.0 kilowatt	1.5 kilowatt at 12.5 metres/second	10 kilowatt at 12.5 metres/second
Number of blades	18	3	3
Rotor diameter	2.44 metres	3.05 metres	7.0 metres
Blade material	Steel	Fibre-reinforced polyester	Fibre-reinforced polyester
Blade construction	N/A	Pultrusion	Pultrusion
Overspeed protection	Horizontal furling	Horizontal furling	Horizontal furling
Generator	N/A	Permanent magnet alternator	Permanent magnet alternator
Pump motor	N/A	1.1 kilowatt 230 volt, 3 phase AC	5.6 kilowatt 230 volt, 3 phase AC
Pump	48 millimetre piston	0.75 kilowatt 10-stage centrifugal	7.5 kilowatt 10-stage centrifugal
Total system cost	US$6,000	US$8,000	US$26,000

N/A = not available.

and (3) to design a controller that allows the use of standard electric motors and pumps when powered by remote wind or solar systems.

Description of Wind-Electric Water Pumping Systems

The two wind-electric water pumping systems described in this study were the 1.5 kilowatt (kW) Bergey Windpower[1] 1500-PD and the 10 kW Bergey Windpower Excel-PD. Information on these wind-electric systems is shown in Table 13.1 along with a mechanical windmill manufactured by Aermotor. For both wind-electric water pumping systems, the electricity produced by the wind turbine is variable-voltage and variable-frequency, three-phase alternating current (AC) electricity. This electricity was used to power off-the-shelf induction motors and pumps without the need of an inverter. Capacitors were added across the three phases of the pump motor, thereby balancing its inductive load and improving the power factor of the wind turbine generator, particularly at high operating frequencies. A controller was necessary to keep the pump motor disconnected from the wind turbine until a certain minimum alternator frequency was reached, otherwise the wind turbine would be braked by the load. Therefore, the controller must have a low frequency cut-in and cut-out. Additional information on the

---
[1] The mention of trade or manufacture names is made for information only and does not imply an endorsement, recommendation, or exclusion by US Department of Agriculture, Agricultural Research Service.

Table 13.2. Specifications of solar-powered water pumps.

Solar pumping system	Solar DC	Solar AC
Rated power	0.1 kilowatts at 1000 watts/metre^2	0.75 kilowatts at 1000 watts/metre^2
Number of panels	2	25
Solar cell material	Crystalline silicon	Amorphous silicon
Controller	Converter	Smart
Motor	24 volt DC	0.56 kilowatts 230 volt AC
Pump	Diaphragm	0.56 kilowatts 12-stage, centrifugal
Total system cost	US$1,500	US$7,000

controllers can be found in Vick and Clark (1995). For a discussion of the instrumentation and experimental setup, see Clark and Vick (1995). The total system cost includes the cost of the wind turbine, controller, capacitors, motor, pump, pipe and tower.

Description of Solar-PV Water Pumping Systems

The two solar-PV water pumping systems discussed in this study are a solar DC (direct current) 0.1 kW pumping system and a solar AC 0.75 kW pumping system. Information on the solar-PV systems is shown in Table 13.2. The solar DC water pumping system used the solar DC-generated electricity (12 volts) and converted it to 24-30 volts through the controller which in turn powered the 24 volt DC motor/diaphragm pump combination. A detailed description of the solar DC pumping system can be found in Clark (1994). The solar AC controller converted the DC solar panel-generated electricity into either one- or three-phase AC. The controller continuously monitored system performance to determine whether the solar radiation was high enough to engage a three-phase AC induction motor which in turn powered a centrifugal pump. The instrumentation and experimental setup for the solar pumping systems was similar to that used for the wind-electric systems. The total system cost includes solar panels, solar panel racks, controller, motor, pump and pipe.

Data and Methods

Each wind-electric water pumping system was tested at several pumping depths and with various pumps. A pumping test consists of operating the system configured for a selected pumping depth. Data were collected and sorted by wind speed to create wind speed bins that were 0.5 meters per second (m/s). Samples were taken at varying rates ranging from 10 samples/s to 1 sample/s and normally averaged to either 1 minute or 2.5 minutes. These averaged values were then placed in the wind speed bins and a test run would be completed when there was a minimum of 1000 samples in wind speed bins between 3 m/s and 10 m/s. Normally we would have over 500 samples in out-lying bins up to 16 m/s. Average values and standard deviations were calculated for each bin. These average values were used in plotting and comparing the pumping data. Finally, performance curves were used to predict monthly and yearly water pumping using 20 years of hourly wind speed data. Wind speed histograms were developed from long-term wind speed data measured at 10 m, 20 m and 30 m heights. Details of data collection methodology and testing procedures for wind water pumping systems can be found in Vick and Clark (1998) and Vick and Clark (1996). Similar analyses were done for the solar pumping systems using radiation intensity as the bin parameter. Details of data collection methodology and testing procedures for solar-PV water pumping systems can be found in Vick *et al.* (2003) and Clark *et al.* (1998).

Results

Performance data are presented from two wind-electric pumping tests and two solar-PV pumping tests. These tests were selected to be representative of tests conducted over the last 15 years, but also represent equipment that is available on today's market. The generating equipment continues to improve and systems sold today are much more reliable and efficient than equipment sold 5 to 10 years ago.

Wind-Electric Systems

Since the traditional windmill has been used for remote water pumping for the last 80 years, all new pumping equipment evaluations have been compared to the performance of a traditional windmill. Figure 13.1 shows the performance of an Aermotor windmill with two different size piston pumps (48 mm and 70 mm) for a 30 m pumping depth. With all mechanical

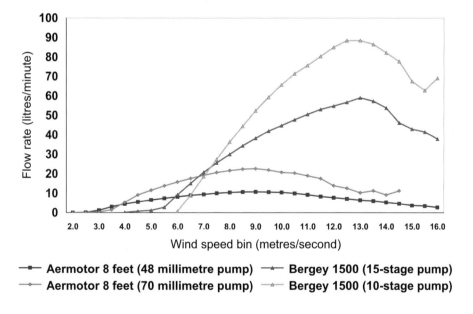

Figure 13.1. Performance curves for mechanical windmill and wind-electric water-pumping systems operating at a 30 metre pumping depth.

windmills, the maximum flow rate occurs at approximately 9 m/s wind speed. Pumping begins at about 3 m/s and will continue to 14 m/s wind speed. The maximum flow rate is 10 litres/minute (L/min) for the small pump and 22 L/min for the larger pump. When comparing this performance curve to the wind-electric system, the results indicate the pump with 15 stages has a peak flow of 58 L/min and the 10-stage pump will deliver 86 L/min. The wind-electric pumping system requires a higher wind speed to start pumping, but continues to pump at higher wind speeds. Because of the totally different design and components of the mechanical windmill and wind-electric system, the efficiency curves are completely different (Fig. 13.2). However, the efficiency curves cross at 7 m/s to 9 m/s, resulting in an average efficiency being about equal when all operational time is considered.

Water needs are normally determined by daily demands; therefore, the average daily water pumped for each month is shown in Fig. 13.3. The mechanical system with the two pump sizes shows an almost constant rate at each month, with the small pump averaging approximately 9,000 L/day and the larger pump 15,000 L/day. The wind-electric water pumping performance varied as the wind speed varied, thus the volume of water for the 15-stage pump ranged from 23,000 L/day in March-May to a low of 11,000 L/day in August. The wind-electric system pumped most of its water when the wind speed was higher. It produced much more water than the mechanical system in all months except August when wind speeds are the lowest.

Performance data for a second wind-electric system is shown in Fig. 13.4. This was a 10 kW system and the maximum flow rate at a 30 m pumping depth was measured at 390 L/min. This system is large enough for small scale irrigation or a large number of cattle. This wind turbine had different aerodynamic characteristics because of its larger size and furling design, which gave it a better electrical load match, resulting in a higher efficiency as shown in Fig. 13.5. The daily water volume by months is similar to the smaller wind-electric pumping systems with the spring providing more water and August being the lowest month. Winter wheat with

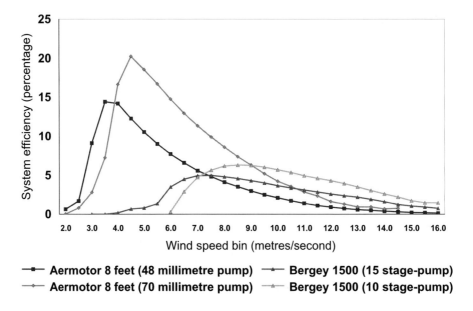

Figure 13.2. System efficiencies for mechanical windmill and wind-electric water-pumping systems.

its highest water use in April and May can easily be irrigated using water provided by a wind-electric system. The use of drip irrigation on fruit trees is another example of using this size wind pumping system for remote locations (Vick *et al.*, 2001). The daily water volumes can

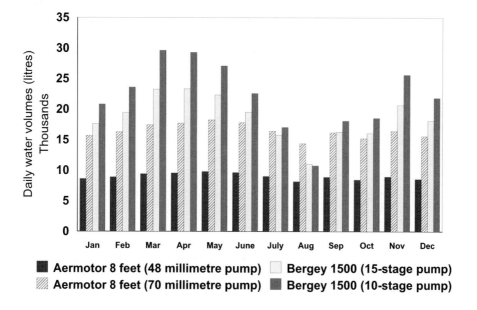

Figure 13.3. Daily water volumes pumped by mechanical windmill and wind-electric water pumping systems (30 metre head; 10 metre hub height; Bushland, Texas).

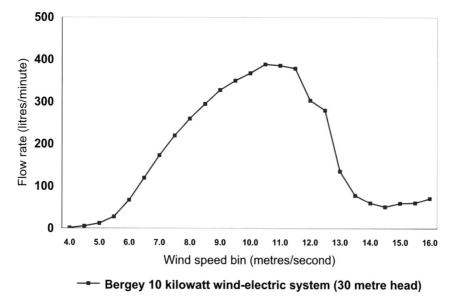

Figure 13.4. Performance curve for a 10 kW wind-electric water-pumping system operating at a 30 metre pumping depth.

be calculated for any location where monthly wind speed data are available. The data in Fig. 13.6 are for the wind resource at Bushland, Texas, but are typical for most of the Great Plains.

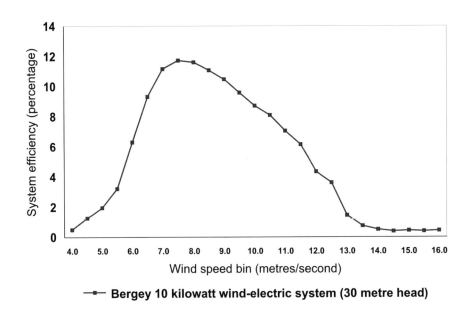

Figure 13.5. System efficiency for a 10 kW, wind-electric water pumping system.

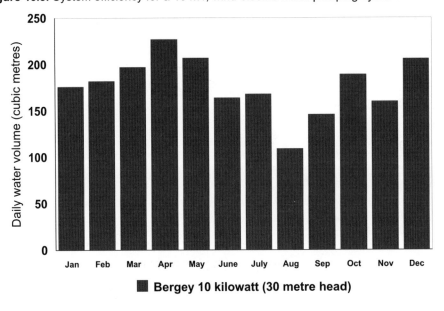

Figure 13.6. Daily water volume pumped by a 10 kW wind-electric system used for irrigation (30 metre head; 18.5 metre hub height; Bushland, Texas).

Solar-PV Systems

The DC electrical output from the solar-PV systems presents several challenges for remote water pumping. DC electric motors are generally small and have limited availability as well as being expensive. Several systems have been manufactured that use a submersible DC motor with a diaphragm pump. The diaphragm pump is a positive-displacement type pump and will operate at variable speed. The major difficulty is the total dynamic lift of the diaphragms. These pumps work well at pumping depths less than 35 m and often have a fixed or maximum speed which yields a fixed pumping rate. Excess incoming energy must be dissipated through excess heat or a dump load. The data shown in Fig. 13.7 for the DC solar pump is a submersible motor powering a diaphragm pump using two 50 watt solar panels. The pump began delivering water at an irradiance of 100 watts/m² and reached its peak flow at 700 watts/m². Additional water could not be pumped, even with additional radiation because the pump reached its maximum speed and the controller had to limit power to the motor.

The other pump shown in Fig. 13.7 was a centrifugal pump operating on AC power provided through a special inverter/controller which converted the DC electric power to a chopped wave form that simulated a sine wave. This system started pumping at an irradiance of 300 watts/m² and continually increased to 65 L/min at 1400 watts/m². Most solar pumps are rated at 1,000 watts/m²; therefore, these pumps would be rated as 4 L/min (1 gal/min) and 52 L/min (13 gal/min).

The system efficiencies of the two solar-PV pumping systems are shown in Fig. 13.8. The DC system is about twice as efficient as the AC system because of the construction materials and method of manufacturing the solar panels. The polycrystalline-silicon construction of the DC panels is much more efficient than the amorphous-silicon-type thin-film construction used on the AC panels. Another factor that contributes to the lower efficiency of the AC system is the conversion of the DC power to AC. Although the AC system has a lower efficiency, it

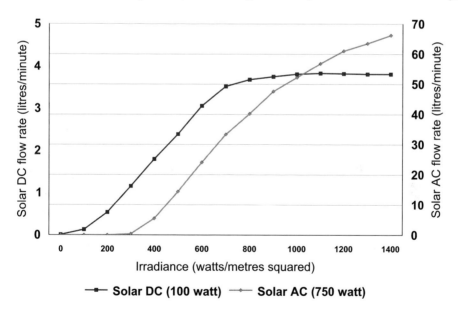

Figure 13.7. Performance curves for two solar-PV water-pumping systems operating at a 30 metre pumping depth. Each is plotted on a different scale.

has many advantages because of the wider range of pumps available and the capability to pump water from greater depths. The diaphragm pump design limits the flexibility and adaptability to meet a variety of pumping needs.

The daily water volume pumped is shown in Fig. 13.9 for these two solar-powered pumps operating in the environment at Bushland, Texas. The solar DC pump with its constant flow will produce almost the same volume of water (1400 L/day) every month of the year. The differences between summer and winter are slight. The volume of water pumped by the AC system shows some variation from month to month. The lower months are either winter months (December, January and February) with shorter days or months with more cloudy days (April and September). The average daily volume was approximately 15,000 L/day.

A typical beef range cow will consume about 60 L of water each day; therefore the volume of water needed for a herd can be easily estimated. If the water is stored in an open tank, then daily water use should be doubled to allow for evaporation from the storage tank. The mechanical windmill with the smaller pump would provide water for approximately 75 animals. The 1.5 kW wind-electric system would provide enough water for 125 animals in all months except August when the number should be reduced to 100. The solar-DC system would provide water for approximately 25 animals and the solar AC system for 110 animals.

The cost of water was determined for each pumping system as a product (m^4) of the unit volume of water (m^3) and the unit pumping head (m). The annual system cost for the wind-electric and solar water pumping systems at a 30 m pumping depth were:

Bergey 1500, 1.5 kW   US$0.036/$m^4$
Bergey XL, 10 kW   US$0.014/$m^4$
Solar DC, 0.1 kW   US$0.085/$m^4$
Solar AC, 0.9 kW   US$0.043/$m^4$

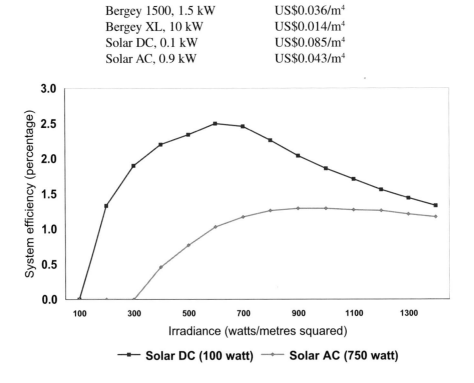

Figure 13.8. System efficiencies for two solar-PV water-pumping systems.

The above numbers assume an availability of 100% for all systems. Additional data needs to be collected to determine reliability, maintenance costs and finally lifetime cost/m^4.

References

Clark, R. (1994) *Photovoltaic Water Pumping for Livestock in the Southern Plains.* Paper No. 94-4529. American Society of Agricultural Engineers, St. Joseph, Michigan. [Accessed 2004.] Available from http://www.cprl.ars.usda.gov/wmru/pdfs/Photovoltaic%20Water%20Pumping%20for%20Livestock%20in%20the%20Southern%20Plains.pdf

Clark, R. and Mulh, K. (1992) Water Pumping for Livestock. *Windpower 92 Conference Proceedings.* Seattle, Washington, October 19-23, 1992. American Wind Energy Association, Washington, DC.

Clark, R. and Vick, B. (1994) *Wind Turbine Centrifugal Water Pump Testing For Watering Livestock.* Paper No. 94-4530, American Society of Agricultural Engineers, St. Joseph, Michigan. [Accessed 2004.] Available from http://www.cprl.ars.usda.gov/wmru/pdfs/Wind%20Turbine%20Centrifugal%20Water%20Pump%20Testing%20for%20Watering%20Li.pdf

Clark, R. and Vick, B. (1995) Determining the Proper Motor Size for Two Wind Turbines Used in Water Pumping. *Wind Energy 1995: Proceedings of the Energy and Environmental Expo 95.* The Energy-Sources Technology Conference and Exhibition, January 29-February 1, 1995, Houston, Texas. American Society of Mechanical Engineers, New York, New York.

Clark, R., Vick, B. and Ling, S. (1998) *Remote Water Pumping using a 1 kilowatt Solar-PV AC System.* Paper No. 98-4087. American Society of Agricultural Engineers, St. Joseph, Michigan.

May, H. (2004) Energy for the South. *New Energy.* 03(June): 19-22.

Postel, S. and Vickers, A. (2004) Boosting Water Productivity. *State of the World 2004.* W.W. Norton & Co., New York, New York, pp. 46-65.

Vick, B. and Clark, R. (1995) Pump Controller Testing on Wind Turbines Used in Water Pumping. *Windpower '95 Conference Proceedings, March 27-30, 1995*, Washington, DC. American Wind Energy Association, Washington, DC.

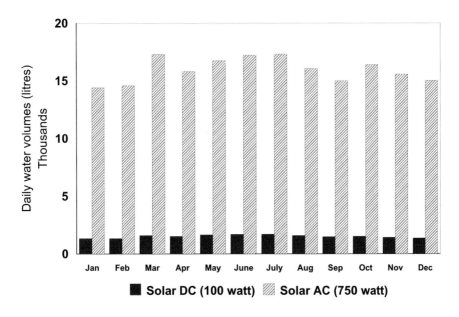

Figure 13.9. Daily water volumes pumped by two solar-PV water-pumping systems.

Vick, B. and Clark, R. (1996) Performance of Wind-electric and Solar-PV Water Pumping Systems for Water Livestock. *Journal of Solar Energy Engineering: Transactions Journal of the American Society of Mechanical Engineers* 118: 212-216.

Vick, B. and Clark, R. (1998) *Ten Years of Testing a 10 kilowatt Wind-electric System for Small Scale Irrigation.* Paper No. 98-4083. American Society of Agricultural Engineers, St. Joseph, Michigan.

Vick, B., Neal, B. and Clark, R. (2001) Wind Powered Irrigation for Selected Crops in the Texas Panhandle and South Plains. *Windpower 2001 Conference Proceedings, June 3-7, 2001.* Washington, DC. American Wind Energy Association, Washington, DC

Vick, B., Neal, B. and Clark, R. (2003) Water Pumping with AC Motors and Thin-film Solar Panels. *Proceedings of SOLAR 2003.* The 32nd American Solar Energy Society Annual Conference and the 28th National Passive Solar Conference, June 21-23, 2003, Austin, Texas. American Solar Energy Society, Boulder, Colorado.

Chapter 14

Trends in US Poultry Housing for Energy Conservation

John W. Worley, Michael M. Czarick and Brian F. Fairchild

Introduction

The last half of the twentieth century saw a tremendous increase in broiler chicken production in the southeastern USA. By 2003, production in the top three broiler-producing states, Georgia, Alabama and Arkansas, was 3.5 billion broilers or 41% of the total US production of 8.5 billion broilers (National Agricultural Statistics Service, 2004). The growth in this region occurred for a number of reasons – among these were mild climate, inexpensive land and available inexpensive labour. Much of the land in the Piedmont areas of northern Georgia, Alabama and other southern states had been depleted by poor agricultural practices and was not suitable for row-crop agriculture. It was suitable for cattle farming if the soil could be restored. Poultry manure is a great agent for restoring depleted soils. The synergy between the poultry industry and cow-calf operations in this region encouraged the growth of the poultry industry. Housing was very low-cost and took advantage of the mild climate by using natural ventilation and minimal insulation to keep costs low.

As genetics and feeding programmes improved, efficiency was increased by producing more meat in a shorter time on less feed; however, the birds that resulted from these increased efficiencies were not as hardy as older breeds of chickens that could survive in the outdoors. As a result, modern broilers do not respond as well to stress as broilers did in the past. The modern broiler requires good, consistent environmental control in order to achieve optimum performance. Housing has evolved, along with bird genetics and improved nutrition, to provide the required environment for broilers to thrive. Much of this change came during a time when energy prices were relatively low and thus, energy conservation was not a major economic driver.

Recent energy price increases and dramatic volatility, especially for natural and liquid petroleum (LP) gas prices, has spurred the poultry industry in the USA to adopt a number of technologies that reduce the amount of energy required to heat and cool broiler chickens. These technologies include: wider production buildings, replacement of curtain-covered openings with solid walls, circulation fans to reduce heat stratification in the building, more energy-efficient ventilation fans and litter treatments that reduce ammonia emissions, thus reducing ventilation and heating requirements. Potential energy savings from most of these technologies is well known, having been documented in previous studies. The extent to which the poultry industry has adopted and is currently utilizing these practices is less understood. The goal of this chapter is to increase the understanding of the current state of

© CAB International 2005. *Agriculture as a Producer and Consumer of Energy* (eds J.L. Outlaw, K.J. Collins and J.A. Duffield)

the broiler industry which would aid in the development of educational programmes for further energy reduction in poultry production.

The objectives of the study were to:

1. Quantify expected energy savings from several technologies being employed in broiler houses including wider houses, more efficient ventilation fans, replacing curtain walls with solid walls, litter treatments to reduce ammonia emissions and circulation fans for heat distribution.
2. Determine the extent to which these new technologies are being adopted.
3. Estimate the potential for energy savings attributable to these new technologies on an industry-wide basis.

Projected Energy Savings from Adoption of New Technologies

This study utilized a survey of the poultry industry to determine the extent of new technology adoption (the polling instrument is shown in the appendix of this chapter). Twenty poultry complexes, located in the southeastern USA were polled to examine the extent of adoption of energy-saving practices at the current time and their plans and projections for the year 2010.

Estimate of Typical Gas Requirements

Worley *et al.* (1999) found that the amount of LP gas used to provide heat for one flock in a typical broiler house (40 feet (ft) by 500 ft, tunnel ventilated) ranged from 95 gallons (gal) in summer to 1,045 gal for cold winter flocks. The majority of winter flock requirements ranged from 600 gal to 900 gal. If a grower produces six flocks per year, with three in cool conditions and three in warm conditions, a reasonable estimate of annual fuel usage would be (650 gal multiplied by 3 flocks) + (250 gal multiplied by 3 flocks) = 2,600 gal. In this study, the authors used this value as a base for gas usage in order to convert percentage savings to gallons of LP gas.

House Tightness

House tightness is important because it allows the producer to control where cold air is introduced into the house. It reduces drafts, makes the bird area more comfortable and reduces overall fuel usage, especially in winter. For the purposes of this study, a 'tight house' is defined as a house where one 48 inch (in) diameter exhaust fan can create a vacuum of 0.1 in static pressure with all inlets closed. If there are cracks, loose curtains, etc., this performance measure is not achievable. Czarick and Lacy (1996) studied the use of curtain pockets to reduce air leakage in broiler houses. Curtain pockets are fabric pockets that cover the edge of the curtain when it is fully closed (see Fig. 14.1). The study compared fuel usage for a house with curtain pockets and a comparable house without curtain pockets. Both houses were well-built and were tight with the only difference being the installation of curtain pockets on one house. The results were 10% to 15% fuel savings on approximately 67% of the days in March. For the flock, savings were approximately 120 gal of LP gas for this one house. Curtain pockets are only one measure of house tightness, but these studies give an indication of the quantity of gas that can be saved by controlling air leaks in the house. For this study, the authors used a 20% reduction in fuel consumption for houses that met the tightness criteria.

Circulation Fans

Circulating fans are small fans, usually hung from the ceiling in the area of the house where brooding occurs. Brooding occurs in the first 10 to 15 days of bird life and requires high temperatures at a time when birds are small and produce very little body heat, thus brooding requires, by far, the greatest fuel usage of broiler grow-out. The purpose of circulating fans is to move hot air from near the ceiling, where it naturally rises, to the bird level on the floor of the house, thus reducing overall heating requirements. The amount of heat that is saved by these fans depends on the heating system used in the house, as well as house tightness. Czarick and Lacy (2000) compared two houses side by side, one with circulating fans installed in the brooding end and the other without circulating fans. Both houses were heated with forced-air furnaces. The circulation fans provided a more comfortable and uniform environment for the broilers and simultaneously resulted in a 30% reduction in fuel usage. The study was repeated (Czarick and Lacy, 2001) in a tightly constructed house with radiant brooders and the circulation fans still saved approximately 10% on fuel usage during brooding. The authors stated that based on these and other studies, an average of 20% savings could be expected using this technology.

In a similar study, Donald *et al.* (2001a) estimated that fuel savings with circulation fans would range from 15% to 40% depending on the age and construction quality of the house. The authors pointed out that the cost for installing fans in a typical house is approximately US$1,000. Based on these studies, the authors used an expected energy savings of 20% for houses with circulation fans installed.

Figure 14.1. Curtain flap installed on a poultry house to reduce air leakage.

Energy-efficient Exhaust Fans

A number of ventilation fans are available with efficiencies that vary greatly. For this study, energy-efficient fans are defined as fans that have exhaust cones (see Fig. 14.2). This definition was chosen for two reasons: 1) Exhaust cones improve fan efficiency significantly, and 2) most fans that have exhaust cones are also relatively energy-efficient. When a customer is willing to spend extra dollars to add the exhaust cone, it indicates concern over energy efficiency and most manufacturers reward that concern by providing a fan package that is highly efficient.

Assuming an electrical cost of US$0.08 per kilowatt-hour (kWh) and an average operating time of 3,000 hours per year, then electricity to operate the tunnel ventilation fans for a typical 40 ft by 500 ft broiler house will cost approximately US$2,600/year (32,500 kWh/year.) If the energy efficiency ratio (EER) is increased from a typical value of 19 cubic feet per minute/watt (cfm/W) to a higher efficiency value of 22 cfm/W, the cost of operating these fans is reduced by approximately 15%. EER's can range from around 12 cfm/W to as high as 31 cfm/W, but most fans marketed for poultry house ventilation typically range from 16 cfm/W to 24 cfm/W. Based on these studies, the authors assumed a 15% savings in energy for high efficiency cone discharge fans.

Solid-walled House

Older broiler houses were constructed with windows that could be opened for natural ventilation or closed by covering them with a plastic curtain. More recently, the curtain openings have not been used for ventilation, but are there as a safety measure in case of

Figure 14.2. Energy-efficient exhaust cone fan.

power failure. The problem with these curtain-covered openings is that they have very limited insulation value and represent a tremendous heat loss in cold conditions. In addition, they typically do not provide adequate ventilation and heat removal during a power failure. By utilizing more reliable backup emergency power equipment and eliminating the curtain openings, substantial energy savings can be achieved.

Expected savings from solid sidewalls versus a curtain-sided house, of course, depends greatly on how the solid side wall is constructed and the width of the curtain opening being replaced. Donald *et al.* (2001b) estimated that energy savings could range from 15% to 40% depending on whether the solid side wall replaced a narrow (24 in wide) opening or a 5 ft wide opening as is sometimes seen on older houses. In addition to this variability, the solid wall resistance to heat flow value (R-factor) can range from around 2 to about 21 (the higher the value the greater the resistance) with modern houses typically being nearer 21. Some houses enclose only the north side (the predominant direction of winter weather) while others are totally enclosed on both sides. The survey tool does not differentiate between any of these variations, but only asks how many houses have one or both side walls solid. The reason for this was to simplify the questionnaire and, hopefully, increase response. Given the uncertainty of these particular points, the authors used a conservative estimate of 15% heat energy savings for solid side wall construction in this study.

Litter Treatments

Litter is a mixture of bedding material (usually pine shavings) and the manure that naturally collects on the floor. Most broilers are raised on previously used litter, often with a top coating of fresh shavings on top of the litter from the last flock of birds. Several litter treatments have been introduced to lower ammonia emissions from the litter. Most litter treatments use the principle of lower litter pH to reduce ammonia gas emissions. Ammonium (NH_4) is produced in the litter from the microbial breakdown of uric acid and faeces. Lowering litter pH reduces microbial growth and the amount of ammonium that converts to the gaseous ammonia (NH_3) form. This strategy is very effective during the first 10 days to 2 weeks of brooding during winter months. During these times, birds are small and the buildings require little air to ventilate except to control ammonia levels. Chicks are unable to maintain their own body temperature after hatching. Approximately 10 to 14 days are required for them to be able to thermoregulate their body temperature. Air temperatures maintained at 90°F to 95°F keep the chicks comfortable. All of the air removed by ventilation must be reheated from cold outside temperatures to 90°F to 95°F; therefore reducing the ventilation rate is very effective at saving energy. Worley *et al.* (1999) showed a 35% fuel savings for winter flocks using alum as a litter treatment. Using the assumptions presented at the beginning of this section (see p. 244), this rate of savings would amount to a 22% savings of year-round gas usage. For the purposes of the current study, the authors used 20% as the expected savings using litter treatments.

Wider Housing

Poultry houses have typically been built 40 ft to 42 ft wide for the past several years. This width has worked well for many years because it provides an optimal area per bird when two feeder lines are installed. Recently, a number of houses have been built 50 ft to 70 ft wide and a number of complexes are looking at this option in order to reduce building cost per bird and energy costs. Wider houses can be built at a lower cost per square foot (and thus per bird) and are more efficient to manage. One of the benefits of wider construction is a reduction in heat loss during winter. Heat is lost primarily through the ceiling and side walls. When a house is wider, the loss through the ceiling per square foot of floor space is practically the same, but the loss through side walls is less because there are fewer square feet of side wall per bird (or per square ft of floor). Losses through side walls would theoretically be 20% less for a 50 ft wide house than a 40 ft wide house with the same wall construction. A 60 ft wide house would lose 34% less energy through the side walls than a 40 ft house. An additional benefit of wider houses is that they are built with thicker walls for greater strength and stability (2x6 boards instead of 2x4 boards[1]) and these thicker walls make it convenient to increase insulation value, further reducing heat loss.

A number of factors make energy savings for wider houses difficult to predict. In order to keep the survey as simple as possible, a wide house was defined as any house 50 ft wide or wider, so the exact width of the house is unknown. Also, wall construction is unknown as well as the type of wall construction (curtain sided, solid wall, etc.). In addition, the portion of the total heating cost attributable to ventilation as opposed to heat loss through the building can vary greatly depending on weather and management. Given these uncertainties, a conservative value of 15% savings in heating energy was used for wider houses.

Electronic Environmental Controller

Electronic controllers use one (or an average of up to six) temperature sensors to determine the temperature in the house and operate fans and heaters in order to maintain a set temperature in the bird environment. Given the accuracy limitations of thermostats, if separate thermostats are used to control heaters and fans, in attempting to maintain the desired temperature, both fans and heaters can be turned on at the same time. This would be analogous to driving a car with one foot on the brake and the other on the accelerator. In theory, one of the advantages of electronic controllers is to keep heaters and ventilation fans from coming on at the same time and competing with each other, thus saving about 15% on energy costs. In reality, most operators using thermostats instead of controllers have learned to set the thermostats for heating and ventilation further apart, thus avoiding this situation. This management, however, results in wider temperature variation, a poorer environment for the birds and reduced bird performance.

For the purpose of this study, the authors did not assign a value for energy savings for controllers, however the number of houses using controllers is considered important information because it is an indication of how well buildings are being controlled and how much improvement can be expected by adding additional technologies. Controllers greatly reduce the amount of time required to manage the environment in the poultry house and thus, allow more time for other duties.

Tunnel Ventilated Houses

The term 'tunnel ventilation' refers to ventilating a building (usually in summer conditions) by causing air to enter at one end, travel down the length of the house and exit at the other end. By doing this, the birds are cooled by convection (breeze) as well as improving conduction (lowering the temperature of the air). The arrangement also makes evaporative cooling systems easy to install and manage. Tunnel ventilation does not necessarily reduce energy consumption because tunnel ventilation fans require a large amount of electrical power to operate. They

---

[1] 2 inch by 6 inch and 2 inch by 4 inch dimensional lumber.

Table 14.1. Adoption of energy-saving technologies by eleven poultry complexes.

Technology	2004		2010 (projected)	
	Number	Percentage of total houses	Number	Percentage of total houses
Total houses	6218	100	6418	100
Tunnel ventilated	4996	80	6280	98
Electronic controllers	3311	53	4948	77
Solid side walls	3130	50	4712	73
Circulating fans	977	16	1916	30
High efficiency exhaust fans	1870	30	3202	50
Litter treatment	3709	60	4372	68
Wide houses	337	5	691	11
Tight houses	3937	63	5641	88

do, however, when properly managed, provide a much more comfortable environment for the birds and thus, allow more efficient production of meat by allowing higher bird densities, fewer bird deaths and greater feed efficiency. Because of the uncertainty of energy efficiency, no value was assigned for tunnel ventilation in this study. It is recognized, however, as an important step in the modernization and increased efficiency of broiler production.

Survey Results

Poultry complexes are typically made up of a number of broiler production farms, a smaller number of breeder hen farms to produce hatching eggs, a hatchery to provide chicks to broiler farms, a feed mill to produce feed for the birds in the complex and a processing plant to process the birds into meat. The survey (see appendix) was distributed to 20 poultry complex broiler service managers in Georgia and surrounding states in March 2004. Eleven surveys were collected by the end of April. All of the surveys received were from Georgia. The 11 complexes represent an annual production of approximately 806 million broilers. This figure represents 64% of the broilers produced in Georgia and 9.5% of all broilers produced in the USA in 2003 (National Agricultural Statistics Board, 2004).

The results of the survey are summarized in Table 14.1. Approximately 6,218 houses were used to produce broilers in these 11 complexes. All of the complexes plan to add new houses and retrofit some older houses, but in most cases, they are not planning expanded production, so the new houses will replace some older existing houses. One of the complexes plans to expand by approximately 200 houses. This would increase the total number of houses to approximately 6,418 by 2010. The data indicate the popularity of tunnel ventilation and electronic controllers. Respondents indicated that their plans are to convert almost totally (98%) to tunnel ventilation with over 75% of houses having electronic controllers by 2010. 'Tightening' (reducing air leaks) of houses is also almost universally seen as important with an expected 88% adoption by 2010. Solid side walls (at least on the north side) are expected to reach 73% adoption by 2010. Circulation fans have a much lower acceptance rate at this point, partially because the research on these fans is fairly new and partially because of the opinion that fixing other problems with the house will reduce the need for adding this technology. The lowest adoption rate, not surprisingly, is that of wider houses. Obviously, wider housing can only be done as new houses are added, so this practice will take much longer to make a large impact on the industry.

Table 14.2. Projected energy savings from adoption of technologies.

	Eleven complexes[a]		Projected industry	
	2004	Projected 2010	2004	Projected 2010
Gas used (million gallons)[b]	16.2	16.7	170.2	175.7
Gas saved (million gallons)[c]				
Solid side walls	1.2	1.8	12.8	19.3
Circulating fans	0.5	1.0	5.4	10.5
Litter treatments	1.9	2.3	20.3	23.9
Wide houses	0.1	0.3	1.4	2.8
Tight houses	2.1	2.9	21.6	30.9
Total gas savings[d]	4.1	5.8	43.0	61.2
Electricity used (million kWh)[e]	202.1	208.6	2127.2	2195.6
Electricity saved (million kWh)[f]				
High-efficiency exhaust fans	9.1	15.6	96.0	164.3

[a] Industry projection based on survey representing 9.5% of total US production.
[b] Gas that would be used if none of the energy saving technologies were adopted.
[c] Gas savings based on the projected savings in the text and the number of houses utilizing the technology.
[d] Total gas savings are adjusted for combination effect of multiple technologies.
[e] Electrical energy that would be used for ventilating houses if high-efficiency fans were not used.
[f] Electrical savings based on the projected savings in the text and the number of houses utilizing the technology.

Projected energy savings from the adoption of these technologies is shown in Table 14.2. Columns 2 and 3 show expected savings for the 11 complexes responding to the survey. These 11 complexes process 9.5% of the broilers processed in the USA. If one assumes that these complexes are representative of 9.5% of the industry, savings can be projected for the entire industry (columns 4 and 5.) The figures for gas and electricity used by the industry are based on the assumption that none of the energy-saving technologies has been adopted. The estimates are based on the annual energy estimates (2,600 gal of propane gas and 32,500 kWh of electrical energy for each house) that were derived earlier in this chapter (see p. 244) and the total number of houses analysed. The electrical energy figure only represents energy for ventilation, but ventilation accounts for a major portion (75% to 85%) of total electrical use and it is the only component affected by the technologies described in this chapter. As shown in Table 14.2, expected gas used by 2010 would be 16.7 million gal of LP gas for the 11 complexes and 175.7 million gal for the USA. Expected electrical consumption would be 209 million kWh by the eleven complexes and 2,196 million kWh for the entire industry.

Expected savings for the adoption of each technology were calculated by multiplying the number of houses utilizing this technology by the percentage savings developed earlier in this chapter (15% for solid side walls, 20% for circulating fans, 20% for litter treatments, 15% for wider houses, 15% for high-efficiency cone fans and 20% for tighter houses). In calculating the expected total savings on gas, the expected savings for each technology should not be added directly. For this reason, the estimated total gas savings were adjusted downward by 30% based on the multiplicative effect of multiple energy-saving technologies. The following example will illustrate. If 20% of gas usage is saved by making the house tighter, then gas usage would be reduced to 80%. If solid walls are then added, 15% of that remaining 80% would be saved so that the total use would be 80% multiplied by 85% = 68% (32% savings) rather than adding 20% and 15% to get 35%. Additional technologies would be

Table 14.3. Potential energy savings if the entire industry adopted technologies.

	Projected industry 2010[a]
Gas used (million gallons)[b]	175.7
Gas saved (million gallons)[c]	
Solid side walls	26.4
Circulating fans	35.1
Litter treatments	35.1
Wide houses	26.4
Tight houses	35.1
Total gas savings[d]	110.7
Electricity used (million kWh)[e]	2195.6
Electricity saved (million kWh)[f]	
High efficiency exhaust fans	329.3

[a] Industry projection based on survey representing 9.5% of total US production.
[b] Gas that would be used if none of the energy saving technologies were adopted.
[c] Gas savings based on 100% adoption of each technology.
[d] Total gas savings are adjusted for combination effect of multiple technologies.
[e] Electrical energy that would be used for ventilating houses if high-efficiency fans were not used.
[f] Electrical savings based on 100% adoption of high-efficiency fans.

calculated similarly so that the total savings for the addition of all technologies would be 63% of the initial usage. If all of the savings are added directly, the savings would be 105% which is impossible. Since only one technology reduces electrical energy use, this adjustment was not necessary for electrical energy.

Given these considerations, the projected annual energy savings for the entire broiler industry would be 61 million gal of LP gas and 329 million kWh of electrical energy by 2010. This represents approximately 35% of total LP gas usage and 75% of total electrical usage.

If these energy-saving technologies were completely adopted by the industry, by the year 2010, even more energy would be saved, as shown in Table 14.3. This table illustrates the potential for future energy-saving measures. Total implementation of all measures would represent a savings of 63% of LP gas and 15% of electrical energy.

Discussion

A number of technologies have been demonstrated to save a substantial amount of energy, for both heating and ventilation, in poultry housing. Because of the large number of poultry houses in the USA and especially in the southeastern USA, the potential for energy savings in this industry is significant. The survey of broiler complexes described in this study illustrates that the industry is already adopting a number of these technologies, but the need still exists for education on additional opportunities for energy savings at both the company and grower levels.

From the survey, it is quite evident that most of the industry acknowledges the advantages of tunnel ventilation, air-tight construction, electronic controllers, litter treatment and solid side walls. These technologies are projected to be adopted at a rate of approximately 70% or more by the year 2010. High-efficiency exhaust fans and circulating fans have been adopted at a lesser rate. Efforts should be made to encourage the industry to install high-efficiency exhaust fans on all new buildings and to replace less efficient fans as opportunities arise.

Circulation fans have not yet been accepted at a high rate, although they are gaining popularity. Opportunity exists to introduce this technology and to demonstrate how much energy can be saved under varying conditions (tighter houses, different types of heating systems, etc.). Wider housing is a relatively new advancement whose benefits have been recognized by a limited number of companies. As experience is gained with these houses, designs can be fine tuned and more companies should see the advantages of moving to wider houses. This technology will obviously have a slow acceptance rate, since it can only happen as new buildings are built, but it can still have a long-term impact on energy use.

Energy conservation is now a driving force in the poultry industry, a trend that has been encouraged by rising energy costs in recent years. Poultry companies and producers are looking for any and all means of conserving energy that will pay for themselves with savings on energy and they are adopting these practices at a high rate. Opportunities still exist for educating consumers on opportunities for adopting new proven technologies and for developing and evaluating new technologies that can save even more.

References

Czarick, M. and Lacy, M. (1996) *Reducing Side Wall Curtain Leakage.* Poultry Housing Tips 8(12). University of Georgia Cooperative Extension Service, Athens, Georgia.

Czarick, M. and Lacy, M. (2000) *Reducing Temperature Stratification in Houses with Forced Air Furnaces.* Poultry Housing Tips 12(4). University of Georgia Cooperative Extension Service, Athens, Georgia.

Czarick, M. and Lacy, M. (2001) *Circulation Fans in Houses with Radiant Brooders.* Poultry Housing Tips 13(1). University of Georgia Cooperative Extension Service, Athens, Georgia.

Donald, J., Eckman, M. and Simpson, G. (2001a) *Paddle* and Recirculating Fans – a Progress Report. *The Alabama Poultry Engineering and Economics Newsletter* 13 (September). Auburn University Cooperative Extension Service, Auburn, Alabama.

Donald, J., Eckman, M. and Simpson, G. (2001b) *Solid Sidewalls for Broiler Houses? The Alabama Poultry Engineering and Economics Newsletter* 12 (July). Auburn University Cooperative Extension Service, Auburn, Alabama.

National Agricultural Statistics Service. (2004) *Poultry – Production and Value, 2003 Summary.* US Department of Agriculture, National Agricultural Statistics Service, Washington, DC. [Accessed 2004.] Available from http://usda.mannlib.cornell.edu/reports/nassr/poultry/pbh-bbp/plva0404.pdf

Worley, J., Risse, L., Cabrerra, M. and Nolan, M. (1999) *Bedding for Broiler Chickens: Two Alternative Systems.* Applied Engineering in Agriculture 15(6): 687-693.

Appendix

Survey instrument.

Energy saving trends in poultry housing

Complex # _____

Total number of birds processed per year _____
Size of birds processed (pounds) _____
Number of houses in complex _____

Technology	Percentage of current houses	Percentage planned (new or retrofit)	Percentage expected in 2010
Enclosed			
Tunnel ventilated (totally enclosed or solid north wall)			
Electronic environmental controller			
Circulation fans			
'High efficiency' (cone exhaust) fans			
Litter treatment (to reduce ammonia emissions)			
'Wide' (50 feet or more) houses			
'Tight houses' (can achieve 0.1 inch static pressure with one 48 inch fan)			

What plans do you have to improve energy efficiency in the houses in your complex over the next five years?

_____
_____
_____

Chapter 15

Experiences Co-firing Grasses in Existing Coal-fired Power Plants

Doug M. Boylan, Jack Eastis, Kathy H. Russell, Steve M. Wilson and Billy R. Zemo

Introduction

In an effort to identify low cost renewable energy options in the southeast USA, Southern Company, a major southeastern US utility, has been researching biomass co-firing in existing coal-fired power plants. This chapter presents handling, emissions, and performance results generated while testing switchgrass (*Panicum virgatum*) and other grasses with two co-firing technologies. Ongoing studies to address issues related to the impact of co-firing on fly ash sales and on catalyst performance in selective catalytic reduction (SCR) nitrogen oxide (NO_x) control systems are also described. Finally, cost estimates associated with generating renewable energy from grass co-firing are presented, based on both current results and expected technology improvements.

Background

With increasing national and world interest in climate change and renewable energy, electric utilities are being presented with both opportunities and challenges. Some utility customers are seeking the opportunity to purchase 'green' or renewable energy as part of their home energy use. These customers are willing to pay a premium price for energy which meets their standard. However, the quantity of renewable energy needed for the green market is currently very small compared with what might be required if a national renewable portfolio standard (RPS) is put in place. The omnibus energy bill proposed during 2003 required that by 2020, 10% of all electric power sold by certain classes of electric utilities be generated from renewable sources. For Southern Company, this would require a substantial investment of approximately 3,000 to 7,000 megawatts (MW) of renewable generation.

To address these opportunities and challenges and to reduce costs to their customers, utilities are seeking the lowest cost renewable energy options. In the southeast USA, these options are limited. Photovoltaic power (solar cell) has high capital cost, and the lack of resource in this area makes wind power impractical. Therefore, Southern Company's renewable energy interest has focused on biomass, and in particular, on co-firing biomass in existing pulverized coal-fired power plants. About 70% of Southern Company's generation is from pulverized coal-fired plants, and if existing units could be used to produce renewable energy, then capital and operating costs might be reduced.

However, these units are designed to run on coal, and biomass differs greatly from coal. Therefore, in establishing cost effectiveness, it is important to determine all factors which contribute to both the costs and benefits of co-firing biomass in a coal-fired unit. Biomass has a fuel energy cost that can be compared with coal, but the biomass will also impact emissions, efficiency, and operations, and these costs and benefits must be included.

Southern Company has been considering a number of biomass co-firing options. This chapter describes studies conducted by Southern Company to use switchgrass and other grasses as fuel. Grasses offer advantages related to high productivity on marginal land and the low moisture content of the product fuel. Grasses also appear to have the best local opportunity to take advantage of possible federal tax credits for closed-loop biomass, which is biomass grown solely for energy production. Two co-firing technologies with grasses were studied in actual power plants, and measurements were made of the impact on unit efficiency, emissions, and operations. Also described are ongoing studies to address issues related to the impact of co-firing on fly ash sales and on catalyst performance in SCR NO_x control systems. Finally, an economic model is used to predict costs of generating renewable energy from grass co-firing based on both current results and expected technology improvements.

Co-firing Technologies

There are two approaches to co-firing biomass in a coal-fired power plant. In co-milling, the most straight forward approach, the biomass material is treated as if it were coal. It is mixed at a low concentration with coal and introduced into the plant through the coal handling system. This approach has very low capital and operating costs, but is limited by handling issues to co-firing only low percentages of biomass. Further, the biomass must be processed to a form and size that will allow it to pass through the coal handling and coal pulverization systems.

With direct injection, a biomass processing, handling, and burner system completely separate from the coal is installed. The biomass is pneumatically conveyed to the boiler through dedicated burners, and the biomass must be ground or shredded fine enough to burn in suspension in the furnace. This approach requires higher capital costs than co-milling, but also allows higher percentages of biomass to be co-fired.

Co-milling Evaluations

The initial approach to co-firing grasses was co-milling. Low percentages of chopped switchgrass would be blended with coal, and introduced into the coal handling system. The process has been successful with sawdust, and was expected to be straight forward with grass as well. The project goal was to determine what percentage of chopped (less than one inch in length) grass could be handled, pulverized and burned without major impact on the unit performance.

Loose Grass Tests

Tests showed that the bulk densities of crushed coal (45-55 pounds/cubic foot) (lb/ft^3) and chopped grass (5-7 lb/ft^3) were greatly different. A mix of 10% grass with 90% coal by weight would require approximately equal volumes of each. After observing trial mixtures of grass and coal, it was decided to conduct laboratory-scale handling tests before filling the power plant bunkers with the mixture.

This proved to be an astute decision. The coal bunker at a power plant is a large storage silo which holds approximately one day's supply of crushed coal for feeding by gravity into the unit. A model of the bunker was constructed for the project by Jenike and Johanson, bulk solids handling specialists in California, simulating the bunker angles and bunker wall material (Bush *et al.*, 2001). They found that crushed coal flowed through the simulated bunker as well as it flowed through the actual power plant bunker. However, a mixture of 5% by weight (about 3% by energy content) grass with coal caused bridging and would not flow at all. The low bulk density of the grass as well as its tendency to thatch contributed to its failure to flow. Based on further measurements and analysis by Jenike and Johanson, it was determined that co-milling loose switchgrass with coal, even at very low percentages, would not be viable.

Cubing Tests

However, because of the relatively low costs of co-milling operations and capital, another approach was tried that involved making the grass into cubes that have more coal-like handling properties (Boylan *et al.*, 2003). The goal was to convert the grass into dense chunks that would avoid the thatching of loose grass and handle more like coal in the coal bunkers. The denser cubes should also improve the transportation costs of the grass.

A test programme was established to evaluate the potential of cubing grass for renewable power plant fuel purposes. In particular, the project sought to determine the parameters for manufacturing quality cubes, and to determine how they would handle and burn in a power plant. A Warren and Baerg Manufacturing Company grass cuber, a low energy system widely used in farm applications, was selected for the trial. The cubing system mixes the grass with a binder (if necessary), and extrudes the grass radially through a set of dies, yielding a product 1.25 inch by 1.25 inch in cross section and roughly 1 inch to 3 inches long. The cubes have a bulk density of roughly 23-24 lb/ft^3, and in bulk, have two to three times the energy content of equal volumes of loose or baled hay. The trial was to include cubes made from both the highly productive switchgrass, and also from bermudagrass (*Cynodon dactylon*) grown in Georgia. A portable cubing system was assembled that was operated for these tests at farms in Alabama and Georgia.

From these tests, moisture limits, binder requirements and feed rates were determined for producing cubes from switchgrass and bermudagrass. As part of the field cubing test programme, 170 tons of switchgrass and bermudagrass cubes were manufactured and transported to Plant Mitchell, a Georgia Power Company facility located south of Albany, Georgia, for field co-firing tests. Georgia Power is one of five regulated retail electric utility subsidiaries of Southern Company.

Grass Cube Co-firing Tests

Following the cubing tests, one week of field testing was conducted evaluating the performance of the cubes in an actual boiler at Plant Mitchell. Plant Mitchell Unit 3 is a 165 MW tangentially-fired combustion engineering unit, normally fuelled with central Appalachian bituminous coal. Coal is pulverized with four bowl mills, with the fine powdered coal introduced into the furnace through four levels of coal nozzles at each corner of the boiler.

To avoid putting the entire unit at risk, a preliminary combustion test was conducted with cubes introduced into a single pulverizer. The coal bunker level above only one of the pulverizers was drawn down, and then refilled with a mixture of approximately 10% by weight (about 20% by volume) cubes with coal. The mix passed through the coal handling belts and crusher to the bunker. The cubes which reached the pulverizer appeared for the most part to pulverize and burn well, although some relatively large burning cube flakes were noted during video observation of the boiler interior.

Later observation of the coal levels, however, showed that the coal-cube mixture was not flowing appropriately in the bunker. The mixture tended to adhere to itself and to the walls, resulting in a 40 feet (ft) deep rat-hole from the top to the bottom of the bunker. A small percentage of the cubes were crushed in fuel handling or by being run over by the coal handling bulldozer (Fig. 15.1). These cubes reverted to grass, and this relatively small amount of grass stems caused the mixture to not flow. A retest with only 2.5% cubes by weight resulted in fewer problems but still unacceptable fuel flow characteristics in the bunker. At this point, it was determined that grass cubes of this design would be unacceptable as a power plant co-milling fuel (Boylan *et al.*, 2003).

Full Scale Direct Injection Tests

Co-firing System and Test Plan

Direct injection tests were conducted at Alabama Power Company's Plant Gadsden, located about 60 miles northeast of Birmingham (Zemo *et al.*, 2002). Alabama Power is a subsidiary of Southern Company. Plant Gadsden is a relatively small power plant, consisting of two

Figure 15.1. Mixing cubes and coal at Plant Mitchell, near Albany, Georgia.

identical 70 MW tangentially-fired pulverized coal units. These units are fuelled with Synfuel, basically Alabama bituminous coal and coal fines with an asphalt binder. Synfuel is a little more difficult to handle than its parent coal, but has essentially identical combustion characteristics.

The biomass direct injection co-firing system was constructed for Plant Gadsden's Unit 2 as part of a co-firing project with Southern Research Institute, Electric Power Research Institute (EPRI), the US Department of Energy, and others, and is shown schematically in Fig. 15.2. Switchgrass was stored as round 1,600 pound bales on a section of the plant coal pile. The bales were transported by tractor and hay spear to a Haybuster Agricultural Products tub grinder, which ground the bales, including twine or net covering, to approximately one inch or smaller pieces. (The Haybuster was later replaced with a larger Vermeer Manufacturing Company TG525 tub grinder.) The ground grass was conveyed to a metering bin, in which augers at the bottom provided a controlled rate of grass feed. The grass was augered into the inlet of a 'dirty' transport fan. On passing through the fan, the grass was pneumatically conveyed to two dedicated biomass burners in the furnace. The system could provide up to 7,000 lb of switchgrass per hour, which would produce roughly 4.3 MW of renewable generation. One thousand bales of switchgrass, each weighing about 1,600 lb, were delivered for the tests.

Six weeks of testing were conducted co-firing switchgrass at rates of up to 10% of the heat input to the furnace. Forty performance tests were conducted at various operating conditions, measuring the effect of co-firing on boiler performance, unit emissions, and operations.

These detailed tests were followed by longer term testing in which the remaining several hundred tons of switchgrass were consumed. Operation sustainability and longer term impact on boiler tube corrosion and tube deposits were examined.

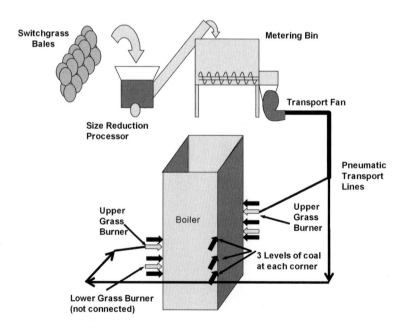

Source: Boylan *et al.* (2003).

Figure 15.2. Schematic of switchgrass direct injection system.

Test Results – Emissions

Emissions were generally lower when co-firing switchgrass with coal. Compared with coal, switchgrass and most other biomass materials have very little sulphur. As a result, plant emission monitors showed that sulphur emissions were reduced by approximately the percentage of coal displaced by grass. Similarly, mercury content of the grass is very low compared with coal, resulting in reduced mercury emissions. Pilot scale testing had suggested that NO_x would be reduced with co-firing (Bush *et al.*, 2001), but NO_x levels appeared to be unaffected, neither increasing nor decreasing when switchgrass was introduced.

A major goal with renewable energy is the reduction of carbon dioxide (CO_2) emissions. Renewable energy generation from biomass is usually considered to be CO_2 neutral because while the grass or tree used as fuel is growing, it absorbs carbon from the atmosphere. When that biomass is burned, it is the same carbon returned to the air, with essentially no net CO_2 increase. Therefore, in the testing, when the unit load was held steady, then the reduction of coal flow with the introduction of biomass was a measure of reduction of new CO_2 to the atmosphere. Tests demonstrated reductions of up to 10% of the coal flow rate.

Test Results – Efficiency

It was found in the series of boiler efficiency tests that efficiency when co-firing switchgrass was slightly less than when firing coal alone. One major cause was the introduction of fairly large volumes of cold transport air into the furnace when the switchgrass system was operated. This transport air had to be heated, using some of the fuel energy that would otherwise have gone to making steam and power. This loss, called the dry gas loss, represents energy from the fuel that escapes up the stack.

Some of the additional loss was due to unburned fuel which fell into the bottom of the furnace. Typically, this material consisted of the nodes in the grass, which resisted being ground in the tub grinder, increasing the unburned carbon content of the bottom ash. In contrast, fly ash, the ash carried in the exhaust gases toward the collection systems near the stack, actually had reduced unburned carbon. This was an unexpected result, and offset some of the other losses.

Boiler efficiency is calculated by summing these and other losses, and subtracting the total from 100%. Figure 15.3 shows full load boiler efficiency plotted with furnace exit oxygen (O_2) levels for co-firing and for coal alone. The two co-firing curves represent introduction of the grass at the upper (u65) and lower (l65) burner locations shown in Fig. 15.2. The third curve (zero) represents boiler efficiency without co-firing, which is between 0.3% and 1.0% higher than with co-firing. The most efficient switchgrass co-firing was measured at about 2.7% furnace exit O_2.

Although these efficiency losses appear small, they take on a different perspective when the entire loss is credited to the grass. These losses occur only when the grass is fired, so it is proper that the loss be calculated as a percentage of the input biomass fuel. In a typical case in which co-firing 7% by energy input resulted in about 0.5% boiler efficiency loss, this was equivalent to a loss of about 7% of the input biomass energy.

Future studies will evaluate the potential for using a higher efficiency transport fan and a single burner. This will reduce the amount of transport air and resulting dry gas loss. If the grass burns well in the furnace in this single burner scenario, the net result may be to substantially reduce the boiler efficiency loss and associated costs to biomass co-firing.

Test Results – Handling and Grinding

Dust was a very real consideration in grinding and handling switchgrass. To reduce dust emissions, the top of the metering bin was covered and the conveyor belt from the tub grinder was enclosed, with a fan drawing dust from the two. The original Haybuster 8 ft diameter tub grinder gave a fairly tight fit with the 6 ft diameter bales, the bale thereby reducing much of the dust from the grinder. However, when the Haybuster was later replaced with the 12 ft diameter Vermeer tub grinder, the fit was not nearly as close, and considerable dust resulted. A low-water-use fogger system was installed over the tub that was reasonably successful in reducing dust.

Both grinders worked well in processing the bales with synthetic and natural twine or netting in place. This was important to the manpower and safety aspects of this project. Pre-grinding removal of the twine would have required exposing additional personnel to added risk, or the redesign of the handling system to automatically remove the strings.

The smaller 150 horsepower (hp) Haybuster grinder worked well with switchgrass. The stems of this grass were large and brittle, fracturing fairly easily into pieces. Wheat straw also ground up easily. However, finer grasses such as fescue (*Festuca* sp.), bermudagrass, and mixed grasses were more wiry and pliable. These grasses tended to form nests and entanglements in the Haybuster grinder, sometimes seizing up and causing damage to the equipment. For the finer grasses, the larger (525 hp) Vermeer tub grinder was obtained. The Vermeer screens were modified with additional stationary teeth for the hammers to pass between, providing additional cutting edges. This system has worked well for the finer grasses, but consumes more energy in processing.

Records were kept of the fuel consumption of grinder and bale handling tractor, together with electrical power and number of bales processed. From these data, it was determined that the bale handling tractor consumed 0.5 gallons of diesel per ton, while the Haybuster

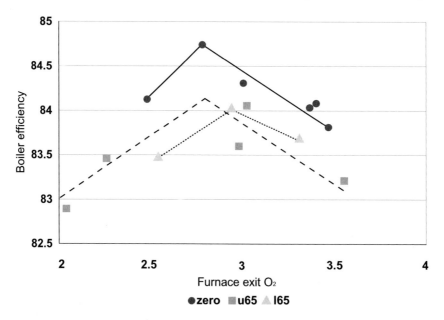

Source: Boylan *et al.* (2003).

Figure 15.3. Switchgrass direct injection boiler efficiency.

consumed 2.25 gallons per ton. With the larger Vermeer grinder, between 3.8 gallons and 4.4 gallons of diesel per ton of grass were needed, depending on the grass being ground. Electrical power consumption for fan and augers was about 12 kilowatt-hours (kWh) per ton. Overall, grinding costs were estimated to be about US$3.75 and US$5.90 per ton of grass depending on the grinder and the type of grass. It is believed that this cost could be reduced by powering the grinder with an electric motor.

Test Results – Longer Term

Following the detailed performance tests, 2 months of longer term evaluation were conducted in which switchgrass was co-fired on a routine basis. Standard plant data were collected and assessed, and during an outage which followed this longer term testing, an inspection was made of the boiler interior. For this time period, no indications of unusual corrosion or deposits were noted, either in operations data or in outage inspection. However, the co-firing unit operation was intermittent, the system used only in daylight and on weekdays. Because the fuel contains relatively high concentrations of potassium and chlorine, further investigations are recommended in this study area.

Test Results – Overall

Co-firing of switchgrass as well as other grasses by direct injection in Plant Gadsden's coal-fired unit was considered successful. Overall emissions decreased measurably, and efficiency penalties associated with co-firing the switchgrass were considered acceptable. However, several other factors remain to be investigated that may raise issues for applying this technology commercially at some plants.

Additional Co-firing Issues

Some cost issues associated with biomass co-firing are plant specific and related to how the units control NO_x emissions and whether they sell their ash.

Ash Sales and ASTM International

Roughly 10% of coal consists of ash, and large power plants can produce ash at rates up to 128 tons per hour. Unless a beneficial use is found for the ash, proper disposal is required, which can incur significant costs.

Typically, about 80% of the ash occurs in the form of fine fly ash which is removed from the flue gas before it enters the stack. Fly ash is commonly used to replace part of the cement in a concrete mixture, or as an additive to enhance the concrete properties. The addition of fly ash to concrete helps improve the physical and mechanical properties of both fresh and hardened concrete, and as a result, there is a good market for the ash. This market can change fly ash from a liability to a revenue source, provided that the fly ash meets ASTM International (formerly American Society for Testing and Materials) standards. However, ASTM International standard C618 essentially defines fly ash for concrete applications as ash deriving solely from coal combustion. As a result, fly ash marketers are hesitant to market a 'non-spec' fly ash containing even small amounts of ash from wood, grass, or other biomass. For power plants that sell their ash, this can be a significant issue.

The addition of small amounts of biomass can jeopardize utilization and sale of an otherwise waste product, and at the same time create significant disposal costs for very large amounts of ash.

Southern Company and others are attempting to address this issue with ASTM International. In 1998, EPRI issued a report (Fouad *et al.*, 1998) on testing by Southern Company and the University of Alabama at Birmingham to consider the effect of wood co-fire ash on concrete properties. Wood itself has little ash content. As a result, it was not surprising that the study indicated that wood ash had no detrimental effect, and could potentially have beneficial effects, at least up to a co-firing level of 30% by mass wood fuel content.

Switchgrass and other grasses can be more of a concern as they contain a higher ash percentage than wood, and that ash has significantly higher alkali metal content – especially potassium. The ASTM International C618 standard limits the amount of allowable sodium and potassium for some applications. Studies are on-going (Amos, 2002) to consider ash from grasses such as switchgrass. These tests should provide a basis for establishing an acceptable level of grass co-firing, which would then be presented for consideration to ASTM International.

SCR Catalyst Effects

In order to decrease NO_x emissions at some power plants, selective catalytic reduction (SCR) systems have been installed. To reduce NO_x, a controlled amount of ammonia is introduced into the boiler exhaust gas, and the mixture passed over a catalyst material. Substantial surface area is required to provide good contact with the flue gas, so the SCR and its catalyst content are physically quite large.

Studies by others (Zheng *et al.*, 2002) have suggested that the presence of alkali metals in fly ash can result in catalyst poisoning and, as a result, significant degradation of the catalyst performance. Because potassium is such a significant component in fly ash from grasses, grass co-firing may present a particular problem for plants equipped with SCR.

Several groups are currently investigating the impact of grass co-firing on SCR, including Southern Company. This project, co-sponsored by EPRI, involves exposing coupons of catalyst material to flue gas while switchgrass is being co-fired. Two probes are used, one of which is a baseline probe, operated only when coal alone is fired. At 250-hour intervals, samples will be removed from each probe to measure changes in their chemical makeup, deposits, and ability to catalyse the NO_x-ammonia reaction. Initial samples are planned to be taken in June 2004.

When the results of these tests and those by others are complete, a better assessment of the impact of grass co-fire ash on SCR catalyst can be made. Until then, due to the risk to a major investment and to the plant performance, Southern Company power plants equipped with SCR will not be co-firing biomass.

Costs

Based on available test and economic data, a model was constructed to predict costs and benefits for a commercial co-firing application at a typical power plant. The system modelled here generates 11 MW of renewable energy, co-firing about 54,000 tons of grass annually. Figure 15.4 presents the model results, showing the increased cost of power due to co-firing compared with coal-alone operation, and breaks down the individual components of cost and benefit. It should be kept in mind that these values are estimates based on a single set of assumptions, and an actual application will be quite site specific. However, they give an indication of approximate overall cost, and of the relative importance of different cost and benefit components. The first three columns on the chart represent predictions of costs and benefits and net costs associated with direct injection co-firing of US$45 per delivered ton of switchgrass. Coal energy value for this calculation was assumed to be US$1.72 per million British thermal units (MBtu) and the heating value of the grass was taken to be 6,900 Btu per pound. The model predicts the largest cost component to be the cost of the biomass fuel compared with coal. As a result, relatively small changes in grass cost or in coal cost could have the most impact on power cost. The cost of renewable energy increases 0.08 cents/kWh for every dollar per ton increase in fuel cost. Labour and capital costs are the next largest components, approximately equal, and each about a third of the fuel cost difference. Capital costs were determined from an engineering assessment for this study by McBurney (Cantrell, 2004). Sulphur benefits reduce the co-firing cost by about 0.34 cents/kWh.

The total column is the total environmental benefits subtracted from the costs column, and represents the additional costs of generating renewable energy by co-firing, above the normal costs of generating coal-based power. The model predicts that at present, the overall additional cost above baseline of producing renewable energy from co-firing grasses is about 3.2 cents/kWh. The baseline cost of power from coal alone is roughly 3 cents/kWh.

The study results also suggest that cost reduction may be possible in the future, and an example of expected improvements is shown in the last three bars of Fig. 15.4. It is not expected that fuel cost will be reduced substantially over time. However, efforts to improve the combustion efficiency by reducing the amount of cold transport air appear promising to cut biomass efficiency costs by half. Capital costs may be reduced slightly and labour costs

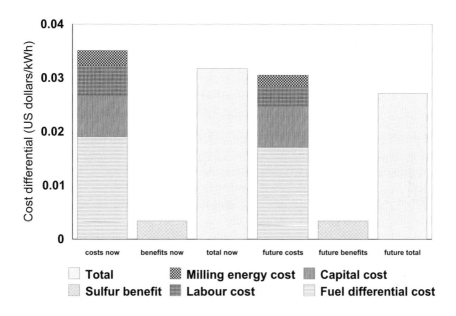

Figure 15.4. Power costs above coal baselines predicted for co-firing grasses now and with expected future improvements.

for handling may be improved by designing for non-power plant personnel off site. Overall, these improvements suggest that the cost of co-firing above that of coal-alone operation might be reduced to about 2.7 cents/kWh.

Renewable energy credits in the proposed 2003 omnibus energy bill were established at 1.5 cents/kWh. This represents an upper cost bound on the added cost of generating renewable energy for meeting the energy bill requirements. If this energy credit cost is what appears in a final RPS, then the results of this study suggest that to be competitive, co-firing of grasses will require price support in the form of tax credits or subsidies.

To help address some of the on-going issues related to reducing costs of co-firing grasses and other biomass, Southern Company is planning to construct a new system at Georgia Power Company's Plant Mitchell. Based on lessons learned in testing on the Plant Gadsden system, a larger, more efficient, and less man-power intensive direct injection system is being planned. The Plant Mitchell system, with up to 10 MW co-firing capability, will be designed to co-fire grasses as well as sawdust. It will provide a semi-commercial platform for better evaluation of operations, maintenance, emissions, and performance costs, and how to reduce them. Construction of the system is planned for spring 2005.

Conclusions

Based on the results of these studies, the following conclusions are made regarding co-firing of grasses in pulverized coal boilers:

- Switchgrass and other grasses can be successfully co-fired with coal in a power plant by direct injection. Up to 10% of energy from biomass was achieved in these tests. Co-milling of grasses with coal was unsuccessful.
- CO_2, sulphur dioxide (SO_2), and metals emissions were reduced with co-firing. NO_x emissions were unchanged.
- Boiler efficiency was slightly reduced when co-firing, due to cold transport air entering the furnace. This loss was offset in part by reduced unburned carbon in fly ash while co-firing.
- Different grasses can have very different grinding characteristics.
- Research continues to address other issues such as SCR catalyst degradation and ash utilization and sales.
- Cost predictions for co-firing grasses were about 3.2 cents/kWh higher than for coal-alone generation. Expected future improvements in these costs suggest that overall cost of co-firing above that of coal-alone operation might be reduced to about 2.7 cents/kWh.
- Based on these results, a subsidy will probably be needed to make grass co-firing competitive with other renewable options.
- Future studies include a planned grass co-firing facility at Georgia Power's Plant Mitchell.

Acknowledgements

The authors wish to express their appreciation to Evan Hughes, Dave O'Connor, David Bransby, Vann Bush, David and Doug Wilson, Ann Jones, David Tillman, Burea Vinson, and Chip Blalock for their support of the projects. They also acknowledge the contributions of

Wayne Edwards, Bill Mashburn and the staffs of Plant Mitchell and Plant Gadsden. Project funding came from Southern Company, Electric Power Research Institute, and US Department of Energy.

References

Amos, W. (2002) Summary of Chariton Valley Switchgrass Co-fire Testing at the Ottumwa Generating Station in Chillicothe, Iowa. Technical Report NREL/TP-510-32424. National Renewable Energy Laboratory, Golden, Colorado.

Boylan, D., Eastis, J., Jones, A. and Felix, R. (2003) *Manufacturing and Co-firing Switchgrass and Coastal Bermudagrass Cubes for Generating Renewable Energy.* EPRI Technical Update 1004814. Electric Power Research Institute, Palo Alto, California.

Bush, P., Smith, H., Taylor, C., Bransby, D. and Boylan, D. (2001) *Evaluation of Switchgrass as a Co-firing Fuel in the Southeast.* Final Technical Report, US Department of Energy Cooperative Agreement DE-FC36-98GO10349, Southern Research Institute, Birmingham, Alabama.

Cantrell, M. (2004) *Biomass Co-Firing System – Final Engineering Report.* McBurney Project 604010, McBurney, Norcross, Georgia.

Fouad, F., Copham, C. and Donovan, J. (1998) *Evaluation of Concrete Containing Fly Ash with High Carbon Content and/or Small Amounts of Wood.* EPRI TR-110633. Electric Power Research Institute, Palo Alto, California, and Southern Company Services, Atlanta, Georgia.

Zemo, B., Boylan, D. and Eastis, J. (2002) Experiences Co-firing Switchgrass at Alabama Power's Plant Gadsden. *Proceedings of the 27th International Technical Conference on Coal Utilization and Fuel Systems.* Clearwater, Florida, March 4-7. Coal Technology Association, Gaithersburg, Maryland.

Zheng, Y., Jensen, A. and Johnsson, J. (2002) *Deactivation of SCR Catalysts in Biofuel (Co)-Combustion – Literature Study.* Department of Chemical Engineering, Technical University of Denmark, Lyngby, Denmark.

Chapter 16

Animal Waste as a Source of Renewable Energy

Soyuz Priyadarsan, Kalyan Annamalai, Ben Thien, John M. Sweeten and Saqib Mukhtar

Introduction

Since World War II, agriculture has vastly expanded, become more specialized and more intensive in both the developed and much of the developing world. Similarly, animal agriculture has undergone substantial shifts in structure, size and productivity in pursuit of higher efficiencies and better economics of operations. Such advances have resulted in consolidation and concentration in animal feeding operations, resulting in significant declines in the overall number of operations, while improving productivity. Animal (livestock and poultry) feeding operations (AFOs) and confined animal feeding operations (CAFOs) are the natural outcome of such consolidation and intensification.

In Texas, feedlots are primarily located in the Panhandle area (the northernmost 26 counties of the state) near Amarillo. There are approximately 70 feed yards in the Texas Panhandle area, which includes border areas of Oklahoma and New Mexico, feeding almost 7 million head (30% of cattle on feed in the USA). Some feedlots have capacities greater than 20,000 cattle, while larger operations have capacities of 50,000 to 85,000 heads. Cattle feeding in the Panhandle area is a US$5.5 billion industry with a total economic impact of about US$15.5 billion. Most poultry in the USA is produced in integrated operations, which raise broilers in houses containing flocks of 10,000 to 30,000 birds per house. On average, each house produces five to six flocks per year. Poultry broiler production in the USA has increased dramatically with total broiler meat production increasing from nearly 3.64 billion kilograms (kg) in 1963 to more than 20.45 billion kg produced by 8.5 billion birds in 2003. Texas alone processes more than 601 million broilers per year and ranks sixth in the nation (National Agricultural Statistics Service, 2003).

Not surprisingly, one by-product of such large concentrated operations is the generation of substantial amounts of animal wastes. Each feeder steer or heifer in the feedlot produces about 28.12 kg or 62 pounds (lb) of wet manure per day containing 88% moisture and 12% solids (Sweeten, 1979). Typically, the cattle waste production is estimated as dry tons per year. It is estimated that six million dry tons of waste are produced per year in the Texas Panhandle area and 75 million dry tons of waste in the USA as a whole. On the other hand, manure excreted by broilers is mixed with feathers, feed, water and bedding material, such as saw dust, rice, or peanut hulls. An average of 1.25 tons of litter is produced per 1,000 birds sold (Collins *et al.*, 1999).

Approximately 10.5 million tons of broiler litter was produced in the USA in 2001, with more than 750,000 tons being produced in Texas. In some cases, large operations may not

© CAB International 2005. *Agriculture as a Producer and Consumer of Energy* (eds J.L. Outlaw, K.J. Collins and J.A. Duffield)

have sufficient land available for spreading the manure as fertilizer in accordance to their manure management plan. This often leads to the extended storage of manure which poses economic and environmental liabilities. US animal agriculture has increasingly come under scrutiny for the sector's potential impact on the country's natural resources. Today, concerns are primarily: 1) nutrient pollution of water resources; 2) greenhouse gas emissions such as methane (CH_4) and carbon dioxide (CO_2) and 3) air quality (e.g., odours and particulates).

Across the USA, the animal industry is faced with regulations or anticipated stringent land application requirements that will impact how animal waste is managed. Scientists are in search for uses or technologies that will not only solve the manure disposal problem, but also generate revenue. Although the options are numerous, the more practical animal waste recycling applications are as a fertilizer, feed, or fuel source. An old concept with a new application is the use of animal manure as a non-conventional, renewable energy source for generating heat and power.

Alternative Technologies

Animal manure has been considered as a potential energy feedstock for more than three decades through: a) bioconversion (anaerobic digestion or gasification in absence of air) and b) non-biological or thermal conversion (gasification at stoichiometric oxygen (O_2) content, pyrolysis at O_2-starved conditions, or combustion with excess O_2). Anaerobic digestion leads to production of CH_4 (60%) and CO_2 (40%) in the absence of O_2 using syntrophic bacteria and therefore there is no need for aeration. However, the production of CH_4 is affected by high ammonia-based nitrogen (total ammonia) levels (3000 mg/litre to 6000 mg/litre) which affects the population of syntrophic bacteria causing failure of the digestion process. For a perfectly stirred reactor, the failure rate is as high as 70%, while with the plug flow reactor, the failure rate is 65%. The biogas (CH_4 + CO_2) cannot be stored or transported economically due to low heat content. Further, only about 40% to 60% of the organic matter is converted into biogas. Also, digestion is a slow process that is very water intensive and generates a digested slurry which is difficult to transport. These two factors make this a less attractive technology. Annamalai *et al.* (1987) and Sami *et al.* (2001) summarized the non-biological technologies which utilize feedlot manure as a sole energy source. Prior research with feedlot biomass combustion in the 1980s and 1990s was conducted in circulating and conventional fluidized bed combustors (Sweeten *et al.*, 1986; Annamalai *et al.*, 1987). Some of these technologies have met with limited success due to the highly variable properties of cattle manure and the associated flame stability problem in combustion systems. Furthermore, while ash is 2% to 10% in woody biomass and 10% in coal, the ash content in animal manure can range from 20% to 50%, leading to low heating values. Since feedlot manure and poultry litter manure have uses as both a fuel source and as a fertilizer, they shall be properly referred to as feedlot biomass (FB) and litter biomass (LB), respectively.

Co-firing

Blending FB or LB with coal and firing it in existing boiler burners can eliminate most of the problem associated with firing FB or LB alone. Currently, coal-fired utilities are already exploring the possibility of supplementing fossil fuels through the utilization of biomass fuels, a renewable CO_2 neutral fuel, in order to reduce fossil fuel generated CO_2 emissions. For a 90:10 coal and FB blend firing, the CO_2 reduction is of the order of 5% to 8% depending upon the FB quality. Co-firing studies have also been done on wood waste (Gold and Tillman, 1996), switchgrass (*Panicum virgatum*) (Aerts *et al.*, 1997), sawdust (Abbas *et al.*, 1994), straw and wood chips blends (Hansen *et al.*, 1998; Sampson *et al.*, 1991), sewage sludge (Spliethoff and Hein, 1998) and refuse-derived fuel (RDF) (Ohlsson, 1994).

Gasification

The co-firing approach in existing suspension fired burners requires grinding of high ash, high moisture and fibrous biomass almost to the same fineness as coal (in the range of 50 micrometres (μm) to 100 μm). The gasification approach does not require fine grinding. Once the solid fuel is converted into a gaseous energy carrier, it can be easily transported and directly burned using conventional gas burners to generate heat and power. Fixed bed gasification studies on various biomass fuels like beech wood, nutshells, olive husks (Blasi *et al.*, 1999), cotton stalks (Patil and Rao, 1993), pigeon pea stalk (Katyal and Iyer, 2000) and bundled jute sticks (Kayal and Chakravarty, 1994) have been conducted. Krishnamoorthy *et al.* (1989) studied the gasification of cattle waste briquettes in 6.0 kilograms/hour (kg/h) or 13.2 lb/h and 25.0 kg/h (55.1 lb/h) fuel feed rate updraft gasifiers. Recently, Priyadarsan *et al.* (2004a) reported studies on fixed-bed co-gasification of coal blended with LB and low-ash FB for two nominal particle size distributions, 9.4±3.1 and 5.15±1.15 millimetres (mm) and two airflow rates, 1.3 m^3/h and 1.7 m^3/h (45 standard cubic feet per hour (SCFH)) and 60 SCFH to yield a low British thermal unit (Btu) product gas, 4.78±0.31 megajoule/kg (MJ/kg) on a dry basis. However, before co-firing or gasification of FB and LB reach a commercial stage, laboratory-scale and pilot-scale studies need to be performed to generate fundamental data evaluating the performance of such units and also economic analyses of FB and LB as a coal supplement.

Economic Analysis

Animal manure cannot sustain much investment in processing and distribution owing to its initially low nutritional content. At the same time, in order to keep biomass fuel costs competitive with coal, the collection and transportation costs must be kept to a minimum. The large variability in FB and LB makes it difficult to generalize the costs of manure handling, storage and transportation. From an economic standpoint, manure can be looked at as a waste to be disposed of at the least cost.

Combustion Studies

In this section, the thermo-physical and chemical properties of FB and LB, such as proximate[1], ultimate[2] and thermo-gravimetric[3] (TGA) analyses are discussed. In addition, performance studies on co-firing and gasification are also presented.

Fuel Properties

Feedlot biomass for the TGA, co-firing and gasification experiments was harvested from a feedlot in Amarillo, Texas, composted, dried and then ground (Sweeten *et al.*, 2003). The manure was collected from feedlots after being deposited for 180 days in a feedlot pen. The

---

[1] The fuel composition as a percentage of ash, moisture, fixed carbon and volatile matter.
[2] The fuel elemental composition as a percentage of carbon, oxygen, nitrogen and sulphur.
[3] An indicator of the temperature at which the solid fuel will start releasing the volatile matter and also the rate at which the volatile matter is released from the fuel.

manure was then placed in windrows and allowed to compost for 45 days. Next, the manure was placed on a tarp and allowed to dry in a greenhouse until the moisture content was less than 8%. Finally, FB used for TGA and co-firing studies was placed in sealed barrels, shipped to California[4] and ground, while the FB for gasification was manually hammered and sized to get a coarser particle size distribution. Table 16.1 shows the proximate and ultimate analyses of all the fuels used for the experimental studies. Coal-A, FB-A and LB-A were used in the TGA experiments, while Coal-B, FB-B and LB-A were used in co-firing experiments and Coal-C, FB-C and LB-B were used in gasification experiments. The LB-B analysis is reported elsewhere (Priyadarsan et al., 2004b). All of the blends were analysed on a mass basis and all the blend properties were obtained through the law of mixtures from the coal and biomass analyses. A 90:10 blend is a mixture of nine parts coal and one part of the respective biomass on a mass basis.

In LB, ash is around 25%; in FB it is around 40%; and in coal it is around 5% (Table 16.1). The higher ash percentage can cause problems in boiler burners by causing fouling and boiler tube corrosion. Additional calculated fuel parameters, including ash in kilograms/gigajoule (kg/GJ), nitrogen (N) in kg/GJ and sulphur (S) in kg/GJ are also shown in Table 16.1. The higher heating values of FB are approximately 34% to 45% of those of coal on heat basis, while those of LB are 56% of coal on heat basis. This requires the mass flow of fuel be increased in order to maintain the same heat throughput when firing blends. Even more troubling is the increased S and N in FB. For 90:10 FB blends, the increase in N (kg/GJ) is approximately 15% to 25% and increase in S (kg/GJ) is approximately 12% to 20% as compared to coal on heat basis. When the fuel burns, N and S in the fuel will combine with O_2 from the air to form nitric oxide (NO) and sulphur dioxide (SO_2) which are recognized air pollutants. On a dry, ash-free basis, both FB and LB contain approximately 80% volatile matter (VM), while coal contains approximately 45% VM. The results of the ultimate and proximate analyses show that FB and LB are lower quality fuels as compared to coal.

Size Distribution

Figure 16.1 shows the cumulative mass size distribution of the fuels used for TGA and co-firing studies. It is seen that 90% mass of Coal-A < 50 μm, 90% mass of Coal-B < 75 μm, 90% mass of FB-A < 150 μm, 90% mass of FB-B < 190 μm and 90% mass of LB-A < 200 μm. The fibrous nature of FB and LB results in a coarser size distribution as compared to coal.

Characteristics of Pyrolysis

After being sieve classified, TGA experiments were conducted to determine the kinetics of pyrolysis for Coal-A, FB-A, LB-A and 90:10 FB blend-A. TGA studies involve heating approximately 25 milligrams (55x10^{-6} lb) of the fuel sample in an inert atmosphere at a rate of 10°C/minute (min) or 18°F/min, with continuous monitoring of the mass loss from the sample. Figure 16.2 shows the comparative TGA traces of the above-mentioned fuels. All the fuels exhibit a moisture loss region, followed by the VM loss region, finally yielding carbon (called fixed carbon, FC) and ash residue, as clearly marked in Fig. 16.2. As observed in Fig. 16.2, for all the fuels, moisture is completely released by 400 Kelvin (K) (261°F). However, the

---

[4] For suspension firing, the manure has to be pulverized to a fine size similar to that of coal (almost 70% of mass less than 100 μm). A typical coal-fired plant will have the pulverized unit on site for grinding the coal, so the manure grinding can also be done on site in the same pulverizer unit. However, since the experiments at Texas A&M University were conducted at a laboratory scale which does not have a pulverizer unit on site, a company in California was identified to grind the dried manure to the desired specification and ship it to Texas A&M University.

Table 16.1. Fuel properties.

Parameter	Coal-A	FB-A	90:10 FB Blend-A	LB-A	90:10 LB Blend-A	Coal-B	FB-B	90:10 FB Blend-B	Coal-C	FB-C	50:50 FB Blend-C
Dry loss	22.8	6.8	21.2	11.6	21.7	15.1	7.7	14.4	21.2	7.6	14.4
Ash	5.4	42.3	9.1	26.8	7.6	5.3	44.2	9.2	4.3	43.9	24.1
Carbon	54.1	23.9	51.0	28.4	51.5	60.3	23.6	56.6	43.8	23.9	33.9
Hydrogen	3.4	20.3	5.1	3.7	3.5	3.6	2.9	3.5	4.6	3.0	3.8
Oxygen	13.1	3.6	12.1	22.8	14.0	14.5	19.1	15.0	25.2	18.8	22.0
Nitrogen	0.8	2.3	1.0	3.0	1.0	1.0	1.8	1.0	0.7	2.3	1.5
Sulphur	0.4	0.9	0.4	0.7	0.4	0.2	0.5	0.3	0.2	0.5	0.4
Chlorine	N/A	N/A	N/A	0.93	N/A	<0.1	1.9	N/A	N/A	N/A	N/A
Phosphorus	N/A	N/A	N/A	1.97	N/A	N/A	N/A	N/A	N/A	N/A	N/A
Gross heating value (GHV)[a]	21.38	9.56	20.20	12.07	20.45	23.71	9.42	22.28	27.70	9.35	18.53
Fixed carbon (FC)	37.3	10.5	37.6	10.9	34.6	42.4	6.5	38.8	41.9	10.3	26.1
Volatile matter	34.5	40.4	32.1	50.7	36.1	37.2	41.4	37.6	32.6	38.2	35.4
Ash (kilograms/gigajoule)[b]	2.5	44.2	4.5	22.2	3.7	2.2	46.9	4.1	1.6	46.9	13.0
Nitrogen (kilograms/gigajoule)[b]	0.38	2.41	0.47	2.52	0.50	0.40	1.89	0.47	0.24	2.47	0.80
Sulphur (kilograms/gigajoule)[b]	0.18	0.94	0.22	0.55	0.20	0.10	0.53	0.12	0.08	0.51	0.19
Carbon dioxide (kilograms/gigajoule)[b]	29.50	29.16	29.48	27.50	29.38	29.67	29.22	29.65	18.46	29.84	21.33
Carbon dioxide from coal (kilograms/gigajoule)[b]	29.50	0.00	28.10	0.00	27.76	29.67	0.00	28.42	18.46	0.00	13.80
Carbon dioxide from biomass (kilograms/gigajoule)[b]	0.00	29.16	1.38	27.50	1.62	0.00	29.22	1.24	0.00	29.84	7.53

[a] As received, megajoule/kilogram (MJ/kg); multiply MJ/kg by 433 to get Btu/pound.
[b] Multiply kilograms/gigajoule by 2.33 to get pounds/million Btu.
N/A = not available.

onset of pyrolysis takes place at different temperatures for different fuels. It approximately begins at 520 K (476°F) for both LB-A and FB-A and 650 K (710°F) for both Coal-A and 90:10 FB Blend-A. The pyrolysis region for the biomass fuels is steeper, suggesting that their mass loss due to pyrolysis is much faster as compared to coal. For the blend, the behaviour is

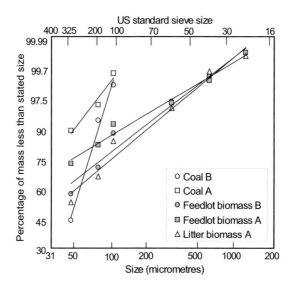

Figure 16.1. Cumulative mass size distribution of Coal-A, Coal-B, FB-A, FB-B, and LB-A.

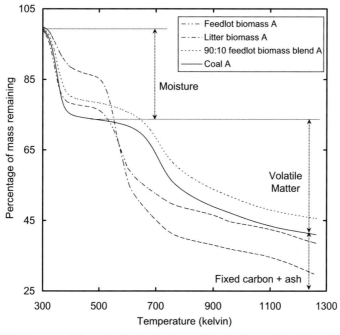

Figure 16.2. TGA traces of Coal-A, FB-A, LB-A, and 90:10 Coal-A FB-A blend.

dominated by the behaviour of coal. Thus under combustion conditions, FB and LB fuels will not only pyrolyse at a lower temperature, but also at a faster rate as compared to coal or blended fuels.

Co-firing Studies

Once the basic fuel properties are understood, the co-firing results can be analysed. The experiments were conducted with pure Coal-B, a 90:10 blend of Coal-B and FB-B and a 90:10 blend of Coal-B and LB-A. Coal-B at 10% excess air is considered the base case for the co-firing experiments. The co-firing setup is a 30-kilowatt (kW) or 100,000 Btu/h downward fired boiler burner setup, with the fuel being supplied in suspension form by the primary air and the secondary air being supplied to complete the combustion process. Annamalai *et al.* (2003) have reported a detailed description of the co-firing setup.

Figure 16.3 shows burnt mass fraction or loosely called combustion efficiency achieved for various fuels under different excess air percentages. The results show that all of the fuels had a similar burnt mass fraction. The lower values for coal are due to limited residence of the 30 kW (100,000 Btu/h) bench scale facility. The greater volatile content of the biomass fuels, lower pyrolysis temperature and lower activation energy of the biomass fuels make up for the decrease in heating value, increased ash in the biomass fuel and larger biomass size to achieve a slightly higher burnt mass fraction compared to coal.

In addition to how biomass changes the degree of combustion or burnt mass fraction, the quantities of air pollutants in the form of nitrogen oxides (NO_x) and SO_2 that are released are also important. When the FB and LB are mixed with coal, the amount of N in the resulting fuel is increased by 8% for FB and 20% for LB. If the entire additional N is converted to NO_x, there may be significant increases in air pollution emissions. Figure 16.4 shows the NO_x emissions on a measured parts per million (ppm) basis and on a 3% O_2 normalized basis. The

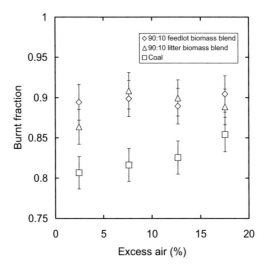

Source: adapted from Thien (2002).

Figure 16.3. Burnt mass fraction versus excess air percentage.

results show that the addition of more nitrogen with the biomass fuels does not lead to increased NO_x emissions when blended fuel is fired. It is assumed that most of the NO_x present is formed from the fuel, as the temperatures in coal combustion usually are not high enough to produce large amounts of thermal NO_x (temperature >1800 K or 2780°F). Although the type of fuel fired does not have an effect on the NO_x emissions, the excess air percentage does have an effect on the NO_x emissions, with greater excess air percentages producing greater NO_x emissions. The greater availability of O_2 at higher excess air percentages leads to greater formation of NO_x.

Gasification Studies

A 10 kW (34,000 Btu/h), updraft fixed-bed gasifier (reactor internal diameter 0.15 m or 0.5 ft, reactor height 0.30 m or 1 ft) facility was built and fired with a 50:50 Coal-C FB-C blend of two nominal particle sizes, 9.4±3.1 or 0.37 inch (in) and 5.15±1.15 mm (0.20 in), at a primary airflow rate 1.3 m³/h (45 SCFH). A detailed description of the setup and method of operation is described by Priyadarsan et al. (2004a). Figure 16.5 shows a comparative temperature profile in the fuel-bed at two different times. For the experiments, there was an initial time period of 0.5 h which was required to achieve the required bed height of 0.17 m (0.56 ft). After this initial time period, the clock was reset and the temperature and gas composition were measured. Since ash was not discharged from the bottom of the bed, ash accumulated and resulted in transient conditions in the bed, which is reflected in the peak-temperature shift towards the free board zone of the gasifier. In Fig. 16.5 it is observed that the peak-temperature shift is slower for 5.15 mm (0.2 in)-sized particles as compared to the 9.4 (0.37 in)-sized particles. Smaller particles result in a smaller void fraction, leading to a higher amount of fuel (combustible) per unit volume of the bed. This leads to a longer burn time per unit volume of the fuel bed, thus resulting in slower peak-temperature propagation. Figure 16.6 shows the

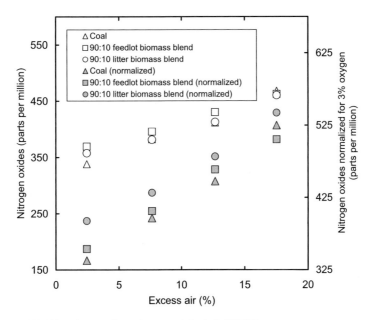

Source: Thien (2002); adapted from Annamalai et al. (2003).

Figure 16.4. NO_X and normalized NO_X emissions versus excess air percentage.

gas species composition along the fuel bed. It can be seen that the product gas composition is similar for both the particle sizes, with the product gas composition at the top of the fuel bed being approximately 27% carbon monoxide (CO), 7% hydrogen (H_2), 2% CH_4 and 5% CO_2 and the remaining is nitrogen (N_2). The higher heating value of the product gas composition is

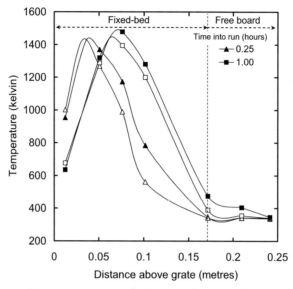

Figure 16.5. Comparative temperature profiles in the fuel bed (solid symbol: 9.4 millimetre-sized particle, hollow symbol: 5.15 millimetre-sized particle). Fuel: 50:50 Coal-C:FB-C blend.

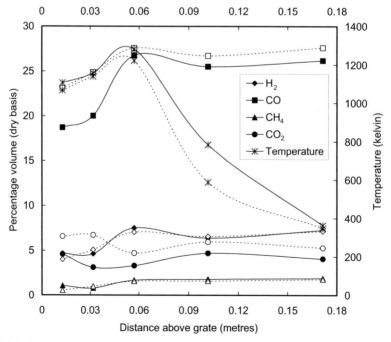

Figure 16.6. Comparative gas species profile in the fuel bed (solid symbol and solid line: 9.4 millimetre-sized particle, hollow symbol and dashed line: 5.15 millimetre-sized particle). Fuel: 50:50 Coal-C:FB-C blend.

4.67 MJ/m³ (125 Btu/SCF) for the 5.15 mm (0.2 in)-sized particles and 4.56 MJ/m³ (122 Btu/SCF) for the 9.4 mm (0.37 in)-sized particles. Thus fixed-bed gasification can be used as a feasible methodology to gasify coarse sized FB by blending it with high-energy content fuels like coal to generate a low-Btu gas.

The above-mentioned results for co-firing and gasification suggest that both of these technologies have the potential to provide an alternative environmentally sound method for disposing of animal waste. But before these techniques reach commercial-scale testing, economic analyses need to be done to determine the feasibility of such technologies.

Economic Analyses for Feedlot Biomass (FB)

Figure 16.7 shows the proposed 'coal:biomass blend energy conversion technology'. It can be observed that fly ash from power plants is collected and recycled and used as fertilizer or a paving surface for feedlot/poultry houses. Amosson *et al.* (1999) conducted an economic survey of cattle feedlots in the Texas Panhandle to determine the current cost of manure handling and delivery to farmers. In all, 47 feedlots were represented in the survey. Manure handling costs for feedlots larger than 15,000 head averaged US$0.05/ton paid by the feedlots to the contract manure collectors/haulers and in turn the manure contractors charged the farmers an average of US$2.28/ton plus US$0.11/ton-mile (one-way haul distance basis) for hauling and spreading. Haul distances to farmland for manure use as fertilizers reportedly averaged 14.5 kilometres (km) or 9 miles, with a range of 0.4 km to 80.5 km (0.25 mile to 50 miles). Of the feedlots surveyed, 77% reported that most of the manure is applied to irrigated farmland. The remaining feedlots indicated that a majority of manure was applied to predominately non-irrigated cropland (Amosson *et al.*, 1995). However, the region suffers from a declining water table, which indicates a reduction of irrigated acreage in the future, which will probably lessen the farmer demand for manure.

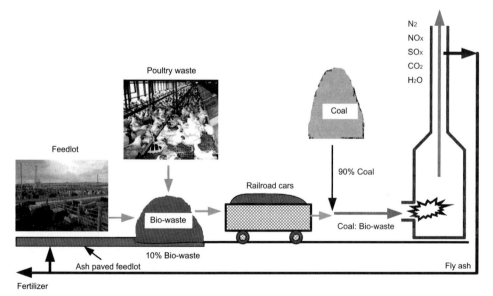

Figure 16.7. Coal-biomass blend energy conversion technology. Biomass is collected from feedlots, transported via trains/trucks to power plants, mixed with coal, fired in existing burners, fly ash collected and recycled back to feedlot/poultry house and excess used for land reclamation and/or as fertilizer.

An economic analysis of feedlot biomass management at both conventional unpaved and paved feedlots was conducted and the CO_2 and SO_2 emissions for a 90% coal:10% manure blend fuel for a 2,146 MW electric power plant operating at 40% efficiency were calculated (I. Harahap, Department of Mechanical Engineering, Texas A&M University, College Station, Texas, 2000, personal communication). The economic analysis included the cost of feedlot biomass collection and transportation, cost savings for the plant and ash produced for disposal or utilization. Collected manure from unpaved and from fly ash-paved feedlots was assumed to have moisture contents of 6.8% and 10% and ash contents of 42.3% and 25%, respectively. Table 16.2 shows the details of the comparative energy cost (US$/MJ) for both paved and unpaved feedlots as compared to coal. The calculations have been performed for a total hauling distance of 30 miles by both trains and trucks. For trains, the number of cars per train was assumed constant at 110, with each car having a hauling capacity of 700 tons. For hauling by truck, each truck was assumed to have a hauling capacity of 20 tons. The transportation cost per mile was assumed to be US$0.12/mile and US$0.11/mile for trains and trucks, respectively. A miscellaneous overhead cost of US$1.00/ton was also included to take into account maintenance, testing and other hidden costs.

Results for the 2,146 MW unit showed that using both unpaved feedlot biomass (UFB) and paved feedlot biomass (PFB) as 10% fuel in the coal:FB blend resulted in a fuel cost reduction of 1.8% and 4.0% as compared to pure coal firing, assuming that the FB is supplied by train. If the same biomass fuels are supplied by trucks then the savings are 2.0% for UFB and 4.1% for PFB compared to pure coal firing. A concern is the increased ash generation requiring marketing for disposal utilization. The ash generation increases by 77.5% for UFB blend and 21.7% for PFB blend as compared to pure coal firing. CO_2 emissions were projected to be 4.7% less for UFB blend and 6.8% less for PFB blend than for pure coal firing, while SO_2 emissions would increase by 20.0% and 28.8% for UFB blends and PFB blends,

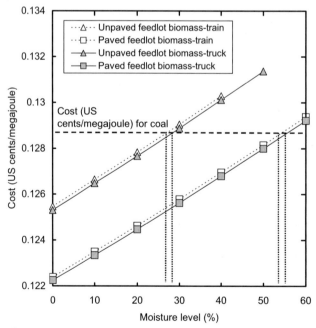

Note: Multiply US$/gigajoules by 1.06 to get US$/million Btu.

Figure 16.8. Variation of fuel cost with biomass moisture level, for a hauling distance of 30 miles.

Table 16.2: Comparative economic analysis for unpaved feedlot biomass (UFB) and paved feedlot biomass (PFB) blends.

Parameter	Coal	90:10 UFB blend	90:10 PFB blend
Electrical capacity (megawatts)	2146	2146	2146
Operating efficiency (%)	0.4	0.4	0.4
Moisture in fuel (%)	22.8	21.2	21.5
Ash in fuel (%, as received basis)	5.4	9.1	6.4
Higher heating value (megajoules/kilogram, as received basis)	21.38	20.20	20.65
Fuel feed rate (kilogram/second, as received basis)	250.89	265.57	259.76
Fuel consumption (million tons/year, as received basis)	8.70	9.21	9.01
Coal consumption (million tons/year, as received basis)	8.70	8.29	8.11
Manure consumption (million tons/year, as received basis)	0.00	0.92	0.90
Ash content (kilogram/kilogram, as received basis)	0.05	0.09	0.06
Ash from coal (million tons/year, as received basis)	0.47	0.45	0.44
Ash from manure (million tons/year, as received basis)	0.00	0.39	0.14
Total ash generated (million tons/year)	0.47	0.84	0.58
Fuel supply cost (train only)			
Million US$/year	217.58	213.62	208.94
Fuel cost/ton (US$/ton)	25.00	23.19	23.19
Cost/gigajoule (US cents/gigajoule) – train only	128.60	126.26	123.50
Fuel cost savings (million US$/year) – train only	N/A	3.96	8.64
Number of trains/year	N/A	12	12
Number of empty cars per train (from plant to feedlot, including coal ash also)	N/A	10	40
Daily frequency of train (train/day)	N/A	0.03	0.03
Fuel supply cost (truck only)			
Million US$/year	217.58	213.34	208.67
Fuel cost/ton (US$/ton)	25.00	23.16	23.16
Cost/gigajoule (US cents/gigajoule) – truck only	128.60	126.10	123.34
Fuel cost savings (million US$/year) – truck only	N/A	4.24	8.91
Number of trucks /year	N/A	46063	45054
Number of empty trucks/year (from plant to feedlot, including coal ash)	N/A	41857	42171
Daily frequency of trucks (trucks/day)	N/A	126	123
Total annual moisture release (million tons/year, as received basis)	1.98	1.95	1.94
CO_2 emission (kilogram/gigajoule)	92.70	92.65	92.63
Reduction in CO_2 emission (%), assuming biomass to be CO_2 neutral	0	4.73	6.82
SO_2 emission (kilogram/gigajoule)	0.36	0.43	0.46
Increase in SO_2 emission (%)	N/A	20.01	28.83

Note: multiply megajoules/kilogram by 433 to get Btu/pound; multiply kilogram/gigajoule by 2.33 to get pounds/million Btu; multiply US$ per gigajoule by 1.06 to get US$/million Btu. N/A = not applicable.

respectively. For the case of hauling by train, the breakeven moisture level is 26% for UFB blend and 54% for the PFB blend for a total hauling distance of 30 miles (Fig. 16.8). To be economically competitive with coal, the economical hauling distance by train for the UFB blend is roughly 70 miles as compared to around 110 miles for PFB blends (Fig. 16.9). The cost per GJ for the 10% PFB blend is US$123.5, for the 10% UFB blend is US$126.26 and US$128.60 for pure coal. Overall, the calculations estimate that use of a 10% blend of PFB with coal is the most economically advantageous compared to both the 10% UFB blend and pure coal cases.

Economic Analysis for Poultry Litter Biomass (LB)

Figure 16.10 shows contract growers' farms (located within the 25-mile radius of a power plant) for the firm, feed mill location and location of power plants in and around Brazos County, Texas. Based upon the information provided by the firms on litter production and removal from the broiler houses, it was assumed that each farm had an average of 6.6 broiler houses, producing 200 tons of litter/house/year at a litter removal cost of US$10/ton. In addition, the cost of transportation to the plants was assumed to be US$2.5/mile for trucks with a litter biomass carrying capacity of 20 tons (i.e., US$0.125/ton-mile).

From Table 16.1, the heating value of the combustibles in LB-A is calculated to be 19.60 MJ/kg. Assuming the heating value of the combustibles to be constant and a typical ash content of 25% and moisture content of 15% and including an additional overhead cost of US$1.00/ton of LB, along with the other costs mentioned previously, the fuel cost analysis is reported in Table 16.3. From Table 16.3 it can be seen that the cost of both coal and blends are almost similar on a US cents/MJ basis; the broiler house operator will just recover the cost of the LB supply cost. It can also be observed that there is a reduction of 5.8% in CO_2 emissions and an increase in 12.6% and 42.3% in SO_2 and ash.

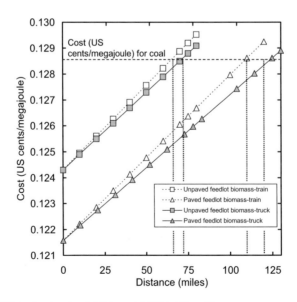

Note: Multiply US$/gigajoules by 1.06 to get US$/million Btu.

Figure 16.9. Variation of fuel cost with hauling distance.

Table 16.3. Comparative analysis for 90:10 LB blend

Parameter	Coal	90:10 LB blend
Fuel cost/megajoule (US cents/megajoule)	0.13	0.13
CO_2 emission (kilogram/gigajoule)	9.27	9.26
Reduction in CO_2 emission (%)[a]	N/A	5.76
SO_2 emission (kilogram/gigajoule)	0.36	0.41
Increase in SO_2 emission (%)	N/A	12.61
Increase in ash (%)	N/A	42.32

[a] Assuming manure to be CO_2 neutral.
N/A = not applicable.
Multiply cents/megajoules by 1060 to get US cents/million Btu; multiply kilogram/gigajoule by 2.33 to get pound/million Btu.

Figure 16.11 shows the potential profit for a broiler house operator, assuming that the LB is supplied at the same cost as coal, which is US$1.28/GJ. It is also assumed that the ash in LB is 25% and the heating value of pure combustibles is fixed at 19.60 MJ/kg. As expected, an increase in hauling distance decreases the maximum moisture content in the fuel which can be transported at profit for the broiler house.

Conclusion

LB and FB have greater volatile matter (80%) on a dry ash-free basis as compared to coal (45%). Both biomass fuels lose volatiles more rapidly and at lower temperatures (520 K) than coal (670 K). The heating values of FB and LB are much lower than coal, so the mass flow of

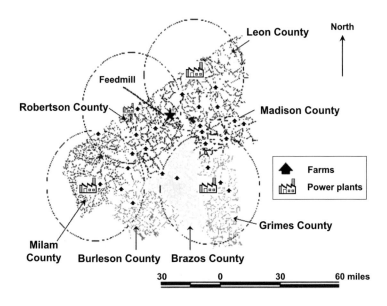

Note: Circles in figure have 25 mile radii.

Figure 16.10. Location of Texas poultry farms relative to coal combustion power plants.

fuel has to be increased in order to maintain the same heat throughput when firing blends. Co-firing results suggest that biomass blends and coal have a similar burnt mass fraction (0.95). During co-firing, even though the fuel nitrogen throughput increased, there was no appreciable change in NO_x emissions as compared to coal. However for all the fuels tested, NO_x increased as the excess air percentage was increased. Gasification of coal and FB blends suggest that FB can be co-gasified with coal to generate a low-Btu product gas which in turn can be used for direct burning to generate heat and power. The economic analysis suggests that it is feasible to co-fire both FB and LB in existing coal power plants. In addition, co-firing results in a decrease in CO_2 emissions, but an increase in SO_2 emissions. To make animal waste an alternative non-conventional source of renewable energy, the handling and transportation cost for these fuels has to be kept minimal. This can be achieved by decreasing the moisture and ash content of the biomass fuel.

Acknowledgements

We gratefully acknowledge the following sponsors: US Department of Agriculture through the National Center for Manure and Animal Waste Management at North Carolina, State of Texas Higher Education Coordinating Board-Advanced Technology Research Program and US Department of Energy contract DE-FG26-00NT40810.

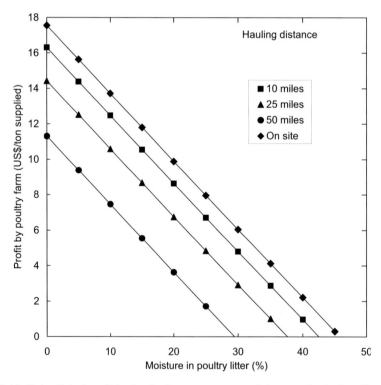

Figure 16.11. Potential of profit by broiler house versus moisture content of poultry litter.

References

Abbas, T., Costen, P., Kandamby, N., Lockwood, F. and Ou, J. (1994) The Influence of Injection Mode on Pulverized Coal and Biomass Co-Fired Flames. *Combustion and Flame* 99: 617-625.

Aerts, D., Bryden, K., Hoerning, J. and Ragland, K. (1997) Co-Firing Switch Grass in a 50 MW Pulverized Coal Boiler. *Proceedings of the 59th Annual American Power Conference, Chicago, Illinois* 59(2): 1180-1185.

Amosson, S., Smith, J. and Rauh, W. (1995) *Texas Crop and Livestock Enterprise Budgets: Texas High Plains*. Texas Agricultural Extension Service, College Station, Texas.

Amosson, S., Sweeten, J., Weinheimer, B. and Camarata, S. (1999) *Feedlot Manure Survey*. Report No. 99-35. Texas Agricultural Extension Service, Texas A&M University Agricultural Research and Extension Center at Amarillo and Texas Cattle Feeders Association, Amarillo, Texas.

Annamalai, K., Ibrahim, Y. and Sweeten, J. (1987) Experimental Studies on Combustion of Cattle Manure in a Fluidized Bed Combustor. *Journal of Energy Resources and Technology: Transactions Journal of the American Society of Mechanical Engineers* 109: 49-57.

Annamalai, K., Thien, B. and Sweeten, J. (2003) Co-Firing of Coal and Cattle Feedlot Biomass (FB) Fuels Part II: Performance Results from 100,000 Btu/h Laboratory Scale Boiler Burner. *Fuel* (82)10: 1183-1193.

Blasi, D., Signorelli, G. and Portoricco, G. (1999) Countercurrent Fixed-bed Gasification of Biomass at Laboratory Scale. *Industrial and Engineering Chemistry Research* 38: 2571-2581.

Collins, E., Barker, J., Carr, L., Brodie, H. and Martin, J. (1999) *Poultry Waste Management Handbook*. NRAES-132. National Resource, Agriculture and Engineering Service, Ithaca, New York, p. 64.

Gold B. and Tillman, D. (1996) Wood Cofiring Evaluation at TVA Power Plants. *Biomass Bioenergy* 10: 71-78.

Hansen, P., Anderson, K., Wieck-Hansen, K., Overgaard, P., Rasmussen, L., Frandsen, F., Hansen, L. and Dam-Johansen, L. (1998) Co-Firing Straw and Coal in a 150-MW Utility Boiler: *in situ* Measurements. *Fuel Processing Technology* 54: 207-225.

Katyal, S. and Iyer, P. (2000) Thermo Chemical Characterization of Pigeon Pea Stalk for its Efficient Utilization as an Energy Source. *Energy Sources* 22: 363-375.

Kayal, T. and Chakravarty, M. (1994) Mathematical Modeling of Continuous Updraft Gasification of Bundled Jute Stick - a Low Ash Content Woody Biomass. *Bioresource Technology* 49: 61-73.

Krishnamoorthy, P., Seetharamu, S. and Bhatt, M. (1989) Comparative Study of Biomass Fuels in an Updraft Gasifier. In: *Proceedings of the 24th Intersociety Energy Conversion Engineering Conference* 4: 1959-1964. [Accessed 2004.] Available from http://ieeexplore.ieee.org/iel5/852/2490/00074740.pdf?isNumber=2490&prod=CNF&arnumber=74740&arSt=1959&ared=1964+vol.4&arAuthor=Krishnamoorthy%2C+P.R.%3B+Seetharamu%2C+S.%3B+Siddhartha+Bhatt%2C+M

National Agricultural Statistics Service. (2003) *U.S. Poultry Charts*. US Department of Agriculture, National Agricultural Statistics Service, Washington, DC. [Accessed 2004.] Available from http://www.usda.gov/nass/aggraphs/poultry.htm

Ohlsson, O. (1994) *Results of Combustion and Emissions Testing when Co-Firing Blends of Binder-enhanced Densified Refuse-derived Fuel (b-dRDF) Pellets and Coal in a 440 MWe Cyclone Fired Combustor. Vol. 1, Test Methodology and Results*. Subcontract Report No. DE94000283. US Department of Energy, Argonne National Laboratory, Argonne, Illinois.

Patil, K. and Rao, C. (1993) Updraft Gasification of Agricultural Residues for Thermal Applications. In: Paul, P. and Mukunda, H. (eds.) *Proceedings of the Fourth National Meet on Recent Advances in Biomass Gasification and Combustion*. Interline Publishing, Bangalore, India, pp. 198-205.

Priyadarsan, S., Annamalai, K., Sweeten, J., Holtzapple, M. and Mukhtar, S. (2004a) Co-Gasification of Blended Coal with Feedlot and Chicken Litter Biomass. In: *Proceedings of the 30th International Symposium on Combustion*. The Combustion Institute, Chicago, Illinois, pp. 2973-2980.

Priyadarsan, S., Annamalai, K., Sweeten, J., Holtzapple, M. and Mukhtar, S. (2004b) Fixed Bed Gasification of Feedlot Manure and Poultry Litter Biomass. *Transactions Journal of the American Society of Agricultural Engineers* 47(5): 1689-1696.

Sami, M., Annamalai, K. and Wooldridge, M. (2001) A Review of Co-Firing of Coal: Bio-solid Fuels Cofiring. *Progress in Energy and Combustion Science* 27: 171-214.

Sampson, G., Richmond, A., Brewster, G. and Gasbarro, A. (1991) Co-firing of Wood Chips with Coal in Interior Alaska. *Forest Products Journal* 41(5): 53-56.

Spliethoff, H. and Hein, K. (1998) Effect of Co-Combustion of Biomass on Emissions in Pulverized Fuel Furnaces. *Fuel Processing Technology* 54(1-3): 189-205.

Sweeten, J. (1979) *Manure Management for Cattle Feedlots*. L-1094. Texas Agricultural Extension Service, Texas A&M University, College Station, Texas, p. 6.

Sweeten, J., Korenberg, J., LePori, W., Annamalai, K. and Parnell, C. (1986) Combustion of Cattle Feedlot Manure for Energy Production. *Energy in Agriculture* 5: 55-72.

Sweeten, J., Annamalai, K., Thien, B. and McDonald, L. (2003) Co-firing of Coal and Cattle Feedlot Biomass (FB) fuels. Part I. Feedlot Biomass (Cattle Manure) Fuel Quality and Characteristics. *Fuel* 82(10): 1167-1182.

Thien, B. (2002) *Cofiring with Coal-Feedlot Biomass Blends*. PhD dissertation, Texas A&M University, Department of Mechanical Engineering, College Station, Texas.

Chapter 17

Development of Genetically Engineered Stress Tolerant Ethanologenic Yeasts using Integrated Functional Genomics for Effective Biomass Conversion to Ethanol

Z. Lewis Liu and Patricia J. Slininger

Introduction

Biomass Pre-treatment Generates Fermentation Inhibitors

As interest in alternative energy sources increases, the significance of agriculture as an energy producer has been recognized. Renewable biomass including lignocellulosic materials and agricultural residues has become attractive as potential low cost feedstocks for bioethanol production. For economic reasons, dilute acid hydrolysis is commonly used to prepare the biomass degradation for enzymatic saccharification (conversion into sugar) and fermentation (Bothast and Saha, 1997; Saha, 2003). However, numerous side-products are generated by this pre-treatment, many of which inhibit microbial metabolism. More than 100 compounds have been detected to have potential inhibitory effects to microbial fermentation (Luo *et al.*, 2002). Among commonly recognized inhibitors, furfural and 5-hydroxymethylfurfural (HMF) are considered as representative inhibitors based on their overall effects to fermentation (Taherzadeh *et al.*, 2000). Other common inhibitors include acetic acid, cinnamic acid, coniferyl aldehyde, ethanol, ferulic acid, formic acid, levulinc acid and phenolics.

During biomass degradation with dilute acid treatment, furfural is mainly derived from pentose dehydration and HMF formed from dehydration of hexoses (Larsson *et al.*, 1999; Lewkowski, 2001). These compounds reduce enzymatic and biological activities, break down DNA and inhibit protein and RNA synthesis (Sanchez and Bautista, 1988; Khan and Hadi, 1994; Modig *et al.*, 2002). Most yeasts, including industrial strains, are susceptible to these inhibitors generated from acid hydrolysis pre-treatment and especially susceptible to the presence of multiple inhibitory compounds (Taherzadeh *et al.*, 2000; Liu *et al.*, 2003; Martin and Jonsson, 2003). To facilitate fermentation processes, additional treatments are often needed to remediate these inhibitory compounds including physical, chemical, or biochemical detoxification procedures. However, these additional steps add cost and complexity to the process and generate additional waste products.

© CAB International 2005. *Agriculture as a Producer and Consumer of Energy* (eds J.L. Outlaw, K.J. Collins and J.A. Duffield)

Microbial Performance is the Key for Improvement

The economics of fermentation-based bioprocesses rely extensively on the performance of microbial biocatalysts available for industrial application. This is true for bioethanol production as well and the improvement of microbial performance will be the key to future commercial sustainability of a biomass-to-ethanol industry. Fermentation is among the oldest microbial applications in human history. Although a tremendous amount of knowledge has been accumulated through years of experience and development of modern technology, each alternative fermentation process remains a 'black box' that challenges scientists to look deeper inside and develop a more thorough understanding of how it works and how to make it work better. The fast life cycle and genetic diversity of microorganisms are invaluable resources that will lead to more cost effective processes in the future. The more one understands about the genetic blueprint of microbial process, the more effectively yeast can be genetically engineered for improvement that will enhance profitability.

Genetically manipulated yeast strains have shown enhanced properties for ethanol fermentation through improved utilization of starch, lactose (milk sugar) and xylose (a five-carbon sugar), as well as enzyme production (Ho *et al.*, 1998; Jeffries and Shi, 1999; Ostergaard *et al.*, 2000; Hahn-Hägerdal *et al.*, 2001). Development of genetically engineered strains with greater inhibitor tolerance, especially to furfural and HMF, is a promising alternative to the additional detoxification steps (Liu and Slininger, 2004; Liu *et al.*, 2004a). However, development of such strains is hindered by a lack of understanding of the basic mechanisms underlying stress tolerance in microorganisms. Few yeast strains genetically improved for inhibitor tolerance are available.

Why Integrated Functional Genomics?

Traditional single-gene studies have contributed significantly to the understanding of gene functions and gene regulatory rules in the past and will continue to do so in the future. However, a biological process often cannot be explained by a single gene's function. Life, as a complex dynamic system, functions under an integrated control programme rather than as an isolated event. Considering thousands of genes in a genome are required to maintain a living yeast system, a significant single-gene alteration impacts other genes in the system. Gene interactions and intermediate metabolic products and constant adaptation to environmental changes add complexity to the metabolic network. For a better understanding of fermentation inhibitory stress tolerance, it is necessary to learn more about functions of key genes of interest as well as the regulatory elements and control system involved in integrated pathways at the genome level. Therefore, to better understand gene functions, it is impossible to have a complete interpretation without insight into integrated life events governed by the entire genome.

Advances in gene microarray technology have revolutionized the understanding of biology. Using this integrated technology as a tool, one can study global transcriptome profiling and gain insight into a living response to inhibitory stimulants such as furfural, HMF and other inhibitory complexes encountered in bioethanol fermentation.

In short, improvement of stress tolerant ethanologenic yeasts using integrated functional genomics is a necessary and practical approach for effective biomass conversion to ethanol. In order to successfully accomplish the mission objectives, it is necessary to clarify and address the basics and uncertainties concerning the conversion of the inhibitors, to evaluate yeast tolerance, to assess feasibility potential and to develop the proper methodology to be applied.

Metabolic Conversion Pathways of Inhibitors

Furfural

Furfural conversion to furfuryl alcohol by yeasts has been well established (Morimoto and Murakami, 1967; Palmqvist *et al.*, 1999). A recently demonstrated biochemical pathway suggested that furfural was first converted to furfryl alcohol and further reduced to pyromucic acid (Nemirovskii *et al.*, 1989). Under anaerobic conditions, furfural is reduced to furfuryl alcohol and oxidizing NADH (an electron carrier in oxidation-reduction reaction) formed in biosynthesis (Palmqvist *et al.*, 1999). During the furfural reduction phase, acetaldehyde and pyruvate are produced, but ATP (a molecule which carries energy in cells) is reduced and acetate production is delayed. Cell replication is inactivated and no glycerol produced. Although it has not been fully proven, the reduction of furfural to furfuryl alcohol is believed to be catalysed by NADH-dependent dehydrogenase.

HMF

Unlike furfural which is well studied, knowledge on HMF conversion is more limited. A commercial source for an HMF conversion product is not readily available. However, HMF and furfural are chemically related since both have a furan ring and an aldehyde group in their chemical structure. Following the furfural conversion route, HMF has been assumed to convert to HMF alcohol (Nemirovskii and Kostenko, 1991). Although most studies assumed this conversion, HMF alcohol has not been isolated from cultures nor its structure identified. Nemirovskii *et al.* (1989) proposed a general reduction model for furfural and HMF that the aldehyde is first reduced to an alcohol and then oxidized to an acid. However, a different metabolic pathway of HMF other than that analogous to the furfural route was suggested when a conversion product of HMF was not recovered (Sanchez and Bautista, 1988). The mechanism of HMF metabolism and its involvement in the fermentation pathway are not clear and need to be clarified.

Identification of 2,5-bis-hydroxymethylfuran

A comparative study of HMF tolerance for ethanologenic yeasts led to consistent observations of a metabolite of HMF using high-performance liquid chromatography (HPLC) analysis. The accumulation of this HMF-associated compound in the medium appeared not to inhibit cell growth and final ethanol production. Once formed, the product was persistent throughout the fermentation. Analysis of fermentation broth supernatant by reverse phase HPLC with UV detection showed that HMF absorbed at 282 nanometers (nm) at the beginning of incubation. At 48 hours, the end of the fermentation, HMF was no longer detected in the broth, but a new peak, the HMF-associated product, was detected at 222 nm. This indicated that HMF was completely converted to another compound 48 hours after incubation.

When samples were analysed using a Biorad Fast Acid HPLC column and measured with a refractive index detector, HMF had a retention time of 11.0 minutes (min) and the conversion product a retention time of 6.2 min. The metabolite was purified from lyophilized medium by preparative HPLC. The pure material formed light crystals after drying by lyophilization. An aliquot was resuspended in methanol and analysed by gas chromatography-mass spectrometry (GC-MS). The resulting chromatogram showed a single peak with a molecular mass ion of 128 (Liu et al., 2004b). The signals for the aldehyde proton and the asymmetric spectra of HMF were absent when the metabolite was analysed using nuclear magnetic resonance (NMR). The NMR spectra are consistent with that of a symmetrical molecule, with a furan ring.

The metabolite from HMF converted by yeasts was isolated and purified. Its chemical structure was characterized using mass and NMR spectra analysis and identified as 2,5-bis-hydroxymethylfuran (also termed as furan-2,5-dimethanol, FDM) (Liu et al., 2004b). The chemical structure of the metabolite was identified as a compound with a composition of $C_6H_8O_3$ and a molecular weight of 128 (Fig. 17.1). This rigorous chemical identification of 2,5-bis-hydroxymethylfuran is the first chemical proof of the structure of the metabolic reduction product of HMF. It clarifies the literature and provides basic evidence for a fermentation pathway that yeasts can utilize to respond to the presence of HMF.

Biotransformation of HMF

With the clarification of an end conversion product of HMF, it can be concluded that in the presence of HMF, ethanologenic yeasts reduce the aldehyde group on the furan ring of the HMF into an alcohol. This is analogous to the reduction of furfural into furfuryl alcohol. It was further demonstrated that this transformation is a yeast-catalysed process and requires active cell growth and metabolism (Liu et al., 2004a). In culture medium amended with a lethal high dose of 120 millimole (mM) HMF, no cell growth occurred even after 7 days incubation; and as measured by HPLC, glucose and HMF remained at the original concentrations and no HMF conversion ever happened.

It is interesting to note that after a prolonged lag phase caused by exposure to HMF at sublethal doses, cell growth resumed and glucose consumption and ethanol production finally reached the maximum. This period of bioactivity was relatively shorter than that for the control culture. It appeared that yeasts were able to adapt to inhibitors at the tolerable concentrations.

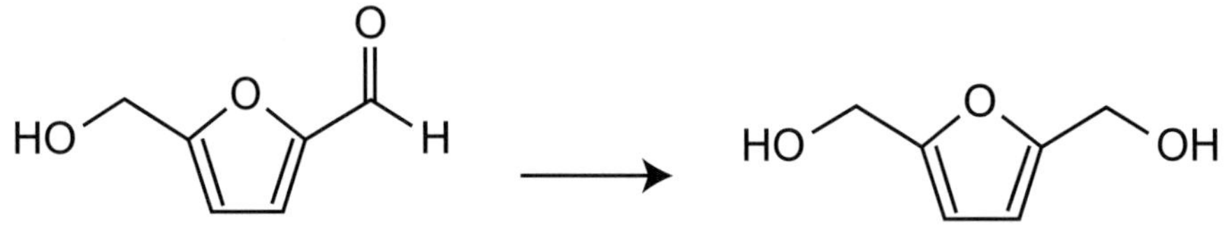

Source: Liu *et al.* (2004b).

Figure 17.1. A schematic diagram showing conversion of 5-hydroxymethylfurfural (1) to 2,5-bis-hydroxymethylfuran (2) by yeasts.

Adaptive Response of Ethanologenic Yeasts to the Inhibitors

In a preliminary study, responses to numerous inhibitors for several ethanologenic yeasts were evaluated including strains of *Candida shehatae, C. tropicalis, C. wickerhamii, Clavispora lusitaniae, Kluyveromyces marxianus, Metschnikowia pulcherrima, Pachysolen tannophilus, Pichia farinose, P. membranifaciens, P. stipitis, Saccharomyces bayanus, S. cerevisiae, Schizosaccharomyces pombe* and *Torulaspora delbrueckii*. Based on results of this study, selected strains of *S. cerevisiae* and *P. stipitis* were identified as more tolerant to inhibitors and worthy of further investigations to study their responses to furfural, to HMF and to the combination of the two inhibitors. *P. stipitis* is a natural pentose-fermenting yeast of potential importance to efficient biomass conversion to ethanol. *S. cerevisiae* is a well accepted hexose-fermenting yeast and in general, shows more tolerance to a broad range of inhibitors than other strains. For an economic biomass conversion to ethanol process, the utilization of both pentoses and hexoses is required. Therefore, both yeasts are potential candidates for further improvement on the inhibitor tolerance from biomass hydrolysates.

Using a series of concentrations of inhibitors incorporated into a synthetic medium, this research demonstrated and characterized a dose-dependent response to furfural and HMF at concentrations from 10 mM to 120 mM (Liu *et al.*, 2003). Given a tolerable concentration, yeast strains were able to recover from a prolonged lag phase during the initial stage of the incubation. The lag phase lasted from a few hours to several days depending upon the amount of inhibitor added to the cultures. Once the cell growth recovered, cultures inoculated with all strains were able to consume glucose and thereafter produce ethanol. This demonstrated a clear dose-dependent inhibition of the yeasts by furfural and HMF. In a synthetic medium, complete suppression of cell growth was observed at 60 mM and 120 mM for furfural and HMF, respectively. When both inhibitors were applied in combination, cell growth was only recovered at 10 mM of each inhibitor, which indicated these inhibitors acted in a negative synergistic fashion even at low concentrations. This negative synergy suggests that the inhibitors may act by different mechanisms or that yeast strains use different processes to adjust to the presence of the combined inhibitors. Kinetic inhibition effects of furfural and HMF are generally known to yeast fermentation (Palmqvist *et al.*, 1999; Taherzadeh *et al.*, 2000; Martin and Jonsson, 2003). These results also demonstrated that the length of the lag phase in batch culture fermentation of both *S. cerevisiae* and *P. stipitis* was dependent on the dosages of furfural and HMF.

The duration of the lag phase may be interpreted as a measurement of varied levels of tolerance to furfural and/or HMF. Cells that did not survive the inhibitor stress simply died and no metabolic activity was detected. Only adapted cells survived and functioned further to produce ethanol. This interpretation further suggests that some yeast strains have more effective mechanisms to withstand these inhibitors than others. The prolonged lag phase before the recovery of the cell growth could also reflect a shift in the physiology of the cells adapting to the chemical stress. Enzymatic induction was suggested during the lag phase adaptation (Kang and Okada, 1973; Liu *et al.*, 2004b). Important metabolic enzymes including alcohol dehydrogenase, aldehyde dehydrogenase and pyruvate dehydrogenase have been reported to be inhibited *in vitro* by furfural and HMF (Modig *et al.*, 2002). The persistence of specific altered gene expression over time supports the hypothesis that yeasts are stimulated to undergo an adaptive process in response to HMF (Liu and Slininger, 2004). Using a directed adaptation strategy, recent research developed several adapted strains with increased tolerance to furfural and HMF (Liu *et al.*, 2004a).

Potential for Yeast Tolerance Improvement

Enhanced Tolerance and Inhibitor Detoxification by Biotransformation

Based on observations of a dose-dependent yeast response to furfural and HMF, a directed adaptation method was developed and several strains more tolerant to furfural and HMF each individually and in combination were generated. These adapted ethanologenic yeast strains were evaluated, demonstrating their significantly higher levels of tolerance to furfural and HMF compared with the parental strains under controlled conditions. These adapted, more tolerant strains showed no significant delay in cell growth and glucose consumption. They produced normal ethanol yield in the presence of furfural or HMF. More significantly, these adapted strains were more tolerant to the inhibitors and showed enhanced biotransformation ability to convert furfural into furfuryl alcohol and HMF into 2,5-bis-hydroxymethylfuran compared with that of the parental strains.

The adapted strains had a nearly normal cell growth in contrast to a prolonged lag phase of the parental strains under the challenge of furfural and HMF. This indicated a qualitative change in cell response to the inhibitors. The adapted strains were observed to be different in metabolic profiles as measured by HPLC compared with their parental strains. *S. cerevisiae* 307-12H60 (Fig. 17.2D), 307-12H120 and *P. stipitis* 307-10H60 showed enhanced biotransformation ability to reduce HMF to 2,5-bis-hydroxymethylfuran at 30 mM (Fig. 17.2D) and 60 mM; and *S. cerevisiae* 307-12-F40 (Fig. 17.2B) converted furfural into furfuryl

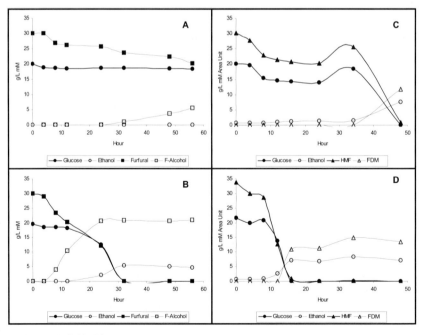

Source: Liu *et al.* (2004b).

Figure 17.2. Comparison of inhibitor tolerance and biotransformation of furfural (A, B) and 5-hydroxymethylfurfural (C, D) between wild type (A, C) and improved tolerant strains (B, D) of *Saccharomyces cerevisiae*.

alcohol at significantly higher rates compared to the parental strains (Fig. 17.2A and Fig. 17.2C). Strains of *S. cerevisiae* converted 100% of HMF at 60 mM into 2,5-bis-hydroxymethylfuran and 307-12-F40 converted 100% of furfural into furfuryl alcohol at 30 mM. Such significant enhancement of biotransformation to reduce these inhibitors by these adapted tolerant strains suggests the high potential for *in situ* detoxification using more tolerant yeasts as a means for more efficient bioethanol production.

Tolerance Improvement Potential

Microorganisms including yeasts live in an ever-changing environment and have to constantly adapt to a specific environmental change for survival. As documented in numerous reports, yeast adaptation to a stress condition is common and accomplished via a variety of molecular mechanisms (Gasch and Werner-Washburne, 2002; Erasmus *et al.*, 2003). Global gene expression analysis supports the existence of yeast adaptation responses to stress conditions (Liu and Slininger, 2004). An adapted *P. stipitis* was also previously reported to have improved ethanol production from hemicellulose hydrolysate (Nigam, 2001). It appeared that adaptation can be an alternative means to improve microbial strain improvement. The application of the directed adaptation method that was developed provided strains of *S. cerevisiae* and *P. stipitis* which were more tolerant to furfural and HMF inhibitor by-products of lignocellulosic acid hydrolysis. As a conclusive remark, yeasts have internal genetic potential for further tolerance improvement against inhibitors such as furfural and HMF.

These adapted, more tolerant strains pose relatively high levels of tolerance to single inhibitors. However, they have not been tested against inhibitor complexes such as those in biomass hydrolysates and may require further directed adaptation where inhibitor complexes provide the selection pressure for adaptation instead of only individual inhibitors. However, single inhibitor-tolerant strains are necessary for studies in dissection of mechanisms of stress tolerance to the multiple inhibitor complexes. These adapted, more tolerant strains will be valuable resources in our upcoming studies of molecular mechanisms of stress tolerance using functional genomics.

By-product Utilization Issues

When using HMF-containing medium for fermentation, it was found that the metabolic profiles of the yeast fermentation also changed. In a 30 mM HMF-treated culture, the yield of acetic acid was found to be significantly higher than in cultures without HMF (unpublished data). The ethanol yield did not appear to be adversely affected by the presence of HMF or by the acetic acid production. The current yield of acetic acid is relatively low compared with that of the main stream production for the acid. However, if it can be separated as a by-product, in addition to the ethanol main product, the utilization of by-products can help to reduce the total cost of the bioethanol production. Biomass hydrolysates bring new ingredients including inhibitors into the fermentation. This might raise additional new challenges in the future development. However, for the potential acetic acid by-product utilization, an economic feasibility evaluation is needed once the acetic acid yield and its kinetic potentials are known.

Understanding Tolerance Mechanisms using Functional Genomics

No information is available for genomic expression studies in response to this chemical inhibitory stress for bioethanol fermentation. A research programme has been initiated to study the genetic basis in metabolic pathways underlying the inhibitor tolerance using functional genomics for ethanologenic yeasts. Building upon the current understanding of molecular mechanisms of the tolerance, plans are underway to genetically engineer more robust novel strains to be suitable for use in bioethanol production from biomass hydrolysates.

Quality Control Development for Microarray Studies

Currently, the most up-to-date 70-mer DNA oligomers are used representing 6,388 genes for the complete genome of *S. cerevisiae* (QIAGEN Operon, Alameda, California). Optimized 70-mer oligonucleotides with 5′ amino-linked sequence were designed which were synthesized for three soybean genes, MSG (major latex protein), CAB (photosystem I chlorophyll ab binding protein) and RBS1 (ribulose bisphosphate carboxylase small chain 1 precursor). These sequences have no homologous similarity in the yeast genome and therefore avoid interference with the yeast expression signal. Each synthesized oligo control gene was included in a sub-block of the microarray gene chip with a total of 32 spots serving as exogenous nucleic acid controls. *In vitro* transcribed mRNA of the three exogenous control genes was spiked into each RNA probe labelling reaction at 10 picograms (pg), 100 pg and 1000 pg. The linearity of signal intensity of the control genes allowed them to serve as a quantitative reference for gene expression measurement (Fig. 17.3). A slide gel electrophoresis system, called sGel, was developed to evaluate the quality of labelled probes in addition to the quantification of the labelled probe by NanoDrop spectrophotometer. Such quality control

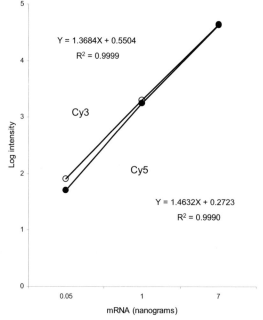

Source: Liu and Slininger (2004).

Figure 17.3. Signal linearity of synthesized control genes for microarray.

measurements provided a normalization reference and allowed an estimate of variability, reduced variation and increased reliability and reproducibility of the microarray data.

Genomic Expression in Response to the Inhibitors

Fermentation profiles and genomic expression profiles of wild type with and without HMF treatment over time were compared. This provided a view of potential candidate genes responsible for the inhibitory HMF stimulant. Microarray experiments were conducted comparing data of HMF-treated cultures between wild type and the adapted, more tolerant strains over time. From the second set of data, the tests were able to distinguish genes responsible for the HMF tolerance relative to the wild type under the same suppressive challenge by HMF. The adapted, more tolerant strains showed distinct expression profiles of selected genes compared with that of the parental strain (Fig. 17.4) by the preliminary analysis.

Actually, tested strains responded in more complex patterns of transcriptional expression, which involved a significantly large number of genes. The expression levels of at least several hundred genes were significantly different between wild type and the newly developed strains at various times after the HMF challenge (Liu and Slininger, 2004). Genes in all categories of biological process, cellular component and molecular function were involved, among which, some genes appeared to be HMF-specific while others were shared with those in a core set of stress genes. Yet, interpretation of some genes was limited by incomplete annotations or lack of known functions. In addition, for those enhanced or repressed genes, some were constantly expressed as up- or down-regulated, but some of them were able to resume their original functions over time. These dynamic expression patterns reflected the complexity of the yeast response to the HMF stress, which involved multiple layers of pathways.

Among the abundance of transcriptionally expressed genes, genes with limited annotation and unknown functions were commonly found. The process of data analysis is on-going. It remains challenging to assign complete functions for these genes. Thanks to the comprehensive database available for analysis of yeast genomic studies, the global expression was explored enabling additional insight into the complexity of HMF stress response. However, a great deal of detailed knowledge is unknown. Challenges remain to draw meaningful conclusions from the vast amount of data collected.

Research is currently underway studying tolerance mechanisms for individual inhibitors and will progressively edge into more challenging studies of yeast tolerance to combined inhibitors during bioethanol fermentation. The dynamics of the gene transcriptional activities underlying the tolerance mechanisms to HMF, furfural and inhibitor complex will be illustrated. Newly induced functions, repressed functions and genetically altered metabolic events are expected. Key genes and regulatory components responsible for the biological conversion of the inhibitor and the major tolerance mechanisms will be elucidated.

Novel Strain Design and Genetic Engineering

A better understanding of the genetic mechanisms and biochemical pathways responsible for the tolerance in yeasts will provide a sound foundation to develop genetically engineered novel strains to withstand major inhibitors generated from the biomass pre-treatment and fermentation process. With the progress of our functional genomic studies and in-depth understanding of the mechanisms of inhibitory stress tolerance involved in fermentation, additional research will be able to develop new hypotheses to design genetically novel more

1. Wild type, HMF_0 mM 0.5 hour
2. Wild type, HMF_30 mM 0.5 hour
3. 307-12H60, HMF_30 mM 0.5 hour
4. Wild type, HMF_30 mM 2 hour
5. 307-10H60, HMF_30 mM 0.5 hour
6. 307-10H60, HMF_30 mM 2hour
7. 307-12H60, HMF_30 mM 2 hour
8. Wild type, HMF_0 mM 2 hour
9. Wild type, HMF_0 mM, 0 hour

Source: Liu and Slininger (2004).

Figure 17.4. Relationship of selected gene expression between wild type and tolerant strains in response to HMF.

robust yeast strains for more efficient ethanol production. A series of biological experiments will be conducted to confirm the new findings as a validation step during the hypothesis development. Novel strains which withstand and detoxify the inhibitors *in situ* during fermentation will be generated using recombinant DNA technologies. Working towards a new biology millennium, current expectations are to use integrated approaches and extend these efforts through multidisciplinary collaborations to accomplish the mission of more efficient bioethanol production.

References

Bothast, R. and Saha, B. (1997) Ethanol production from Agricultural Biomass Substrate. *Advances in Applied Microbiology* 44: 261-286.

Erasmus, D., van der Merwe, G., and van Vuure, H. (2003) Genome-Wide Expression Analysis: Metabolic Adaptation of *Saccharomyces cerevisiae* to High Sugar Stress. *Yeast Research* 3: 375-399.

Gasch, A. and Werner-Washburne, M. (2002) The Genomics of Yeast Response to Environmental Stress and Starvation. *Functional & Integrative Genomics* 2 (4-5): 181-192.

Hahn-Hägerdal, B., Wahlbom, F., Gárdony, M., Van Zyl, W., Otero, R. and Jönsson, L. (2001) Metabolic Engineering of *Saccharomyces cerevisiae* for Xylose Utilization. *Advances in Biochemical Engineering/Biotechnology* 73: 53-84.

Ho, N., Chen, Z. and Brainard, A. (1998) Genetically Engineered *Saccharomyces* Yeast Capable of Effective Cofermentation of Glucose and Xylose. *Applied and Environmental Microbiology* 64 (5): 1852-1859.

Jeffries, T. and Shi, Q. (1999) Genetic Engineering for Improved Fermentation by Yeasts. *Advances in Biochemical Engineering/Biotechnology* 65: 117-161.

Kang, S. and Okada, H. (1973) Alcohol Dehydrogenase of *Cephalosporium* sp. Induced by Furfural Alcohol. *Journal of Fermentation Technology* 51 (2): 118-124.

Khan, Q. and Hadi, S. (1994) Inactivation and Repair of Bacteriophage Lambda by Furfural. *Biochemistry and Molecular Biology International* 32: 379-385.

Larsson, S., Palmqvist, E., Hahn-Hägerdal, B., Tengborg, C., Stenberg, K., Zacchi, G. and Nilvebrant, N. (1999) The Generation of Inhibitors During Dilute Acid Hydrolysis of Softwood. *Enzyme and Microbial Technology* 24: 151-159.

Lewkowski, J. (2001) Synthesis, Chemistry and Applications of 5-Hydroxymethylfurfural and Its Derivatives. *Arkivoc* 1: 17-54.

Liu, L. and Slininger, P. (2004) *Global Gene Expression Analysis of Ethanologenic Yeasts in Adaptation to Bioethanol Fermentation Inhibitory Stress*. Presented paper, Genome 2004: International Conference on the Analysis of Microbial and Other Genomes, April 14-17, Hinxton, Cambridge, UK.

Liu, L., Slininger, P., Dien, B., Kurtzman, C. and Gorsich, S. (2003) *Comparative Tolerance Studies Reveal Dose-dependent Inhibition of Ethanologenic Yeasts by Furfural and 5-Hydroxymethylfurfural*. Presented paper, 2003 Society of Industrial Microbiology Annual Meeting, August 10-14, Minneapolis, Minnesota.

Liu, L., Slininger, P. and Gorsich, S. (2004a) *Enhanced Biotransformation of Furfural and 5-Hydroxymethylfurfural by Newly Developed Ethanologenic Yeast Strains*. Presented paper, 26th Symposium on Biotechnology for Fuels and Chemicals, May 9-12, Chattanooga, Tennessee.

Liu, L., Slininger, P., Dien, B., Berhow, M., Kurtzman, C. and Gorsich, S. (2004b) Adaptive Response of Yeasts to Furfural and 5-Hydroxymethylfurfural and New Chemical Evidence for HMF Conversion to 2,5-Bis-hydroxymethylfuran. *Journal of Industrial Microbiology and Biotechnology* 31(8): 345-352.

Luo, C., Brink, D. and Blanch, H. (2002) Identification of Potential Fermentation Inhibitors in Conversion of Hybrid Poplar Hydrolyzate to Ethanol. *Biomass and Bioenergy* 22: 125-138.

Martin, C. and Jonsson, L. (2003) Comparison of the Resistance of Industrial and Laboratory Strains of *Saccharomyces* and *Zygosaccharomyces* to Lignocellulose-Derived Fermentation Inhibitors. *Enzyme and Microbial Technology* 32: 386-395.

Modig, T., Liden, G. and Taherzadeh, M. (2002) Inhibition Effects of Furfural on Alcohol Dehydrogenase, Aldehyde Dehydrogenase and Pyruvate Dehydrogenase. *The Biochemical Journal* 363 (3): 769-776.

Morimoto, S. and Murakami, M. (1967) Studies on Fermentation Products from Aldehyde by Microorganisms: the Fermentative Production of Furfural Alcohol from Furfural by Yeasts (part I). *Journal of Fermentation Technology* 45: 442-446.

Nemirovskii, V. and Kostenko, V. (1991) Transformation of Yeast Growth Inhibitors Which Occurs During Biochemical Processing of Wood Hydrolysates. *Gidroliz Lesokhimm Prom-st* 1: 16-17.

Nemirovskii, V., Gusarova, L., Rakhmilevich, Y., Sizov, A. and Kostenko, V. (1989) Pathways of Furfurol and Oxymethyl Furfurol Conversion in the Process of Fodder Yeast Cultivation. *Biotekhnologiya* 5: 285-289.

Nigam, J. (2001) Ethanol Production from Wheat Straw Hemicellulose Hydrolysate by *Pichia stipitis*. *Journal of Biotechnology* 87: 17-27.

Ostergaard, S., Olsson, L. and Nielsen, J. (2000) Metabolic Engineering of *Saccharomyces cerevisiae*. *Microbiology and Molecular Biology Reviews* 64: 34-50.

Palmqvist, E., Almeida, J. and Hahn-Hägerdal, B. (1999) Influence of Furfural on Anaerobic Glycolytic Kinetics of *Saccharomyces cerevisiae* in Batch Culture. *Biotechnology and Bioengineering* 62(4): 447-454.

Saha, B. (2003) Hemicellulose Bioconversion. *Journal of Industrial Microbiology and Biotechnology* 30: 279-291.

Sanchez, B. and Bautista, J. (1988) Effects of Furfural and 5-Hydroxymethylfurfrual on the Fermentation of *Saccharomyces cerevisiae* and Biomass Production from *Candida guilliermondii*. *Enzyme and Microbial Technology* 10: 315-318.

Taherzadeh, M., Gustafsson, L. and Niklasson, C. (2000) Physiological Effects of 5-Hydroxymethylfurfural on *Saccharomyces cerevisiae*. *Applied Microbiology and Biotechnology* 53: 701-708.

Chapter 18

Case Studies of Rural Electric Cooperatives' Experiences with Bioenergy

Carol E. Whitman

Introduction

Rural electric cooperatives, not-for-profit utilities which are owned by those they serve, are in the business of generating and delivering electric energy and therefore have significant interests in bioenergy as both producers and consumers. Often rural utilities are viewed as a geographically well-situated market for energy from agriculture, whether it is from ethanol, biodiesel, electricity, or related products. While many electric cooperatives have taken a lead in using bioenergy and have a vested interest in its commercial success, technical and market barriers to its use remain. As businesses, cooperatives apply the same considerations in their purchasing decisions in terms of cost, availability and performance as with other business structures. So, to supplant established energy products, new bioenergy products must be fully competitive. This chapter will first introduce electric cooperatives and their use of bioenergy and then share the practical experience of two cooperatives with biofuels and biopower.

Rural Electric Cooperatives

Rural electric cooperatives are part of a complex system of electricity generation and delivery serving the USA, from densely populated cities to remote rural areas. Electric cooperatives arose in rural areas in the 1930s where there were few customers per mile of line and for-profit utilities considered electricity delivery uneconomical. They are not-for-profit, owned by the consumer-members they serve.

Cooperatives both deliver and generate electricity. Distribution cooperatives deliver electricity to the end-user. In some areas, distribution cooperatives have formed generation and transmission cooperatives (G&Ts) that generate and transmit electricity to serve them.

Today, there are 930 cooperatives located in 47 states. They serve 37 million people or 12% of the population and over three-quarters of the continental US land mass. Each year in the USA, electric cooperatives generate 5% of the electricity produced, purchase the remainder of their needs from federal utilities and the wholesale power market and deliver 10% of the total kilowatt-hours sold. They own and maintain 43% or 2.4 million miles of the distribution lines in the USA.

© CAB International 2005. *Agriculture as a Producer and Consumer of Energy* (eds J.L. Outlaw, K.J. Collins and J.A. Duffield)

Bioenergy and Cooperatives

Bioenergy is a natural area of interest to rural electric cooperatives since it combines energy and agriculture – energy being the cooperatives' business and agriculture being the business of many of their consumer-members. Cooperatives currently use bioenergy as fuel for both transportation and generation and invest in it. The reasons why cooperatives support bioenergy vary and include legal requirements, commitment to community, environmental stewardship, interest by cooperative members and that it is simply good business (National Rural Electric Cooperative Association, 2004).

In 1993, members of the National Rural Electric Cooperative Association passed a national resolution urging electric cooperatives to use and promote the use of biofuels to address energy and environmental issues. Since that time, cooperative electric utilities have increased their use of biofuels. For example, one-fourth of the distribution cooperatives in Illinois currently use biodiesel fuels to support local farmers. Bioenergy is also a major focus for cooperatives' economic development programmes. In South Dakota, East River Electric Power Cooperative – the area's G&T – and its distribution cooperatives aided the financing of Dakota Ethanol, a maize ethanol cooperative in East River Electric Power Cooperative's service territory producing 40 million gallons of ethanol a year.

Renewable generation from biomass such as manure and other agricultural residues has also received increasing attention from cooperatives. Over the last 10 years, the Cooperative Research Network, the research arm of the electric cooperatives, has devoted a considerable amount of resources to develop biopower and enabling technologies such as a low-energy microturbine. Over 200 distribution cooperatives now offer green power programmes and the number is increasing. Although the cost of biopower continues to be higher than coal-based generation, advances in technology and production incentives can make electric power from manure digesters cost-competitive with central station generation (Electric Power Research Institute, 2000).

Bioenergy holds a great deal of promise and its costs are decreasing, but constraints to its use remain. Two cooperatives that have used bioenergy share their experiences below and identify the significant barriers.

Biodiesel Case Study

Southern Maryland Electric Cooperative (SMECO) is a large distribution cooperative serving about 300,000 people in a historically rural agricultural area near the Chesapeake Bay. Major crops of the southern Maryland peninsula are tobacco, maize and soybeans. The cooperative began using biodiesel fuel when it fell under the US Department of Energy's (US DOE) Alternative Fuel Transportation Program (AFTP) in the late 1990s and it had to acquire alternative-fuelled vehicles.

The AFTP is a provision of the *Energy Policy Act of 1992* intended to reduce US dependence on foreign oil and accelerate the development of an alternative fuels infrastructure and the availability of alternative-fuelled vehicles. Under the programme, alternative fuel providers that satisfy certain conditions, including businesses that import and sell electricity, must purchase alternative fuels and alternative-fuelled vehicles. As part of AFTP, US DOE also established a credit trading programme so that businesses that are unable to meet the programme's requirements may acquire needed credits from others.

Cooperative Experience with Biodiesel

The decision by SMECO to use biodiesel fuel under AFTP supported both local businesses and farmers. But SMECO found biodiesel carried substantial up-front costs as well as operational costs. In addition to higher costs for biodiesel fuel compared to diesel, SMECO had to deal with multiple suppliers, diminishing its ability to get better prices for large purchases. They also found that they had reduced fuel economy since biodiesel has 5% to 7% less energy per gallon of fuel than pure diesel (Herman and Associates, 2003). SMECO also had to purchase chemical treatments for the storage tanks to retard bacterial growth and remove water buildup.

Another concern for SMECO was engine warranties. Engine manufacturer information usually states that the company 'neither approves nor prohibits the use of biodiesel fuels' (Herman and Associates, 2003). It then generally states that failures caused by the use of biodiesel fuels would be outside the manufacturer's warranty. Warranties may cover biodiesel blends up to 5%, but distillate fuels are preferred. To add to concerns over engine warranties, SMECO could only buy B20, a 20% biodiesel blend; they had to blend the biodiesel themselves to 5%. They also found vehicles using biodiesel wore out the injectors in their engines.

Credit Trading As an Alternative

In the end, concerns of cost, availability and performance caused SMECO to look for other options to meet its AFTP requirements and SMECO has begun purchasing credits through AFTP's credit trading programme. The cooperative's cost of purchasing credits is almost equal to the additional cost of biodiesel fuel, so the up-front costs essentially cancel out. SMECO has also eliminated worries of invalidating its engine warranties and reduced vehicle maintenance costs and fuel storage management. Through the credit trading programme, SMECO is offsetting the costs of another covered fuel provider capable of exceeding its AFTP requirements.

Biomass-to-Electricity Case Study

Dairyland Power Cooperative (Dairyland) is a G&T that serves 25 distribution cooperatives and 20 municipal utilities in five states (Wisconsin, Iowa, Michigan, Illinois and Minnesota). In turn, the cooperatives and municipal utilities supply the energy needs of 575,000 people. Four of the states in its service territory – Wisconsin, Iowa, Illinois and Minnesota – have renewable energy targets for power sales. So, Dairyland began analysing available renewable resources to help meet its anticipated future energy needs and its members' renewable targets.

Dairyland's farmers also were looking for ways to offset increasing costs of on-farm waste management with the implementation of new federal confined animal feeding operation (CAFO) rules. As a result, Dairyland began getting calls from its distribution cooperative managers expressing interest in anaerobic manure digesters. Since dairy, pig and cattle farms are a significant part of the Dairyland system and their success is important to its own economic viability, Dairyland's engineers decided to look into anaerobic digester systems.

Cooperative Experience with Anaerobic Digesters

Dairyland's engineers soon discovered that every anaerobic digester system was different. Given the unreliability that several unique distributed systems could introduce to the electric grid and to take advantage of the economies of scale that a standard design provides, the engineers decided to take a systems approach.

Last year, Dairyland launched a 25 megawatt manure digester programme to be implemented over 5 years. It uses a proven thermophilic biogas technology to efficiently extract methane from farm waste for the production of renewable electricity. Dairyland has partnered with Microgy Cogeneration Systems, Inc., a subsidiary of Environmental Power Corporation and an expert in biogas technology, to operate and maintain the digester. The dairy or pig farmer will own the manure digester system and will sell the methane to Dairyland. The farmer is responsible for normal manure management, nothing else. Dairyland owns and operates the generator.

Each digester producing methane gas for a 0.775 megawatt generator should produce about 6,500 megawatt-hours of electricity annually from a 1,000 head dairy herd. Manure is a steady and predictable fuel source at a dairy or feedlot. At a target price of 4.5 cents per kilowatt-hour, the production price of generation is equivalent to the national average. The cost for new coal-based generation is comparable to the price of wind power (Ganley, 2004). However, electricity generated from methane digesters still costs 50% more than electricity from the existing fleet of coal-based generation, the benchmark for utilities. Unlike renewable energy sources such as wind or solar, electricity from a methane digester is dispatchable and can be produced on demand.

Technology Pros and Cons

The cookie cutter approach adopted by Dairyland provides economies of scale through consistency of design and equipment, equipment that is transportable between sites, simplified interconnections to the electric grid, discounts on volume purchases and reductions in engineering and legal costs. In addition, the programme is portable and may be copied by others. The cooperative anticipates significant carbon reductions from methane, a greenhouse gas 23 times more potent than carbon dioxide (Intergovernmental Panel on Climate Change, 2001). But it has also found that the capital costs of the project are high and has had permitting issues due to unfamiliarity with the digester technology. In some of Dairyland's service areas, transmission constraints have prevented projects from being considered.

Stability of the farm is critical to a project's success. The revenue from the methane can offset the farmer's manure management costs and provide ancillary benefits to the farmer such as cogeneration of heat for hot water or other on-farm uses, pathogen control and odour mitigation.

Dairyland has developed a solution that will work for it and, hopefully, dairy and pig farmers in the Dairyland service territory.

Conclusions

Cooperatives' experience with bioenergy shows that there are still some practical cost, availability and performance constraints to its widespread use. Biodiesel has high costs and inconveniences due to a poor distribution infrastructure which reduce economies of scale in purchasing and contributes to issues for consumers. Perhaps a more significant issue for vehicle owners is engine warranties. Very few owners will knowingly invalidate their engine warranties, a potentially very expensive proposition. Other engine performance issues and bacterial growth are less serious and will be overcome with technological advances.

In the case of anaerobic digestion, the higher cost of generation remains an issue, but is offset by the fact that it is a reliable electricity source, not intermittent like wind or solar power. More significant issues are methane digesters' high capital costs and the impact of transmission constraints on utilizing distributed generation.

In both cases, it is market barriers that pose the biggest challenges. Sooner or later, continued investment in research and development will yield solutions to the remaining technical issues. But the market barriers are likely to be more difficult. Tax incentives for the use of renewable fuels such as those passed by the US Senate in S. 1637, the *Jumpstart Our Business Strength Act*, can make bioenergy cost-competitive with other fuels and increase its use. Cost-share programmes, such as the Renewable Energy Systems and Energy Efficiency Improvements program (2002 farm bill Sec. 9006) administered by the US Department of Agriculture, can help small businesses and farmers buy down the cost of anaerobic digesters and other renewable technologies. Increased use of bioenergy should also help spur infrastructure improvements and facilitate the enabling technologies necessary to sustain bioenergy without incentives.

The potential for bioenergy remains significant. Electric cooperatives are helping to realize that potential.

References

Electric Power Research Institute. (2000) *Greenhouse Gas Reduction with Renewables 2000 Update*. TR-113785. Electric Power Research Institute, Palo Alto, California.

Ganley, M. (2004) *Electric Utility Industry Overview*. National Rural Electric Cooperative Association, Arlington, Virginia.

Herman and Associates. (2003) *2003 Heavy Duty Diesel Manufacturer Fuel Recommendations*. Renewable Fuels Association, Washington, DC. [Accessed 2004.] Available from http://www.ethanolrfa.org/2003dieselrecommendations.pdf

Intergovernmental Panel on Climate Change. (2001) *Climate Change 2001: The Scientific Basis*. Contribution of Working Group I to the Third Assessment Report of the Intergovernmental Panel on Climate Change. Cambridge University Press, Cambridge, UK.

National Rural Electric Cooperative Association. (2004) *Electric Cooperatives and Alternative Energy: A Snapshot*. National Rural Electric Cooperative Association, Arlington, Virginia. [Accessed 2004.] Available from http://www.nreca.org/nreca/Policy/Regulatory/Documents/alternativeenergyOct04.pdf

Chapter 19

Potential for Biofuel-based Greenhouse Gas Emission Mitigation: Rationale and Potential

Bruce A. McCarl, Dhazn Gillig, Heng-Chi Lee, Mahmoud El-Halwagi, Xiaoyun Qin and Gerald C. Cornforth

Introduction

The Intergovernmental Panel on Climate Change (IPCC) asserts that the Earth's temperature rose by 0.6°C (1°F) during the 20th century and projects that temperature will continue to rise with a forecast increase ranging from 1.5°C to 4.5°C by 2100 (Intergovernmental Panel on Climate Change, 2001a,b). The IPCC also asserts that anthropogenic greenhouse gas (GHG) emissions are the dominant causal factor (Intergovernmental Panel on Climate Change, 2001a,b). In response, society is actively considering options to reduce net GHG emissions.

Emission reductions can be expensive to achieve and take time to develop. According to the US Environmental Protection Agency (US EPA) emissions inventory for 2002, 79% of US emissions arose from the combined emissions from electricity generation (33%), transportation (27%) and industry (19%) (US Environmental Protection Agency, 2004). The US EPA inventory also estimated that carbon dioxide emissions from fossil fuel consumption amounted to 80% of all US GHG emissions. Given such an origin for GHG emissions, a large reduction would require less energy use, development of new technologies, or fuel switching including biofuel usage. Biofuel feedstocks are a source of raw material that can be transformed into petroleum product substitutes in liquid energy usages or used as a direct input to power plants substituting for coal.

In the USA, liquid fuel biofuel production has not proven to be broadly economically feasible with maize-based ethanol being the dominant, but highly subsidized example. Power plant substitution has proven feasible, but only in areas with ready supplies of low cost feedstocks (generally co-generation from wood milling or pulping residues). Widespread production is unlikely under either alternative in the near future in the absence of some form of subsidy.

Schneider and McCarl (2003) list four possible outcomes of expanded biofuel production that could justify subsidization:

- A widespread biofuel market would support agricultural prices and incomes by adding to demand replacing other forms of farm income support.
- Replacement of some fuel components with ethanol has desirable environmental attributes and its widespread use in place of methyl tertiary-butyl ether (MTBE), for instance, would be environmentally desirable. A similar environmental motivation involves

replacement of coal and its inherent mercury content with mercury-free biomass feedstocks.
- Substitution of biofuel-based products for petroleum reduces dependence on imports and contributes to energy security.
- Biofuel combustion substantially offsets net GHG emissions by the recycling of carbon.

Current levels of GHG emissions are beginning to be judged as undesirable. Government pursuit of some mix of these outcomes could justify biofuel-related actions.

In this chapter, we examine the first and last of the four motivations from institutional and quantitative viewpoints. The institutional section discusses the context in which offsets may be needed and the role that agriculture and forestry may play with an emphasis on biofuels. The quantitative portion extends the work of McCarl *et al.* (2000) and Schneider and McCarl (2003) to further consider the potential of biofuel and other possible agriculture/forestry GHG emission offset alternatives in the face of a range of GHG offset prices emphasizing dynamics.

Greenhouse Gas Policy as Motivation for Biofuels

In the recent past, substantial scientific endeavour has been directed toward understanding and forecasting the linkage between GHG emissions, climate change and economic activity. The IPCC scientific-based assessments, as they have arisen over time, report an ever-growing consensus scientific opinion that over the last 100 years a 0.6°C warming has been observed and that in the foreseeable future increasing concentrations of GHGs will cause a substantial degree of climate warming (Intergovernmental Panel on Climate Change, 2001b). Atmospheric concentrations of carbon dioxide (CO_2), the most abundant GHG, are forecast to increase by 90% to 250% over pre-industrial levels by the end of the 21st century (Intergovernmental Panel on Climate Change, 2001b). Such large concentrations coupled with increases in other GHGs are predicted to cause substantial rises in global temperatures (Watson (2000) elaborates) along with accompanying alterations in precipitation patterns.

Numerous scientific efforts have tried to predict the impact of such a climate shift. Predictions to date suggest GHG concentration increases will alter economic and environmental attributes of agriculture, forestry, energy, human diseases, coastal resources, ecosystems and water availability among many other impact categories. One way to partially avoid prospective climate change or climate change risk is by reducing the amount of GHGs in the atmosphere. The IPCC documents present compelling arguments that, while it will be a long time before we know the exact effects of climate change, future reductions in GHG concentrations will take a very long time to be achieved and that perhaps as a precautionary move the USA should begin reduction efforts now. The increase in atmospheric GHG concentrations is largely caused by rising emissions from a diverse set of sources including emissions from fossil fuel combustion, deforestation, agricultural land use changes and land degradation. A reduction in the rate of GHG emissions would reduce future atmospheric concentrations.

In the face of such evidence, society has begun to consider actions to reduce net GHG emissions. The USA, along with most of the developed world, is a member of the United Nations Framework Convention on Climate Change (UNFCCC), which has the stated objective of stabilizing atmospheric GHG concentrations. The USA was also among the parties formulating the UNFCCC-generated international agreement called the Kyoto Protocol

(KP), which mandates actions reducing GHG emissions (United Nations Framework Convention on Climate Change, 1998) but announced in 2002 that it would not ratify the KP. However, in 2002, President George W. Bush announced a US unilateral programme called the Clear Skies Initiative that among other things states a policy goal of reducing future GHG emissions (Executive Office of the President, 2002).

The amounts of GHG emission reduction that are being considered are substantial. If ratified, the KP would have required that the USA achieve a 7% reduction in net GHG emissions relative to 1990 levels by a 2008-2012 commitment period. Considering likely growth levels, the emissions reduction would have risen to roughly a 30% cut in anticipated net GHG emissions as of the commitment period. The Clear Skies Initiative would require an 18% reduction in GHG emissions per dollar of gross domestic product by 2012. However the Pew Center on Global Climate Change estimates place the Clear Skies GHG emission reduction as a 'very modest improvement over the 'business as usual' emissions projections for 2012..., it appears to continue the same trend of GHG-intensity reductions and GHG emissions increases experienced over the last two decades' and asserts that 'Emissions in 2012 would be 30 percent above 1990 levels' as opposed to the Kyoto commitment that would have been 7% below 1990 (Pew Center on Global Climate Change, 2003).

Greenhouse gas emissions reductions in a biofuel or other context may also be motivated for several other reasons:

- Adoption of a precautionary stance toward climate forcing – the exact incidence of climate change, the degree of forcing caused by incremental emissions, the economic implications thereof and the reversibility of deleterious effects are highly uncertain. It may be desirable to adopt a go slow, precautionary stance to slow down potential GHG emission forcing of climate change and preserve societal options motivated by the uncertainties involved with GHG emission forcing.
- International pressures on emissions – the USA has been estimated (depending on accounting stance) to be emitting as much as 40% of global GHGs. Other countries have expressed opinions that US emissions are excessive and have begun to place international pressure on the USA to reduce GHG emission. Actions to reduce US emissions may be desirable to facilitate international affairs, trade negotiations and other cooperation.
- Domestic policies directed toward pollution – the Clear Skies and other federal governmental initiatives indicate that future actions will be needed to reduce GHG emissions along with sulphur oxides, nitrous oxides and mercury. All of these would be addressed by forms of biofuel usage. A number of states have also adopted regulatory stances or policy goals that also promote changes in emissions.
- Industry planning under uncertainty – industry in the USA is faced with an uncertainty today as to whether GHG emission limits will be imposed in the next 10 to 20 years. Multinational corporations (which are common in the energy component of industry as well as in the other emitting sectors) face a higher degree of certainty of emission caps for operations in Europe and other parts of the industrialized world. This uncertainty places a large part of their business assets at risk. Many firms have started the quest to discover ways to reduce GHG emissions in an economically sound manner.
- Need for cheap emission offsets – the previously stated reasons for GHG emission offsets involve considerations that would cause emitters in the USA to reduce net GHG emissions via choices from a broad set of emission reduction and offset alternatives. Such forces would not necessarily favour agricultural options including biofuels. However, focused

interest on biofuel and other agricultural alternatives has arisen because of anticipated relatively low per metric ton (MT) carbon costs and a feeling that adjustments can be made in a relatively short time frame. A number of biofuel and other agricultural GHG emission-related strategies are currently in use without any GHG emission-related programme in place. Furthermore, some of the strategies involve existing, well-known technologies which could be adopted in the near term as opposed to a number of energy sector strategies which require costly and time-consuming technological innovation and/or engineering efforts to reduce implementation costs. Industrial concerns are interested because the availability of cheap GHG emission offsets may allow them to avoid making large investments to reduce emissions or directly capture and store CO_2.
- Congruence of effects of expanded use of biofuel and other agricultural GHG offsets with other policy desires – many of the potential agricultural GHG emission-reducing practices have been previously encouraged in US government agricultural programmes designed to achieve environmental improvements and agricultural income support. Conservation incentives involving reduced tillage have been present in recent agricultural farm bills and have been justified on the basis of environmental improvement and agricultural income support. Bioenergy provisions have also been present with motivations involving energy security. Many of the potential practices involve direct support payments to farmers and have supply control characteristics which tend to lower aggregate production and raise market prices. Water quality and soil erosion programmes have also been undertaken which involve GHG emission-reducing practices. World Trade Organization (WTO) regulations also contribute to the desirability of GHG-related payments as a form of farm income support.
- Development of another market for farm products – the above motivations largely cover factors which stimulate action or concern on the behalf of general society and policymakers. Agricultural producers are also interested. Proposed GHG policies such as the emission caps allow for direct expanded product markets wherein biofuel feedstocks could be sold along with access to emissions trading markets. Offset producers could sell GHG emission offset credits to those in need of GHG emission reductions or rights both generating income-enhancing opportunities. It is also likely that market demand in those arenas would be much more elastic than conventional agricultural markets. Economically this would generate welfare gains accruing to producers.

Biofuels as a GHG Offset

Biofuel production can provide an important GHG emission offset. Namely biofuels mitigate GHG emissions because their usage displaces coal and oil essentially entering into a carbon recycling operation. As plants grow, they remove CO_2 from the atmosphere via photosynthetic processes. In turn, when the biofeedstocks or their derivative fuels are combusted, the carbon is released into the atmosphere. Fossil fuel use, on the other hand, releases 100% of the contained carbon. The net GHG emission consequences of a biofuel then depend on the amount of fuels from fossil sources used in producing the biofuel energy in the form of petroleum and coal-based electrical energy to raise, transport and process the feedstock into energy. Net carbon emissions from a poplar-fed power plant amount to approximately 5% of the emissions from an energy equivalent amount of coal (Kline et al., 1998). Ethanol replacement relative to petroleum yields a lower offset rate.

Agricultural and Forestry GHG Emission Offset Possibilities

There are a number of opportunities to pursue GHG emission offsets beyond the employment of biofuel practices. These include pursuit of GHG emission offsets in the agricultural sector broadly defined to include forests. Following the arguments in McCarl and Schneider (2000), there are at least three ways agriculture may participate in GHG emission offset enhancement efforts:

- Agriculture may reduce GHG emissions generated during operations.
- Agriculture may enhance absorption of GHG emission by creating or expanding sinks, commonly called carbon sequestration.
- Agriculture may provide products, which substitute for GHG emission-intensive products displacing emissions largely in the form of biofuels but also in the form of other bioproducts.

The first and second of these and the non biofuel aspects of the third are the principal agricultural opportunities we discuss here.

Emission Reductions

In terms of emission reductions, IPCC (1996) estimates that globally agriculture emits about 50% of all methane (CH_4) emissions, 70% of all nitrous oxide (N_2O) and 20% of all CO_2. Sources of CH_4 emissions include rice, ruminants and manure. These sources can be reduced by altering items such as crop mix, livestock herd size, livestock feeding and rearing practices and manure management. N_2O emissions come from manure, legumes and fertilizer use and can be reduced by altering items such as livestock herd size, crop mixes and fertilization practices. CO_2 emissions arise from fossil fuel usage, soil tillage, deforestation, biomass burning and land degradation. CO_2 emissions can be reduced by altering items such as production fuel use; allocation of land between crops, pasture, grass lands and forests; forest harvest rates; crop residue management; forest harvest management and land restoration. McCarl and Schneider (2000) present a further discussion and a review of the types of activities that can be pursued as do the US Environmental Protection Agency (US EPA) documents on methane and gas emission inventories (2004). On the forest side, management practices that offset emissions according to Brown *et al.* (1996) and Brown (1999) include reduced deforestation or logging, protection of forests in reserves and reduced disturbances by managing forest losses through fire and pest outbreaks.

The relative magnitude of emission sources varies substantially across countries, with the greatest differences between developing and developed countries. Deforestation and land degradation mainly occur in developing countries. Developed country agriculture generally uses more energy, intensive tillage systems and fertilizer, resulting in CO_2 fossil fuel-based emissions, carbon emissions from more intensive tillage, N_2O emissions from fertilizer and N_2O plus CH_4 emissions from animal herds and resultant manure.

Sequestration/Sink Enhancements

There are two major types of actions that can be employed to pursue sequestration. These involve agricultural and forestry management changes and changes in land use.

Changes in Agricultural Land Management

The most commonly discussed agricultural management sequestration-enhancing changes involve tillage and nutrient and residue management. These involve reductions in tillage intensity (adopting conservation tillage or no-till), adding organic manure, altering fertilization and/or somehow leaving behind more crop residues. Strategies have also been mentioned which involve changing rotations, altering crop mixes, employing more perennials, growing crops with more above-ground biomass, using winter cover crops and utilizing erosion control techniques such as terracing, contour ploughing, strip cropping, buffer strips and water management (Lal *et al.* (1998) cover the crop lands topic in much more detail).

Agricultural carbon sequestration (ACS) can also be stimulated by changes in pasture and rangeland management which involve altering plant species on pasture lands, improving grass productivity, reducing grazing intensity, employing fertilization, or otherwise altering management so as to increase the amount of organic manner incorporated into the soil (Follett *et al.* (2000) cover the grazing lands topic in much more detail).

Forest Management Options

Forest management practices that can be altered to increase carbon retention can be classified into several groups:

- Management to retain carbon in forests including longer rotations, reduced deforestation or logging, protection of forests in reserves, reduced impact logging and reduced disturbances by managing forest losses through fire and pest outbreaks (Brown *et al.* (1996); Brown (1999); and Murray (2003) elaborate).
- Management for increased carbon in standing forest biomass or forest soils through use of enhanced silvicultural treatments, natural or artificial regeneration in secondary forests and other degraded forests whose biomass and soil carbon densities are less than their maximum value (Brown *et al.* (1996) and Brown (1999) elaborate).
- Altered management and use of harvested wood products shifting demand into longer-lasting wood products and extending the lifetime of wood products through disposal, recycling and other preservation efforts (Brown *et al.* (1996); Brown (1999); and Skog and Nicholson (2000) elaborate).

Changes in Land Use

Carbon sequestration may also be increased by changing land use. Generally uses that disturb the soil less often enhance sequestration. Thus, the strategies that have been prominently discussed include:

- Afforestation of non-forested agricultural or other lands and increased tree cover on agricultural or pasture lands through agroforestry.
- Conversion of croplands to grasslands, pasture, rangelands or wetlands.
- Restoration of degraded lands, in an effort to reestablish their organic content.

Substitute Product-based GHG Offsets Beyond Biofuels

Agricultural products may be grown which replace fossil fuel-intensive products. One such product category involves biomass for energy generation or transformation into liquid fuels.

Another category is an emerging, wide array of bioproducts made from agricultural materials, ranging from lubricants to cleaning solvents to plastics. Forestry products can also be used to substitute for fossil fuel-intensive use of steel and concrete in construction (Brown *et al.* (1996); Marland and Schlamadinger, (1997); and Brown (1999) elaborate). Finally, there may be gains from substituting cotton and other fibres for petroleum-based synthetics.

Factors Affecting the Competitiveness of Biofuels

The amount of subsidy, the extent of biofuel penetration into the energy market and the rate of technical progress are key considerations in future biofuel prospects. In this study, the only form of subsidy considered is in the form of a carbon price which may not be a subsidy at all but rather a reflection of the future externality cost of GHG emission into the atmosphere. Namely, if a prospect offsets 95% of the carbon created from an energy equivalent amount of fossil fuel then at a US$40/MT carbon price this amounts to a US$38/MT subsidy. Because wood is about 50% carbon, this would be a US$19/MT implicit subsidy and relative to a wood price in the vicinity of US$25 to US$30/MT, this substantially offsets the purchase price to a biofeedstock user.

Energy market penetration of biofuels is similarly important. Biofuels can penetrate when new plants are built or older plants are retrofit. New power or petroleum refining plants are needed as existing plants are retired or as demand for energy use grows. Model constraints control the rate of penetration. Technical change is also important. Agricultural productivity has grown over time. Progress in biofeedstock yields whether at the farm level or in the form of conversion ratios to energy products will be important to maintain a competitive role.

Greenhouse Gas Offset Potential Analysis Requirements

The agricultural and forestry (AF) sectors are complex and highly interrelated. A number of features of these sectors influence the needed form of analytical approach to assessment of AF GHG emission mitigation potential. In particular, the features that we take note of involve dynamics, multiple gas implications, mitigation alternative interrelatedness, co-benefits, market/welfare implications and differential offset rates. The importance of each factor will be addressed below:

Dynamics

Agricultural and forestry activities develop over time. Forest and longer term perennial usage diverts resources from other usages. Carbon is sequestered in AF ecosystems as long as carbon addition exceeds decomposition. However, as carbon accumulates, decomposition rates rise. Eventually sequestration will stop. Forest biofuel feedstocks take years to mature. Examination of sequestration and biofuels requires attention to these dynamics.

Multiple Gas Implications

Greenhouse gas emission mitigation strategies impact emissions of CO_2, N_2O and CH_4. These gasses have different effects on climate change radiative forcing. Equivalency rates have been established. This chapter uses 100-year global warming potentials as suggested in Reilly *et al.* (1999).

Mitigation Alternative Interrelatedness

The AF mitigation alternatives are linked in many ways. Modelling must be complex, depicting intermediate products, product substitution and competition for land, among other factors across the AF sectors.

Co-Benefits

The AF mitigation alternatives directly affect environmental pollutant loadings, water usage, animal habitat and many other environmental processes.

Market/Welfare Implications

The USA exhibits large production relative to domestic needs and total global market volume for many commodities. Therefore, the commodity production and effects of US GHG emission mitigation policies will also affect prices and welfare in domestic and world markets.

Differential Offset Rates

The AF GHG emission-related strategies exhibit substantially different GHG offset rates. For example, tillage changes produce about 0.625 MT of carbon offsets per hectare while still producing crops. Forests or biofuels can produce offset rates above 2.5 MT but with no complementary crop production. Furthermore, these offset production rates vary over time. At low offset prices, complementary production is likely to be favoured. Under discounting, near term offsets will be favoured over longer term offsets. This implies a need to look at offset price and mitigation timing characteristics when assessing a strategy.

Modelling

In order to investigate the dynamic role of agriculture and forest carbon offsets including biofuels and other alternatives, an analytical framework is needed that can depict the time path of offsets from AF possibilities. To accomplish this, Lee's (2002) GHG version of the Forest and Agricultural Sector Optimization Model (hereafter referred to as FASOMGHG) will be used (Adams *et al.*, 1996). This model has forest carbon accounting unified with a detailed representation of the possible mitigation strategies in the agricultural sector adapted from Schneider (2000) and McCarl and Schneider (2001).

FASOMGHG is a 100-year inter-temporal, price-endogenous, mathematical programming model depicting land transfers between the AF sectors in the USA. The model solution portrays a multi-period equilibrium on a decadal basis along with a dynamic simulation of prices, production, management and consumption within these two sectors under the scenario depicted in the model data.

Geographic Scope

FASOMGHG divides the USA into 11 regions, nine of which produce forest products and 10 of which produce agricultural products.

Product Scope

FASOMGHG simulates production of 50 crop and livestock commodities and 56 processed commodities along with 10 forestry commodities.

Land Transfers

There are period by period land transfer possibilities involving land from: (1) forestry to agriculture in the pasture or cropland categories; (2) agriculture to forestry in either the pasture or cropland categories; (3) cropland to pasture; and (4) pasture to cropland. Many forested tracts are not suitable for agriculture, thus, the model accounts for land that is not mobile between uses. Costs for converting forestland reflect stump removal, land grading and other factors.

Agricultural Management

The agricultural component depicts activity during each of 10 decades. Agricultural output is produced using land, labour, grazing and irrigation water. Once commodities enter the market, they can go to livestock use, feed mixing, processing, domestic consumption, or export. Imports are also represented. The model structure incorporates the agricultural sector model described by Chang *et al.* (1992) with Schneider (2000) added GHG features. Demand and supply components are updated between decades. The model uses constant elasticity functions for domestic and export demand as well as factor and import supply. In the first two decades, the solution is required to be within a convex combination of historical crop mixes, following McCarl (1982), but is free to change thereafter. Possibilities for GHG emission management are included by incorporating the strategies summarized in Table 19.1.

Forest Management

The forest sector model allows multiple harvest age possibilities. Multiple-decade forest production processes are represented by periodic regional timber yields from the Aggregate Timber Land Analysis System (ATLAS) (Mills and Kincaid, 1992). Logs are differentiated into sawlogs, pulpwood and fuelwood for both hardwoods and softwoods, yielding six product classes. Upon harvest, forestlands may be regenerated into forestry or may migrate into agriculture. Forested land is differentiated by region, ownership class, age cohort of trees, forest cover type, site productivity class, timber management regime and suitability of forestland for agriculture use.

Terminal Conditions

Given that the model is defined for a finite period, there will be immature trees at the end. Terminal conditions are imposed that value ending immature trees and land remaining in agriculture.

Table 19.1. Mitigation strategies in FASOMGHG.

Mitigation strategy	Strategy nature	Greenhouse gas affected		
		Carbon dioxide	Methane	Nitrous oxide
Afforestation	Sequestration	Yes	No	No
Rotation length	Sequestration	Yes	No	No
Timberland management	Sequestration	Yes	No	No
Defforestation (avoided)	Sequestration	Yes	No	No
Biofuel production	Offset	Yes	Yes	Yes
Crop mix alteration	Emission, sequestration	Yes	No	Yes
Rice acreage reduction	Emission	No	Yes	No
Crop fertilizer rate reduction	Emission	Yes	No	Yes
Other crop input alteration	Emission	Yes	No	No
Crop tillage alteration	Sequestration	Yes	No	No
Grassland conversion	Sequestration	Yes	No	No
Irrigated/dry land conversion	Emission	Yes	No	Yes
Livestock management	Emission	No	Yes	No
Livestock herd size alteration	Emission	No	Yes	Yes
Livestock system change	Emission	No	Yes	Yes
Liquid manure management	Emission	No	Yes	Yes

Soil and Ecosystem Saturation

Terrestrial carbon sinks are limited by ecosystem capability in interaction with the management system. In particular, carbon only accumulates until a new equilibrium is reached. FASOMGHG assumes tillage practice change-induced carbon gains/losses stop after 30 years based on West and Post (2002).

Forest carbon accounting is based on the Forest Carbon Budget (FORCARB) model as developed by Birdsey (1992) and Birdsey and Heath (1995) and the Harvested Carbon HARVCARB model of Rowe (1992). Forest carbon is accounted in four basic pools, soil, ecosystem, standing trees and products after harvest.

Biofuel Production and Use

In FASOMGHG, the potential for biofuel penetration into the energy sector coupled with the cost of production is represented in several ways. Biomass input is modelled to the point in the power plant energy generation process where fuel is fed into the burners. Two biomass production and power plant use scenarios are used: (1) diverting milling residues from traditional pulp and paper or other uses and (2) producing switchgrass and short rotation woody crops for biomass. The model also allows the possible use of wood chips from short rotation woody crops for pulp and paper production. Biomass hauling costs are from McCarl et al. (2000). Switchgrass and short rotation woody crop production information is from yields and costs of production studies at the Oakridge National Laboratories (Walsh and Graham, 1999, Graham et al., 1996). Maximum potential by year for biomass penetration into the electrical energy market data is based on information provided by Z. Haq (US Department of Energy, Energy Information Administration, Washington, DC, 2003, personal communication). Cellulosic and maize starch-based ethanol were used in the model.

Table 19.2. Emission reduction in million metric tons of carbon dioxide equivalent.

	Price in US$ per metric ton carbon dioxide equivalent				
	5	15	30	50	80
Afforestation	2	110	450	845	1264
Soil sequestration	120	153	147	130	105
Biomass offsets	17	844	952	957	960
Methane and nitrous oxide	13	34	65	107	159
Forest management	106	216	313	385	442
Crop management	29	56	74	91	106
All strategies	288	1413	2001	2514	3037

Results and Implications

The fundamental analysis is an examination of AF mitigation strategies that arise under different CO_2 equivalent (CO2E) prices. The CO2E price is applied to CO_2, CH_4 and N_2O emissions/offsets adjusted for their GHG warming potential. CO2E prices are varied from US$0 to US$80 per MT. Offset estimates are computed on a total US basis relative to a zero CO2E scenario.

Static Mitigation Quantity

The strategies employed vary over time. The annuity equivalent offset is used to summarize the strategies with GHG increments discounted at a 4% rate following Richards (1997). Trends appearing in the results (Table 19.2 and Fig. 19.1) are:

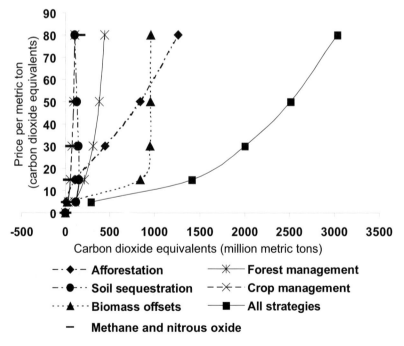

Figure 19.1. Annualized mitigation potential of chosen mitigation tools at different greenhouse gas offset prices.

- At low GHG offset prices the first options chosen are agricultural soil carbon and existing forest stand management largely in the form of longer rotations.
- At higher GHG offset prices biofuel for power plants and afforestation dominate with agricultural soil share reduced from its peak.
- Non-CO_2-related strategies largely in the form of livestock and fertilization management are small but rise with the GHG offset price.
- Liquid fuel replacement biofuels do not enter the solution.

Overall, at lower prices, mitigation involves the use of management alternatives complementary to current land uses that continue traditional production. However, at higher prices, biofuels and afforestation appear, displacing the traditional production but generating larger offsets.

Biofuel-related results show the dominance of power plant usage as opposed to liquid fuel production. This is largely because the power plant replacement uses little energy relative to the offset quantity, but the liquid fuel biofuel replacement uses substantially more energy.

Dynamic GHG Mitigation

One can look at the results over time. Figures 19.2-19.4 present the accumulated GHG mitigation credits from forest sequestration, agricultural soil sequestration, power plant feedstock biofuel offsets and non-CO_2 strategies in different GHG offset prices.

At low prices and in the near term, the carbon stock on agricultural soil and in existing forests initially grows rapidly. However, the offset quantity later diminishes and becomes stable with saturation. Carbon stocks in the forest grow for about 40 years at low prices. Non-CO_2 strategies continually grow throughout the whole model period. Biofuel is not a factor in the near term as it is too expensive to be part of a low carbon price mitigation plan.

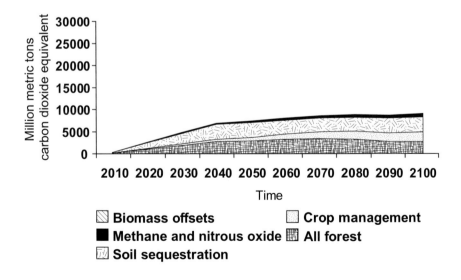

Figure 19.2. Cumulative mitigation contributions from major strategies at a US$5 carbon dioxide equivalent price.

When the prices are higher, the forest carbon stock increases first and then diminishes. The agricultural soil carbon stock is much less important – especially in the later decades.

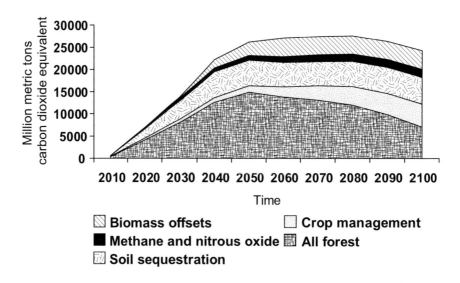

Figure 19.3. Cumulative mitigation contributions from major strategies at a US$15 carbon dioxide equivalent price.

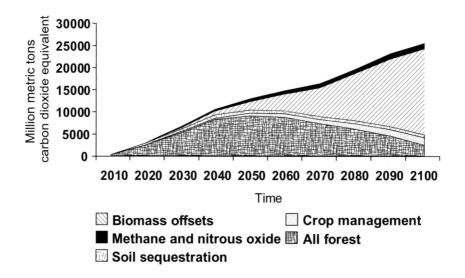

Figure 19.4. Cumulative mitigation contributions from major strategies at a US$50 carbon dioxide equivalent price.

Non-CO_2 mitigation grows but is not a very large contributor to mitigation. Power plant feedstock biofuel potential grows dramatically (largely due to net carbon emissions considerations, the endogenous option of making unsubsidized ethanol as a GHG offset is not found to be desirable) over time and becomes the dominant strategy.

Across the results of the various scenarios, several patterns emerge:

- Carbon sequestration, including agricultural soil and forest carbon and power plant feedstock biofuel offsets are the high quantity mitigation strategies. The importance of these strategies varies by price and time.
- At low prices and in the earlier decades, agricultural soil carbon and existing forest management are dominant. When carbon prices are higher, afforestation and power plant feedstock biofuels are dominant because they have higher per acre carbon offsets.
- The carbon sequestration activities tend to rise over time then stabilize largely due to saturation phenomena. Soils saturate faster than trees.
- The higher the price, the more carbon stored in the forests in the early decades. However, the intensified forest sequestration comes with a price in that CO_2 emissions from forests increase later.

Market Effects and Co-benefits

The introduction of the GHG offset prices causes changes in land use, tillage, fertilization, crop mix and other management practices, commodity production and consumption and trade flows. In turn, this causes changes in market conditions and environmental pollutant loadings. Market-related results found include: (1) a decline in production of traditional agricultural commodities; (2) a rise in agricultural and short-term forest commodity prices; (3) losses in consumer welfare due to higher prices; (4) gains in producer welfare due to higher food prices and GHG-related offset payments; and (5) losses in export earnings. The environmental impacts include: (1) a drop in the amount of traditionally cropped agricultural land; (2) a drop in irrigated area; (3) an increase in forested land; (4) an increase in biofuel land; and (5) a decline in loadings for nitrogen, phosphorus and soil erosion into the environment. An interesting result is that pollutant loadings decline substantially at low prices, but rise at higher prices due to production intensification as land is diverted away from agriculture.

An Analysis of Biofuel Competitiveness

Key factors in biofuel production involve the subsidy level (carbon price), technical progress and market penetration. The results show that the carbon price is quite important. At low CO2E prices, biofuels are not important. However, when carbon prices exceed US$50 per MT (or about US$15 per MT of CO_2), biofuels become very important. At even higher prices, biofuels quickly reach their maximum market potential. Removal of market penetration limits shows a great increase in potential biofuel offsets (see Lee (2002) where such caps were not included). Finally, the effect of shifts in the technical progress rate is from the study by McCarl et al. (2000). In that paper it was found that the competitiveness of the biofuel for power plants depends critically on the rate of technical progress continuing without large increases in costs.

Future Research

Future biofuel investigation will continue under a US Department of Agriculture (USDA) and US Department of Energy (US DOE) supported project focused on: (1) life cycle analysis of biofuel liquid products and biofuel-fired power plant feedstocks in comparison with coal and petroleum in terms of net energy use, GHG balance and other environmental emissions; (2) environmental biocomplexity analysis (El-Halwagi, 2003) of socio-economic, technical and environmental implications of biofuel use; (3) chemical reaction analysis of discharges from coal and biofuel power plants; (4) techno-economic and environmental analysis of co-fed (biomass and fossil fuel) power plants; (5) integrated environmental and economic analysis of biofuel production, including farm level emissions analysis and hydrological pollutant load routing; (6) joint environmental, economic and energy performance modelling of electric turbine and post-combustion treatment for different biomass/coal mixes; (7) the role of fundamental combustion developments for biofuel usage; and (8) technological innovations in farm biomass product yields, handling and storage requirements, energy yields from combustion or processing and control of production and processing costs.

Conclusions

This study discussed the reasons why biofuels and other GHG offsets might be adopted by society then conducted a modelling analysis regarding AF sector GHG emission mitigation strategies in response to GHG offset prices. Focus was placed on the role of biofuel alternatives in general and over time. The results show that the AF sectors offer substantial potential to mitigate GHG emissions. Mitigation strategy choices change dynamically depending on price and time. Tillage-based agricultural soil carbon sequestration and forest rotation length-induced sequestration are the primary mitigation strategies implemented in early decades and at low prices (below US$10 per MT CO_2). However, both saturate and turn into CO_2 sources after 40 to 60 years. On the other hand, power plant feedstock biofuel activities and afforestation become more important in the long run and/or at higher prices. Crop and livestock management are small but steady contributors across the entire spectrum of prices and time periods.

Generally, at high carbon prices, biomass feedstocks can be a way of altering the emission characteristics of US electrical generation but appear to be of limited usefulness in the liquid fuel markets. The competitiveness of biomass depends upon the success of research in developing improved production methods for biofuel crops.

The results support the argument that AF carbon sequestration provides more time to find long-term solutions such as new technologies to halt the increasing ambient GHG concentration as discussed in Marland et al. (2001). It also shows that power plant feedstock biofuels are likely to be an important long-run strategy when GHG offset subsidies are high.

Acknowledgments

The work underlying this chapter was supported by USDA, the US Environmental Protection Agency, US DOE, the Sustainable Agriculture and Natural Resource Management Collaborative Research Support Program and the Texas Agricultural Experiment Station.

References

Adams, D., Alig, R., Callaway, J., McCarl, B. and Winnett, S. (1996) *The Forest and Agricultural Sector Optimization Model (FASOM): Model Structure and Policy Applications*. US Department of Agriculture Forest Service Research Paper PNW-495, Pacific Northwest Research Station, Portland, Oregon.

Birdsey, R. (1992) *Prospective Changes in Forest Carbon Storage from Increasing Forest Area and Timber Growth*. US Department of Agriculture, US Forest Service, Washington, DC.

Birdsey, R. and Heath, L. (1995) Carbon Changes in U.S. Forests. In: Joyce, L (ed.) *Climate Change and the Productivity of America's Forests*. US Department of Agriculture, US Forest Service, Ft. Collins, Colorado, pp. 56-70.

Brown, S. (1999) *Guidelines for Inventorying and Monitoring Carbon Offsets in Forest Based Projects*. Winrock International, Morrilton, Arkansas.

Brown, S., Sathaye, J., Cannell, M. and Kauppi, P. (1996) Management of Forests for Mitigation of Greenhouse Gas Emissions. In: Watson, R., Zinyowera, M. and Moss, R. (eds.), *Climate Change 1995: Impacts, Adaptations and Mitigation of Climate Change: Scientific Analyses. Contribution of Working Group II to the Second Assessment Report of the Intergovernmental Panel on Climate Change*. Cambridge University Press, Cambridge, UK, pp. 773-798.

Chang, C., McCarl, B., Mjelde, J. and Richardson, J. (1992) Sectoral Implications of Farm Program Modifications. *American Journal of Agricultural Economics* 74: 38-49.

El-Halwagi, M. (2003) Industry and Environmental Biocomplexity: Impact, Challenges and Opportunities for Multidisciplinary Research. *Journal of Clean Technologies and Environmental Policy* 4(3): 135.

Executive Office of the President. (2002) *Executive Summary – The Clear Skies Initiative*. The White House, Executive Office of the President, Washington, DC. [Accessed 2004.] Available from http://www.whitehouse.gov/news/releases/2002/02/clearskies.html

Follett, R., Kimble, J. and Lal, R. (2000) *The Potential of U.S. Grazing Land to Sequester Carbon and Mitigate the Greenhouse Effect*. CRC Press, Boca Raton, Florida.

Graham, R., Lichtenberg, E., Roningen, V., Shapouri, H. and Walsh, M. (1996) *The Economics of Biomass Production in the United States*. US Department of Energy, Oak Ridge National Laboratory, BioFuels Feedstock Development Program, Oak Ridge, Tennessee. [Accessed 2004.] Available from http://bioenergy.ornl.gov/papers/bioam95/graham3.html

Intergovernmental Panel on Climate Change. (1996) *Climate Change 1995: The IPCC Second Assessment Report, Volume 2: Scientific-Technical Analyses of Impacts, Adaptations and Mitigation of Climate Change*. Watson, R., Zinyowera, M. and Moss, R. (eds.) Cambridge University Press, Cambridge, UK.

Intergovernmental Panel on Climate Change. (2001a) *Climate Change 2001: Mitigation, IPCC Third Assessment Report: Contributions of IPCC Working Group III*. Metz, B., Davidson. O., Swart, R. and Pan, J. (eds.) Cambridge University Press, Cambridge, UK.

Intergovernmental Panel on Climate Change. (2001b) *IPCC Third Assessment Report: Climate Change 2001: The Scientific Basis: A Report of Working Group I of the Intergovernmental Panel on Climate Change*. Intergovernmental Panel on Climate Change, Geneva, Switzerland. [Accessed 2004.] Available from http://www.grida.no/climate/ipcc_tar/wg1/index.htm.

Kline, D., Hargrove, T. and Vanderlan, C. (1998) *The Treatment of Biomass Fuels in Carbon Emissions Trading Systems*. Center for Clean Air Policy, Washington, DC. [Accessed 2004.] Available from http://www.ccap.org/pdf/biopub.pdf.

Lal, R., Kimble, J., Follett, R. and Cole, C. (1998) *The Potential of U.S. Cropland to Sequester Carbon and Mitigate the Greenhouse Effect*. Sleeping Bear Press Inc., Chelsea, Michigan.

Lee, H. (2002) *The Dynamic Role for Carbon Sequestration by the U.S. Agricultural and Forest Sectors in Greenhouse Gas Emission Mitigation*. PhD dissertation, Texas A&M University, Department of Agricultural Economics, College Station, Texas

Marland, G. and Schlamadinger, B. (1997) Forests for Carbon Sequestration or Fossil Fuel Substitution A Sensitivity Analysis. *Biomass and Bioenergy* 13: 389-397.

Marland, G., McCarl, B. and Schneider, U. (2001) *Soil Carbon: Policy and Economics*. Climatic Change 51: 101-117.

McCarl, B. (1982) Cropping Activities in Agricultural Sector Models: A Methodological Proposal. *American Journal of Agricultural Economics* 64(4): 768-772.

McCarl, B. and Schneider, U. (2000) Agriculture's Role in a Greenhouse Gas Emission Mitigation World: An Economic Perspective. *Review of Agricultural Economics* 22: 134-159.

McCarl, B. and Schneider, U. (2001) The Cost of Greenhouse Gas Mitigation in U.S. Agriculture and Forestry. *Science* 294 (21): 2481-2482.

McCarl, B., Adams, D., Alig, R. and Chmelik, J. (2000) Analysis of Biomass Fueled Electrical Powerplants: Implications in the Agricultural and Forestry Sectors. *Annals of Operations Research* 94: 37-55.

Mills, J. and Kincaid, J. (1992) *The Aggregate Timberland Assessment System—ATLAS: A Comprehensive Timber Projection Model*. Gen. Tech. Rep. PNW-281. US Department of Agriculture, US Forest Service, Pacific Northwest Station, Portland, Oregon.

Murray, B. (2003) Carbon Sequestration: A Jointly Produced Forest Output. In: Sills, E. and Abt, K. (eds.) *Forests in a Market Economy*. Kluwer Academic Publishers, Boston, Massachusetts.

Pew Center on Global Climate Change. (2003) *Analysis of President Bush's Climate Change Plan*. Pew Center on Global Climate Change, Arlington, Virginia. [Accessed 2004.] Available from http://www.pewclimate.org/policy_center/analyses/response_bushpolicy.cfm.

Reilly, J., Prinn, R. Harnisch, J., Fitzmaurice, J., Jacoby, H., Kicklighter, D., Melillo, J. and Stone, P. (1999) Multi-gas Assessment of the Kyoto Protocol. *Nature* 401: 549-55.

Richards, K. (1997) The Time Value of Carbon in Bottom-up Studies. *Critical Reviews in Science and Technology* 27(special): 279-292.

Rowe, C. (1992) *Carbon Sequestration Impacts of Forestry: Using Experience of Conservation Programs*. Presented paper, Symposium on Forest Sector, Trade and Environmental Impact Models: Theory and Applications, April 30-May 1, Seattle, Washington.

Schneider, U. (2000) *Agricultural Sector Analysis on Greenhouse Gas Emission Mitigation in the U.S.* PhD dissertation, Texas A&M University, Department of Agricultural Economics, College Station, Texas.

Schneider, U. and McCarl, B. (2003) Economic Potential of Biomass Based Fuels for Greenhouse Gas Emission Mitigation. *Environmental and Resource Economics* 24: 291-312.

Skog, K. and Nicholson, G. (2000) Carbon Sequestration in Wood and Paper Products. In: Joyce, L. and Birdsey, R. (eds.) *The Impact Of Climate Change On America's Forests: A Technical Document Supporting The 2000 USDA Forest Service RPA Assessment*. RMRS-GTR-59. US Department of Agriculture, US Forest Service Rocky Mountain Experiment Station, Fort Collins, Colorado.

United Nations Framework Convention on Climate Change. (1998) *The Convention and Kyoto Protocol*. United Nations Framework Convention on Climate Change, Bonn, Germany. [Accessed 2004.] Available from http://www.unfccc.de/resource/convkp.html

US Environmental Protection Agency. (2004) *US Emissions Inventory 2004: Inventory Of U.S. Greenhouse Gas Emissions And Sinks: 1990-2002*, EPA 430-R-04-003, April 2004. Available at http://yosemite.epa.gov/oar/globalwarming.nsf/content/ResourceCenterPublicationsGHGEmissionsUSEmissionsInventory2004.html

Walsh, M. and Graham, R. (1999) *Biomass Feedstock Supply Analysis: Production Costs, Land Availability, Yields*. US Department of Energy, Oak Ridge National Laboratory, BioFuels Feedstock Development Program, Oak Ridge, Tennessee.

Watson, R. (2000) *Report to the Sixth Conference of the Parties of the United Nations Framework Convention on Climate Change*. Intergovernmental Panel on Climate Change, Geneva, Switzerland. [Accessed 2004.] Available from http://www.ipcc.ch/press/sp-cop6-2.htm

West, T. and Post, W. (2002) Soil Organic Carbon Sequestration Rates by Tillage and Crop Rotation: A Global Data Analysis. *Soil Science Society of America Journal* 66: 1930-1946.

Chapter 20

Life Cycle Assessment of Integrated Biorefinery-Cropping Systems: All Biomass is Local

Seungdo Kim and Bruce E. Dale

Introduction

Agricultural processes play a crucial role in the environmental performance of fuels, chemicals and materials (biobased products) produced from agricultural raw materials. Thus, more sustainable farm management results in better environmental performance of these biobased products. This chapter compares the environmental performance of different cropping systems providing raw materials to an integrated biorefinery. The base case is continuous maize cultivation under no-tillage conditions. Effects of maize stover removal and planting winter cover crops are also included in the analysis.

Two biobased products, ethanol and polyhydroxyalkanoates (PHA), are considered. Ethanol derived from biomass has the potential to be a renewable transportation fuel that can substitute for petrol. About 3.2 billion gallons (gal) of ethanol are produced annually in the USA, primarily from maize grain (Renewable Fuels Association, 2004). PHA is generally considered an environmentally friendly material in terms of biodegradability and use of renewable resources. A primary market driving force for PHA is the perceived scarcity of petroleum resources. Most conventional polymers currently available are made from crude oil.

Carbohydrates in maize stover are potentially a major feedstock for biobased products. For example, maize stover plays an important projected role in lignocellulose-based bioethanol production (Aden *et al.*, 2002). However, the full utilization of crop residues, including maize stover, might lead to greater soil erosion and lower soil organic carbon levels, thereby undermining the sustainability of the entire system. Winter cover crops, such as winter wheat, can be planted in the late fall after the main crop (usually maize or soybeans) is harvested. The new plants grow for a few weeks in late fall, remain viable through the winter and then grow rapidly again in the spring. Normal agricultural practice is to till under the cover crop or kill it with herbicides prior to planting the primary crop in the late spring. However, some cover crops are harvested for animal consumption. Winter cover crops are known to reduce nitrate leaching and increase soil organic carbon levels (Kuo *et al.*, 1997; Reicosky and Forcella, 1998). This study is interested in the effect of these cover crops on sustainability when maize stover is removed from the land. When maize stover is removed, an integrated production system for each product in which both maize grain and maize stover are utilized as fermentation raw materials for biobased products is also investigated.

Table 20.1. Cropping system scenario (M: maize, w: wheat).

Abbreviation	Crop rotation	Winter cover crop	Residue removal rate (%)
CM	Continuous maize	No	0
CM50	Continuous maize	No	50
CMw70	Continuous maize	Wheat	70

This study implements life cycle assessments on the ethanol production system and the PHA production system, respectively, to identify the environmental performance of the arable land. Ethanol and PHA are produced separately. Sustainable biomass utilization is also investigated using the eco-efficiency concept. The environmental profile is addressed in terms of natural resources used (e.g., coal, crude oil, natural gas), non-renewable energy consumption, global warming, acidification, eutrophication and photochemical smog.

Methods and Data

Washington County, Illinois, is modelled in this study as a rather typical Corn Belt (US Midwest) farming location. In Washington County, the average fraction of clay in the top 20 centimetres of soil is 18%, while the average fraction of sand is 8% (Natural Resources Conservation Service, 2004). The average annual temperature is about 55°F, and average monthly rainfall is about 3.2 inches (National Oceanic and Atmospheric Administration, 2004). A specific location was chosen because the local soil quality strongly depends on site-specific conditions. The cropping systems studied here are summarized in Table 20.1. The effects of maize stover removal and a winter cover crop in the continuous maize systems with 50% residue removal (CM50) and 70% residue removal with wheat as a cover crop (CMw70) are estimated. No-till agricultural practice is applied to all the scenarios. The winter cover crops are not harvested but instead are killed by herbicides before the primary crop is planted.

Dynamics of soil organic carbon and nitrogen are predicted by an agro-ecosystem model. The daily version of the CENTURY model (DAYCENT) estimates soil organic carbon levels, nitrous oxide (N_2O) and nitrogen oxide (NO_x) emissions from soil, and inorganic nitrogen losses due to leaching (Parton *et al.*, 1996; Del Grosso *et al.*, 2000, 2001). The cropping systems are simulated for a 40-year period for soil organic carbon and soil nitrogen dynamics. In each cropping system, crop residues cover more than 60% of the soil surface to keep soil erosion at tolerable levels (Renard *et al.*, 1996; Nelson, 2002).

Ethanol derived from maize grain (CM) is modelled as being produced in a wet-milling system. In the wet-milling process, maize gluten meal (MGM), maize gluten feed (MGF) and maize oil are also produced. When maize stover is utilized (CM50 and CMw70), ethanol derived from maize grain is produced by wet-milling, and ethanol derived from maize stover is produced by a maize stover conversion process. In this conversion process, maize stover is treated by dilute sulphuric acid to convert cellulose and hemicellulose to soluble sugars, which are then fermented to produce ethanol (Sheehan *et al.*, 2002). Lignin-rich residues are utilized to generate electricity and steam. It is assumed that the efficiency of electricity generation from burning lignin-rich residues is 32% (Stahl and Neergaard, 1998) and the efficiency of steam generation (i.e., a ratio of energy output as steam to energy input based on higher heating value of the fuels) is 51% (Sheehan *et al.*, 2002). The ethanol yield is 2.55 gal of ethanol/bushel of maize grain in the maize wet-milling process (Wang, 2000) and 89.7 gal of ethanol/ton of dry maize stover (Sheehan *et al.*, 2002).

PHA is also derived from dextrose produced in wet-milling utilizing maize grain (CM) (J. van Walsem, Metabolix Inc., Cambridge, Massachusetts, 2003, personal communication). When maize stover is utilized (CM50 and CMw70), PHA is derived from maize grain and maize stover (Kurdikar *et al.*, 2001). In the maize stover conversion process (for both ethanol and PHA production systems), lignin-rich residues are utilized to generate electricity and steam. The PHA yield is 10.9 pounds (lb) of PHA/bushel of maize grain (J. van Walsem, Metabolix Inc., Cambridge, Massachusetts, 2003, personal communication) and 294 lb of PHA/ton of dry maize stover (Kurdikar *et al.*, 2001). Products and their applications are summarized in Fig. 20.1.

The functional unit in this study is defined as one acre of farmland. The system boundary in the ethanol production system includes the agricultural process, transportation of biomass, wet-milling and the maize stover conversion process, transportation of ethanol to the user, and driving operations by an ethanol-fuelled compact passenger vehicle. It is assumed that ethanol is used as E10 fuel - a mixture of 10% ethanol and 90% petrol by volume. The system boundary in the PHA production system includes the agricultural process, transportation of biomass, wet-milling, PHA fermentation and recovery, and the maize stover conversion process. Management of PHA wastes is not included in the system boundary because the fate of waste PHA is not clear at this point. PHA will replace conventional polymers, probably polystyrenes.

To rigorously compare the environmental performance of each cropping system, an equivalent function for each cropping system must be defined. Thus, different products from the cropping systems are normalized. The normalization is done by introducing alternative product systems using the system expansion method (International Organization for Standardization, 1998). As seen in Fig. 20.2, the alternative system for a driving operation by an ethanol-fuelled vehicle is a driving operation by a petrol-fuelled vehicle. The alternative

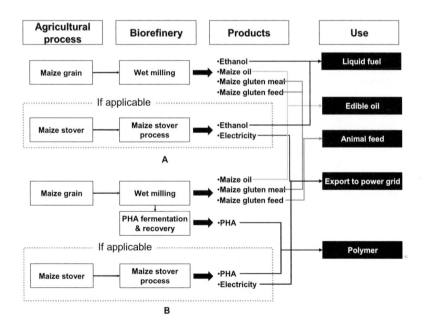

Figure 20.1. Products and their applications in a biorefinery (A: Ethanol production system, B: PHA production system).

products for maize gluten meal and maize gluten feed are maize grain and nitrogen in urea used as animal feeds (Wang, 2000). Surplus electricity from the maize stover conversion process in the ethanol production system is exported to replace electricity generated by fossil energy. In this study, we assume that the alternative electricity is generated in a coal-fired power plant. Lignin-rich residues could be blended with coal in a coal-fired power plant to generate electricity. The alternative product system for PHA is petroleum-based polymers. In this study, polystyrene resin is assumed to be an alternative product for PHA. Surplus electricity from maize stover conversion to PHA is also exported to the power grid.

Information on the agronomic inputs for maize cultivation in Illinois is available (National Agricultural Statistics Service, 2004). About 0.5 lb/acre (active ingredient) of herbicides is assumed sufficient to kill the cover crop (Kim et al., 2003). The lower heating value (LHV or net heating value) is used in the energy analysis. The electricity power grids are fixed for a specific biorefinery location. For example, the agricultural and the biorefinery processes occur in Illinois, in which the electricity power grid of the Mid-America Interconnected Network (MAIN) region is used. For other non-agricultural processes (e.g., fertilizers, agrochemicals, and fuels), which occur outside Illinois, the US average power grid data are applied. The data sources for each process are summarized in Table 20.2.

A compact passenger car is used as a reference vehicle, and its fuel economy is available from the US Environmental Protection Agency (US EPA) (US Environmental Protection Agency, 2004). The combined fuel economy is based on 55% city driving and 45% highway driving. The combined fuel economy of a 10% ethanol (E10)-fuelled vehicle is assumed to be equal to that of a conventional vehicle, which is 24 miles per petrol equivalent volume.

Air emissions from driving vehicles are also available from the US EPA. Carbon dioxide emissions are estimated by the amount of petrol used in the E10 fuel. Sulphur oxide (SO_x) emissions are calculated from the sulphur content in petrol (Wang, 2000). Using E10 fuel in

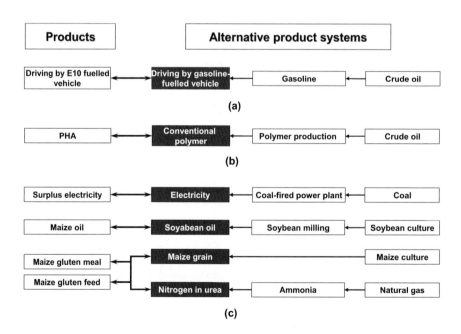

Figure 20.2. Alternative product systems for ethanol production and PHA production. (a) Ethanol production system; (b) PHA production system; (c) Co-product systems in both production systems.

Table 20.2. Primary data sources.

Process	References
Agronomic inputs	National Agricultural Statistics Service (2004).
Fuel requirement in the cropping system	Economic Research Service (2004).
Climate and soil data in the cropping site	National Oceanic and Atmospheric Administration (2004); Natural Resources Conservation Service (2004).
Fuel consumption for harvesting maize stover	Sheehan *et al.* (2002).
Maize wet-milling process/hauling distance of biomass	Wang (2000).
Maize stover conversion to ethanol	Sheehan *et al.* (2002).
Efficiency of electricity generation from burning lignin content in maize stover	Stahl and Neergaard (1998).
Dextrose production (wet-milling)	E. Vink, Cargill Dow LLC, Minnetonka, Minnesota, 2004, personal communication.
PHA fermentation and recovery	J. van Walsem, Metabolix Inc., Cambridge, Massachusetts, 2003, personal communication.
Maize stover conversion to PHA	Kurdikar *et al.* (2001).
Burdens associated with E10 driving/gasoline driving	US Environmental Protection Agency (2004).
US electricity production system	Kim and Dale (2004).
Nitrogen fertilizer/phosphorus fertilizer	International Fertilizer Industry Association (1998, 2004); Energetics, Inc. (2000); Kim and Overcash (2000).
Urea	Pagani and Zardi (1994); Energetics, Inc. (2000).
Other data (e.g., fuels, potassium fertilizer, polystyrene, etc.)	DEAM™ LCA database, see Ecobilan (2004).

the reference vehicle can reduce the air emissions from the vehicle except for non-methane organic gases and NO_x. The data for air emissions from E10 fuelled and petrol fuelled vehicles are presented in Table 20.3.

The potential environmental impact categories considered here are: natural resources used, non-renewable energy, global warming, acidification, eutrophication, and photochemical smog. In the global warming analysis, the carbon contents of biobased products are not taken into account because their carbon would be released to the atmosphere during later

Table 20.3. Air emissions from driving a compact passenger vehicle 50,000 miles.

Air emissions	Unit	Petrol fuelled Average	Petrol fuelled Standard deviation	E10 fuelled Average	E10 fuelled Standard deviation
CO_2[a]	Pound/mile	815	N/A[c]	758	N/A
CO	Pound/mile	1.6619	0.4659	0.8077	0.3366
HCHO	Pound/mile	0.0024	0.0009	0.0022	0
Non-methane organic gases	Pound/mile	0.0967	0.0059	0.0969	0.0159
NO_x	Pound/mile	0.2203	0	0.3304	0.1557
SO_x[b]	Pound/mile	0.1731	N/A	0.1612	N/A

[a] Carbon dioxide emission in driving a petrol-fuelled vehicle is obtained from the GREET model (Wang, 2000).
[b] Sulfur content in petrol is obtained from the GREET model (Wang, 2000).
[c] Not available.
Sources: Wang (2000); US Environmental Protection Agency (2004).

product life phases. Global warming is estimated by the 100-year global warming potential. The characterization factors are adapted from the Tools for the Reduction and Assessment of Chemical and Other Environmental Impacts (TRACI) model developed by the US EPA (Bare et al., 2003).

Results

Cropping System

Maize stover removal (CM50 and CMw70) consumes more fuels than maize culture with only maize grain removal due to the fuels used in harvesting maize stover. Higher energy consumption associated with the agronomic inputs in the CMw70 cropping system also results from the non-renewable energy related to herbicides used in killing winter cover crops. The cropping systems with winter cover crop and maize stover removal (CMw70) therefore consume more energy than continuous maize alone.

Carbon credits due to changes in soil organic carbon levels are 0.9 ton CO_2/acre/year for the CM cropping system, 0.7 ton CO_2/acre/year for the CM50 cropping system, and 1.2 ton CO_2/acre/year for the CMw70 cropping system. For this locality, soil organic carbon increases even with 50% maize stover removal mainly due to no-tillage practice. This result indicates that maize stover removal could lower soil organic matter accumulation rates, but cultivation of winter cover crops with maize stover removal would increase soil organic carbon levels. Environmental burdens due to nitrogen compounds (e.g., N_2O, NO_x and nitrate (NO_3^-)) will be reduced by harvesting maize stover, probably due to two fundamental physical phenomena: 1) the nitrogen content of stover is then not available for nitrification and denitrification processes and 2) the carbohydrates in maize stover are less available to promote nitrous oxide formation when maize stover is removed. The winter cover crops (CMw70) can further reduce the soil nitrogen-related burdens because winter cover crops serve as a nitrogen scavenger, particularly in early spring (Kuo et al., 1997; Reicosky and Forcella, 1998).

Table 20.4. Environmental impacts of the agricultural process (cradle-to-gate).

	Unit	CM	CM50	CMw70
Coal	Pound/acre/year	69	70	76
Crude oil	Pound/acre/year	105	142	188
Natural gas	Pound/acre/year	228	232	240
Non-renewable energy	Million Btu/acre/year	8.1	8.9	10.0
Global warming	Pound CO_2 equivalents/acre/year	743	1255	-195
Photochemical smog	Pound NO_x equivalents/foot/acre/year	0.01	0.02	0.01
Acidification	Moles H^+ equivalents/acre/year	856	864	758
Eutrophication	Pound nitrogen equivalents/acre/year	5.87	5.45	3.27

The potential environmental impacts associated with the cropping systems are summarized in Table 20.4, in which the figures represent cradle-to-gate values, including upstream processes (i.e., fertilizers, agrochemical and fuel manufacturing, etc.), cultivation, and transportation from farm to biorefinery. Maize stover removal requires more natural resources and non-renewable energy due to fuel consumption in harvesting maize stover. The CM50 cropping system has the highest global warming impact because of high fuel consumption and low carbon credits due to changes in soil organic carbon levels. The CMw70 cropping system offers a global warming credit (negative global warming) because of the significant increase in soil organic carbon levels, and the lower nitrous oxide emissions from the soil even though more fuels are required in the harvest. Although winter cover crops and maize stover removal can reduce the soil nitrogen-related burdens, the total impacts associated with acidification, eutrophication, and photochemical smog in the CMw70 cropping system are not significantly lower than the continuous maize culture case (CM) because the impacts associated with fuel consumption in harvesting maize stover and with planting winter cover crops are dominant.

Ethanol Production

The utilization of maize stover as a raw material for ethanol production increases the ethanol production rate per acre by 41% to 65% (depending on the maize stover removal rate) compared to the CM cropping system (continuous maize culture), in which maize grain is the only feedstock. Annual ethanol production per acre in each cropping system is 346 gal in CM, 485 gal in CM50, and 591 gal in CMw70. Ethanol from one acre, when used as an E10 fuel, can drive a compact passenger vehicle up to 79×10^3 miles in CM, 111×10^3 miles in CM50, and 135×10^3 miles in CMw70. The utilization of maize stover also generates surplus electricity that is exported to the power grid and thereby replaces electricity generated by a coal-fired power plant. Surplus electricity exported from the maize stover conversion process is 0.9 megawatt-hours (MWh) per acre per year in the CM50 cropping system and 1.4 MWh/acre/year in the CMw70 cropping system. The ethanol production system also produces animal feeds (e.g., MGM and MGF) at about 0.95 ton/acre/year to 1 ton/acre/year. Table 20.5 summarizes the products from various cropping systems when biomass is utilized for ethanol production.

The cradle-to-grave environmental impact profiles of each cropping system are summarized in Table 20.6. Utilization of ethanol as an E10 fuel from each cropping system has environmental credits in terms of crude oil use, non-renewable energy and global warming. Thus using ethanol as an E10 fuel saves crude oil, conserves non-renewable energy, and reduces greenhouse gases. Utilization of ethanol does not provide environmental credits in

Table 20.5. Products for various cropping systems when biomass is utilized for ethanol production.

	Unit	CM	CM50	CMw70
Biomass products				
Maize grain (dry basis)	Bushel/acre/year	115	114	124
Maize stover (dry basis)	Ton/acre/year	-	1.6	2.4
Final products				
Ethanol from maize grain (A)	Gallon/acre/year	346	342	373
Ethanol from maize stover (B)	Gallon/acre/year	-	143	218
Total ethanol (A+B)	Gallon/acre/year	346	485	591
MGM (dry basis)	Pound/acre/year	361	357	390
MGF (dry basis)	Pound/acre/year	1,556	1,537	1,678
Maize oil (dry basis)	Pound/acre/year	288	285	311
Electricity exported*	Megawatt-hour/acre/year	-	0.94	1.4
Distance driven by an E10-fuelled vehicle	10^3 Miles/acre/year	79	111	135

* Efficiency of electricity generation from burning lignin-rich residues: 32% (Stahl and Neergaard, 1998).

Table 20.6. Cradle-to-grave environmental impacts in the different cropping systems utilized for producing ethanol.

	Unit	CM	CM50	CMw70
Coal	Pound/acre/year	988	-52.2	-506
Crude oil	Pound/acre/year	-2,305	-2,973	-3,231
Natural gas	Pound/acre/year	160	107	78
Non-renewable energy	Million Btu/acre/year	-28.4	-51.0	-65.5
Global warming	Pound CO_2 equivalents/acre/year	-4,592	-8,428	-12,689
Photochemical smog	Pound NO_x equivalents/foot/acre/year	0.022	0.022	0.020
Acidification	Moles H+ equivalents/acre/year	1,462	1,228	1,054
Eutrophication	Pound nitrogen equivalents/acre/year	6.14	5.35	3.00

photochemical smog, acidification and eutrophication because large nitrogen (and phosphorus)-related burdens are released from the soil during maize cultivation. Utilization of maize stover combined with maize grain as raw materials for producing ethanol could reduce all the environmental impacts considered here due to energy recovery in the form of electricity and greater ethanol production. This occurs even though maize stover removal has adverse effects on some environmental indicators at the agricultural step (see Table 20.4). Use of winter cover crops could further reduce the environmental impacts because 1) winter cover crops reduce nitrogen-related environmental burdens, 2) a greater increase in soil organic carbon levels occurs, and 3) more maize stover is available without deteriorating soil quality. The CM50 and the CMw70 cropping systems offer negative coal use (credit) due to surplus electricity exported. Thus the utilization of maize stover for ethanol production would save coal as well as crude oil and non-renewable energy.

Sensitivity Analysis

Two factors are considered in this sensitivity analysis: 1) the avoided electricity production systems for electricity exported from the maize stover conversion process, and 2) the exhaust emissions from vehicle operations.

The avoided electricity system has been assumed to be electricity generated from a coal-fired power plant in the CM50 and the CMw70 cropping systems. It is also possible

Table 20.7. Cradle-to-grave natural resources used in the different avoided electricity systems (pound/acre/year).

	CM50			CMw70		
	Coal-fired power plant	Natural gas-fired power plant	Petroleum oil-fired power plant	Coal-fired power plant	Natural gas-fired power plant	Petroleum oil-fired power plant
Coal	-52	941	930	-506	1011	995
Crude oil	-2,973	-2,964	-3,457	-3,231	-3,218	-3,970
Natural gas	107	-312	76	78	-563	31

Note: Negative values are environmental credits.

that electricity generated from a natural gas-fired power plant or from a petroleum oil-fired power plant can be replaced by electricity exported from the maize stover conversion process. The environmental impacts of fossil-fuel-fired power plants are quite different from each other. However, the differences between the impacts caused by the choice of the avoided electricity systems are less than 7%, except for natural resources used, global warming and acidification. No significant differences in non-renewable energy, photochemical smog, and eutrophication occur because the fractions of the impacts associated with the avoided electricity system compared to the total impacts are small. Greenhouse gas production from the cropping systems is increased by 10% and 7%, respectively, in both CM50 and CMw70 when a natural gas-fired power plant (a petroleum oil-fired power plant) is the avoided electricity generation plant. A coal-fired power plant has the highest greenhouse gas emissions among fossil fuel-fired power plants.

Acidification varies significantly with the choice of the avoided electricity systems. The cradle-to-grave acidification rate is increased by 29% in CM50 and 52% in CMw70 when a natural gas-fired power plant is the avoided electricity generation plant. Likewise, when a petroleum oil-fired power plant is the avoided electricity generation plant, the cradle-to-grave acidification rate is increased by 12% in CM50 and 21% in CMw70. Thus the environmental outcomes of the avoided electricity systems are very sensitive to the natural resources used. Regardless of the avoided electricity system, using ethanol as an E10 fuel can save crude oil and non-renewable energy, and reduce greenhouse gas emissions. Table 20.7 presents the natural resources used in the different avoided electricity systems.

The primary contributing processes are the vehicle operations by E10 fuel and petrol. The upper/lower bound exhaust emissions are applied to identify the sensitivity of the exhaust emissions on photochemical smog, acidification, and eutrophication. The largest changes occur in the CMw70 cropping system due to the large amount of ethanol produced. The upper/lower bound exhaust emissions are changed for photochemical smog by 22% to 42%, for acidification by 15% to 36%, and for eutrophication by 9% to 31%, compared to the results obtained from applying average exhaust emissions. When the upper bound exhaust emissions are applied, the CM cropping system offers the best environmental performance in photochemical smog, while the application of the lower bound exhaust emissions indicates that the best performance in photochemical smog occurs in CMw70. The conclusions on other environmental impacts (i.e., acidification and eutrophication) remain unchanged regardless of the exhaust emissions.

Table 20.8. Products for various cropping systems when biomass is utilized for PHA production.

PHA production	Unit	CM	CM50	CMw70
PHA from maize grain (A)	Pound/acre/year	1,484	1,466	1,600
PHA from maize stover (B)	Pound/acre/year	-	469	716
Total PHA (A+B)	Pound/acre/year	1,484	1,935	2316
MGM (dry basis)	Pound/acre/year	1,174	1,160	1,266
MGF (dry basis)	Pound/acre/year	682	674	735
Maize oil (dry basis)	Pound/acre/year	222	220	240
Electricity exported	Megawatt-hour/acre/year	-	0.32	0.49

Table 20.9. Cradle-to-grave environmental impacts in the different cropping systems utilized for producing PHA.

	Unit	CM	CM50	CMw70
Coal	Pound/acre/year	2,400	1,973	1,978
Crude oil	Pound/acre/year	-2,147	-2,613	-3,051
Natural gas	Pound/acre/year	7.29	-198	-323
Non-renewable energy	Million Btu/acre/year	-0.70	-17.6	-26.8
Global warming	Pound CO_2 equivalents/acre/year	1,756	359	-2,001
Photochemical smog	Pound NO_x equivalents/foot/acre/year	0.013	0.011	0.008
Acidification	Moles H^+ equivalents/acre/year	1,098	840	617
Eutrophication	Pound nitrogen equivalents/acre/year	5.63	4.99	2.67

PHA Production

Annual PHA production per acre in each cropping system is 0.74 ton for CM, 0.97 ton for CM50, and 1.16 ton for CMw70. Utilization of maize stover as raw material for PHA production increases PHA production by 469 lb/acre to 716 lb/acre. Furthermore, the maize stover conversion process exports surplus electricity generated by burning lignin-rich residues. PHA production and electricity exported are summarized in Table 20.8.

Management of PHA wastes is not included in the system boundary because the fate of waste PHA is not clear at this point. Thus, the waste treatment phase of the alternative product (polystyrene) is also excluded. It is assumed that PHA replaces an equal mass of polystyrene (Heyde, 1998). The cradle-to-grave environmental impact profiles of each cropping system for PHA production are summarized in Table 20.9. Using PHA could save crude oil use, and thus conserve non-renewable energy. Utilization of maize stover as a raw material for PHA production combined with grain utilization could reduce many environmental impacts and offer environmental credits in terms of natural gas use. Winter cover crop utilization could also significantly improve the environmental performance of PHA and provide credits toward global warming.

Sensitivity Analysis

Two aspects are considered in this sensitivity analysis: 1) the avoided electricity production systems for electricity exported from the maize stover conversion process and 2) the alternative product system for PHA. Even though the environmental impacts vary with the choice of the avoided electricity systems, the final conclusion that the CMw70 cropping system provides the best overall environmental performance is unchanged.

Table 20.10. Impact categories offering environmental credits when a given petroleum based polymer is replaced by PHA.

Alternative petroleum-based polymer	CM	CM50	CMw70
PS	Crude oil; Non-renewable energy	Crude oil; Natural gas; Non-renewable energy	Crude oil; Natural gas; Non-renewable energy; Global warming
PC	Crude oil; Natural gas; Non-renewable energy; Global warming	Crude oil; Natural gas; Non-renewable energy; Global warming	Crude oil; Natural gas; Non-renewable energy; Global warming
HDPE	Crude oil; Natural gas	Crude oil; Natural gas; Non-renewable energy	Crude oil; Natural gas; Non-renewable energy; Global warming
LDPE	Crude oil; Natural gas; Non-renewable energy	Crude oil; Natural gas; Non-renewable energy	Crude oil; Natural gas; Non-renewable energy; Global warming
PP	Crude oil	Crude oil; Natural gas; Non-renewable energy	Crude oil; Natural gas; Non-renewable energy; Global warming

PHA can replace petroleum-based polymers like high density polyethylene (HDPE), low density polyethylene (LDPE), polycarbonate (PC), polypropylene (PP), and polystyrene (PS). Sensitivity analysis investigates what environmental impact categories change significantly when different petroleum-based polymers are replaced by PHA. When PHA replaces PC polymer, the environmental credits occur in crude oil, natural gas, non-renewable energy and global warming in the CM and the CM50 cropping systems. If PHA replaces HDPE, LDPE or PP, the CM cropping system does not offer environmental credits in global warming because these polymers release fewer greenhouse gases than PC polymers. The impact categories which offer environmental credits in the different alternative petroleum-based polymers for PHA are summarized in Table 20.10.

Eco-efficiency

If ethanol and PHA are produced together from the same unit of arable land, a question arises. What fraction of biomass would be utilized to produce ethanol in the more sustainable practice? To quantify the term sustainable practice, an eco-efficiency is used as defined in Equation (1).

$$\text{Eco-efficiency} = \frac{\text{Economic value added}}{\text{Environmental impact ratio}} \quad (1)$$

Table 20.11. Fractions of maize grain and maize stover utilized for producing ethanol when maximum and minimum eco-efficiencies occur in the CMw70 cropping system producing ethanol and PHA together.

	Maximum eco-efficiency	Minimum eco-efficiency
Coal	(1,1)	(0,0)
Crude oil	(0,0)	(0.56,1)
Natural gas	(0,0)	(0.82,1)
Non-renewable energy	(0,0)	(0.08,1)
Global warming	(1,0)	(0,1)
Photochemical smog	(0,0)	(0.95,1)
Acidification	(1,0)	(0,0)
Eutrophication	(1,0)	(0,0)

(x,y) x: fraction of maize grain utilized for producing ethanol, y: fraction of maize stover utilized for producing ethanol.

In Equation (1), the environmental impact ratio is defined as the ratio of the environmental impacts from the operating processes (e.g., maize production, wet-milling, PHA fermentation and recovery process, etc.) to the environmental credits from the avoided product systems. This term reflects how much environmental impact per environmental credit occurs in the product system. If the ratio is one, the impact and credits are equal. If less than one, the credits exceed the impacts. The economic value added is defined as the ratio of the market value of biobased products (i.e., ethanol and PHA) and co-products (e.g., MGM, electricity, etc.) to the cost of raw materials and fuel (e.g., maize grain, maize stover, coal, crude oil, natural gas, etc.) used in process operations. Thus a practice with a greater eco-efficiency would be more sustainable as defined here.

Since the eco-efficiency is dimensionless, it is possible to add the eco-efficiency of each impact category if all the impact categories are appropriately weighted. However, the weighting factors applied to each impact category vary significantly between decision-makers. Therefore, the eco-efficiency of each category is presented here instead of weighted results. Results are illustrated in Table 20.11. In estimating the eco-efficiency, it is assumed that PHA replaces five types of petroleum-based polymers (HDPE, LDPE, PC, PP, PS) by equal mass fractions. As seen in Table 20.11, the fraction of biomass utilized for ethanol production varies with the impact categories. For example, the maximum eco-efficiency with respect to global warming in the CMw70 cropping system occurs when 100% of maize grain and no maize stover are utilized for producing ethanol (see Fig. 20.3). For crude oil use and non-renewable energy, the maximum eco-efficiency occurs when both maize grain and maize stover are utilized for PHA production.

Conclusions

These results show that the cropping systems play an important role in the environmental performance of biobased products. Even though these analyses predict that maize stover removal reduces the rate of accumulation of soil organic matter (under scenarios simulated here) and consumes more fossil energy for harvesting during the agricultural process step, the utilization of maize stover could nonetheless improve the environmental performance of biobased products because: 1) more biobased products are made and 2) energy is recovered from maize stover. Removing maize stover combined with planting winter cover crops increases soil organic carbon levels and reduces the environmental burdens associated with soil nitrogen losses under the no-tillage systems modelled in this study. Furthermore, utilizing maize stover combined with winter cover crop production (CMw70) is the most environmentally favourable cropping system studied here. This result occurs because more maize stover is available for harvest without lowering soil quality. Additionally, more energy recovery from burning lignin-rich residues from the maize stover conversion process enhances the environmental performance of integrated cropping/biorefinery systems.

Both ethanol and PHA provide environmental credits in terms of crude oil use, non-renewable energy and global warming. Therefore, using ethanol as a liquid fuel and PHA could save crude oil and non-renewable energy and reduce greenhouse gas emissions. The greenhouse gas credit for PHA production in the CM and the CM50 cropping systems depends on which petroleum-based polymer is replaced by PHA. The PHA production system also offers an environmental credit in natural gas consumption, depending on the avoided petroleum-based polymers.

The utilization of biomass for ethanol and PHA would increase photochemical smog, acidification and eutrophication, particularly because of nitrogen- (and phosphorus-) related burdens from the soil during cultivation. Thus, other approaches to reduce these burdens in the agricultural process (e.g., use of buffer strips, etc.) are necessary to achieve better profiles for photochemical smog, acidification and eutrophication associated with the cropping systems.

Considering only sustainable utilization of biomass (i.e., at maximum eco-efficiency), the fractions of maize grain and maize stover utilized for producing ethanol vary with the impact categories. These calculations focus on only biomass utilization, not on societal demands for each product. Each product delivers its own function and therefore market forces

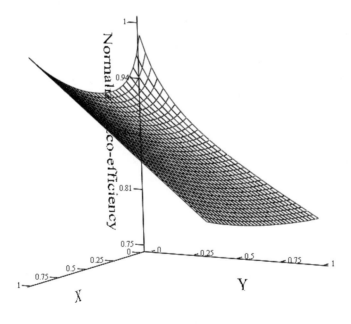

Note: the eco-efficiency is normalized by the maximum eco-efficiency.

Figure 20.3. Normalized eco-efficiency with respect to global warming.

may set the relative amounts of each product (and therefore raw material used), rather than eco-efficiency considerations.

Acknowledgements

The authors gratefully acknowledge support provided by Cargill Dow, LLC (Minnetonka, Minnesota), by DuPont Biobased Materials, Inc. (Wilmington, Delaware) and by the Center for Plant Products and Technologies at Michigan State University (East Lansing, Michigan).

References

Aden, A., Ruth, M., Idsen, K., Jechura, J., Neeves, K., Sheehan, J., Wallace, B., Montague, L., Slayton, A. and Lukas, J. (2002) *Lignocellulosic Biomass to Ethanol Process Design and Economics Utilizing Co-Current Dilute Acid Prehydrolysis and Enzymatic Hydrolysis for Corn Stover.* NREL/TP-510-32438. US Department of Energy, National Renewable Energy Laboratory, Golden, Colorado.

Bare, J., Norris, G. and Pennington, D. (2003) Tools for the Reduction and Assessment of Chemical and Other Environmental Impacts. *Journal of Industrial Ecology* 6(3-4): 49-78.

Del Grosso, S., Parton, W., Mosier, A., Ojima, D., Kulmala, A. and Phongpan, S. (2000) General Model for N_2O and N_2 Gas Emissions from Soils Due to Denitrification. *Global Biogeochemical Cycles* 14(4): 1045-1060.

Del Grosso, S., Parton, W., Mosier, A., Hartman, M., Brenner, J., Ojima, D. and Schimel, D. (2001) Simulated Interaction of Carbon Dynamics and Nitrogen Trace Gas Fluxes using the DAYCENT Model. In: Shaffer, M., Ma, L. and Hansen, S. (Eds.) *Modeling Carbon and Nitrogen Dynamics for Soil Management.* Lewis Publishers, Boca Raton, Florida.

Ecobilan. (2004) *DEAM™ LCA database.* Ecobilan, Paris, France.

Economic Research Service. (2004) Effects of Government Programs on Costs and Returns. *Data: Commodity Costs and Returns.* US Department of Agriculture, Economic Research Service, Washington, DC. [Accessed 2004.] Available from http://www.ers.usda.gov/Data/CostsAndReturns/testpick.htm

Energetics, Inc. (2000) *Energy and Environmental Profile of the U.S. Chemical Industry.* US Department of Energy Office of Industrial Technologies, Washington, DC. [Accessed 2004.] Available from http://www.oit.doe.gov/chemicals/pdfs/profile_chap5.pdf

Heyde, M. (1998) Ecological Considerations on the Use and Production of Biosynthetic and Synthetic Biodegradable Polymers. *Polymer Degradation and Stability* 59(1-3): 3-6.

International Fertilizer Industry Association. (1998) *Mineral Fertilizer Production and the Environment. Part 1. the Fertilizer Industry's Manufacturing Processes and Environmental Issues.* Technical Report No. 26 Part 1. International Fertilizer Industry Association, Paris, France.

International Fertilizer Industry Association. (2004) Nitrogen, Phosphate and Potash Statistics from 1973/74 to 2001/2002. *Statistics.* International Fertilizer Industry Association, Paris, France. [Accessed 2004.] Available from http://www.fertilizer.org/ifa/statistics/IFADATA/summary.asp

International Organization for Standardization. (1998) *ISO 14041: Environmental Management - Life Cycle Assessment - Goal and Scope Definition and Inventory Analysis.* International Organization for Standardization, Geneva, Switzerland.

Kim, S. and Dale, B. (2004) Life Cycle Inventory Information of the United States Electricity System. *International Journal of Life Cycle Assessment.* [Accessed 2004.] Available from http://www.scientificjournals.com/sj/lca/abstract/doi/lca2004.09.176

Kim, S. and Overcash, M. (2000) Allocation Procedure in Multi-Output Process - an Illustration of ISO 14041. *International Journal of Life Cycle Assessment* 5(4): 221-228.

Kim, S., Dale, B. and Hettenhaus, J. (2003) *Cover Crop Effect for Sustainable Stover Removal?* Presented paper, 25th Symposium on Biotechnology for Fuels and Chemicals, May 4-7, Breckenridge, Colorado.

Kuo, S., Sainju, U. and Jellum, E. (1997) Winter Cover Crop Effects on Soil Organic Carbon and Carbohydrate in Soil. *Soil Science Society of America Journal* 61(1): 145-152.

Kurdikar, D., Paster, M., Gruys, K., Fournet, L., Gerngross, T., Slater, S. and Coulon, R. (2001) Greenhouse Gas Profile of a Plastic Material Derived from a Genetically Modified Plant. *Journal of Industrial Ecology* 4(3): 107-122.

National Agricultural Statistics Service. (2004) Corn. *Reports by Commodity-Index of Estimates.* US Department of Agriculture, National Agricultural Statistics Service, Washington, DC. [Accessed 2004.] Available from http://www.usda.gov/nass/pubs/estindx1.htm#agchem

National Oceanic and Atmospheric Administration. (2004) Locate Weather Observation Station Record - Washington Co., Illinois station. *Web Climate Services.* US Department of Commerce, National Oceanic and Atmospheric Administration, Washington, DC. [Accessed 2004.] Available from http://lwf.ncdc.noaa.gov/oa/climate/stationlocator.html

Natural Resources Conservation Service. (2004) NSSC Soil Survey Laboratory Soil Characterization Data Query Interface - Washington Co., Illinois. *NSSC Soil Survey Laboratory Soil Characterization Database.* US Department of Agriculture, Natural Resources Conservation Service, Washington, DC. [Accessed 2004.] Available from http://ssldata.nrcs.usda.gov/querypage.asp

Nelson, R. (2002) Resource Assessment and Removal Analysis for Corn Stover and Wheat Straw in the Eastern and Midwestern United States - Rainfall and Wind-Induced Soil Erosion Methodology. *Biomass and Bioenergy* 22: 349-363.

Pagani, G. and Zardi, U. (1994) *Process and Plant for the Production of Urea.* US Patent 5276183. US Patent and Trademark Office, Washington, DC. [Accessed 2004.] Available from http://patft.uspto.gov/netacgi/nph-Parser?u=/netahtml/srchnum.htm&Sect1=PTO1&Sect2=HITOFF&p=1&r=1&l=50&f=G&d=PALL&s1=5276183.WKU.&OS=PN/5276183&RS=PN/5276183

Parton, W., Mosier, A., Ojima, D., Valentine, D., Schimel, D., Weier, D. and Kulmala, A. (1996) Generalized Model for N_2 and N_2O Production from Nitrification and Denitrification. *Global Biogeochemical Cycles* 10(3): 401-412.

Reicosky, D. and Forcella, F. (1998) Cover Crop and Soil Quality Interactions in Agroecosystems. *Journal of Soil and Water Conservation.* 53(3): 224-229.

Renard, K., Foster, G., Weesies, G., McCool, D. and Yoder, D. (1996) *Predicting Soil Erosion by Water: A Guide to Conservation Planning with the Revised Universal Soil Loss Equation (RUSLE).* Agricultural Handbook 703. US Department of Agriculture, Agriculture Research Service, Washington, DC.

Renewable Fuels Association. (2004) U.S. Fuel Ethanol Production Capacity. *Ethanol Production Facilities.* Renewable Fuels Association, Washington, DC. [Accessed 2004.] Available from http://www.ethanolrfa.org/eth_prod_fac.html

Sheehan, J., Aden, A., Riley, C., Paustian, K., Killian, K., Brenner, J., Lightle, D., Nelson, R., Walsh, M. and Cushman, J. (2002) *Is Ethanol From Corn Stover Sustainable?- Adventures in Cyber-Farming: A Life Cycle Assessment of The Production of Ethanol from Corn Stover for Use in a Flex Fuel Vehicle.* (Draft report for peer review, December 2002.) US Department of Energy, National Renewable Energy Laboratory, Golden, Colorado.

Stahl, K. and Neergaard, M. (1998) IGCC Power Plant for Biomass Utilisation, Varnamo, Sweden. *Biomass and Bioenergy* 15(3): 205-211.

Wang, M. (2000) *Greet 1.5a - Transportation Fuel-Cycle Model.* US Department of Energy, Argonne National Laboratory, Argonne, Illinois.

US Environmental Protection Agency. (2004) Annual Certification Test Results & Data - Certified Vehicle Test Result Report Data. *Transportation and Air Quality.* US Environmental Protection Agency, Washington, DC. [Accessed 2004.] Available from http://www.epa.gov/otaq/crttst.htm

Glossary

Adenosine 5′-triphosphate (ATP) Nucleoside triphosphate compound of adenine, ribose, and three phosphate groups that is the principal carrier of chemical energy in cells. The terminal phosphate groups are highly reactive in the sense that their hydrolysis, or transfer to another molecule, takes place with release of a large amount of free energy.

Barrel (bbl) A unit of measure in the petroleum industry equivalent to 159 litres, 35 imperial gallons, or 42 US gallons. It is different from a standard barrel in US measure which contains 31.5 US gallons or a standard barrel in British and Canadian measure which contains 36 imperial gallons.

Biodiesel A diesel fuel substitute made from vegetable oils or animal fats for use in existing diesel engines. It can be created from soybean oil, other vegetable oils, animal fats, or recycled cooking oil. (see **Soy diesel**)

Biomass Biological materials such as grains, wood, wood residues, animal wastes, and grasses which can be processed into fuels such as ethanol or biodiesel or used to generate electricity.

British thermal unit (Btu) The standard measure of heat energy. It takes one Btu to raise the temperature of one pound of water by one degree Fahrenheit at sea level.

Canola An edible oil derived from any of several varieties of genetically altered rape plant. Canola was originally a trademark in Canada for 'Canadian Oil,' but is now a generic term. (see **Oilseed rape**)

Cellulose The chief constituent of plant cell walls. Cellulose is tough and fibrous and is the principal structural material of plants. (see **Lignocellulose**)

Conservation Reserve Program (CRP) A voluntary US agriculture programme in which landowners receive annual rental payments and cost-share assistance to establish long-term, resource-conserving covers on eligible farmland.

Corn Belt The midwestern United States (mainly Illinois, Iowa, Indiana, Missouri and Ohio); excellent for raising maize and maize-fed livestock.

Corporate average fuel economy (CAFE) standards The sales-weighted average fuel economy, expressed in miles per gallon, that a manufacturer's fleet of passenger cars or light trucks manufactured for sale in the United States must equal or exceed, as required under the *Energy Policy and Conservation Act*.

Contract price The bulk rate paid at a loading terminal for a commodity on the basis of a long-term supply contract. (see **Wholesale price/markets**)

Decatherm (Dth) The quantity of heat energy which is equivalent to 1,000,000 British thermal units (Btu).

Denaturant A bitter or aversive substance which is added to another to prevent accidental or deliberate human consumption, such as petrol added to ethanol.

Distillers dried grains with solubles (DDGS) A co-product of the dry-milling ethyl alcohol distillation process. DDGS is an economical, partial replacement for maize, soybean meal, and dicalcium phosphate in livestock and poultry feeds.

Deoxyribonucleic acid (DNA) Polynucleotide formed from covalently linked deoxyribonucleotide units. It serves as the store of hereditary information within a cell and the carrier of this information from generation to generation.

Dry-mill (dry-grind) An ethanol production process where the grain is ground into flour and then processed without separation of the starch. (see **Wet-mill**)

Ethanol An alcohol that is a colourless, flammable liquid produced by fermentation of sugars. Ethanol can be used as a transportation fuel directly or blended, such as being used as an oxygenate in petrol. It may be produced from grain, sugarcane, sugar beets, leafy or woody plant material, or food and beverage wastes. It is also known as ethyl alcohol or grain alcohol.

Eutrophication A condition in an aquatic ecosystem where high nutrient concentrations stimulate blooms of algae.

Exajoule 10^{18} joules (the metric unit of energy and work). A quadrillion British thermal units equals 1.055 exajoules.

Feedlot A facility where livestock are fattened for market.

Fermentation A process in which an agent causes an organic substance to break down into simpler substances; especially, the anaerobic breakdown of sugar into alcohol.

Functional genomics The study of genes, their resulting proteins, and the role played by the proteins in the body's biochemical processes.

Furan One of a group of colourless, volatile, heterocyclic organic compounds containing a ring of four carbon atoms and one oxygen atom, obtained from wood oils and used in the synthesis of furfural and other organic compounds.

Furfural A liquid aldehyde with a penetrating odour; made from plant hulls and maize cobs; used in making furan and as a solvent. Furfural is also produced as a side-product when dilute acid hydrolysis is used to prepare biomass for enzymatic saccharification and fermentation. Furfural inhibits fermentation.

Gallon A unit of volume used for measuring liquids. One gallon equals 3.78541 litres.

Gas chromatograph-mass spectrometer (GC-MS) An advanced piece of analytical equipment typically used to characterize complex organic mixtures. The gas chromatography stage will separate the mixture into its constituent parts (i.e., substrate and series of active compounds). Each of these stages will then be passed to the mass spectrometer for individual characterization.

Gasohol A blend of petrol and alcohol (generally ethanol, but sometimes methanol) in which 10 per cent or more of the product is alcohol.

Gene A region of DNA that controls a discrete hereditary characteristic, usually corresponding to a single protein or RNA. This definition includes the entire functional unit, encompassing coding DNA sequences, non-coding regulatory DNA sequences, and introns.

Gene chip Also known as DNA chip, DNA microarray or DNA array; includes array of biological molecules such as DNA fragments connected to a matrix.

Gene expression The process by which a gene's coded information is converted into the structures present and operating in the cell. Expressed genes include those that are transcribed into mRNA and then translated into protein and those that are transcribed into RNA but not translated into protein (e.g., transfer and ribosomal RNAs).

Genetic engineering Altering the genetic material of cells or organisms to enable them to make new substances or perform new functions.

Genome The totality of genetic information belonging to a cell or an organism; in particular, the DNA that carries this information.

Gigajoule (GJ) 10^9 joules (the metric unit of energy and work) or approximately 0.95 million Btu.

Great Plains In the USA, the elevated plains which lie east of the Rocky Mountains and generally west of the 100th meridian. The Great Plains includes all or portions of the states of New Mexico, Texas, Oklahoma, Colorado, Kansas, Nebraska, Wyoming, Montana, South Dakota and North Dakota. The Great Plains is the USA's premier livestock and wheat-producing region.

Hexose Any of a group of simple sugars containing six carbon atoms in each molecule, as dextrose or fructose.

High Fructose Corn Syrup (HFCS) A sweetener produced by processing maize starch to fructose.

High performance liquid chromatography (HPLC) Type of chromatography that uses columns packed with tiny beads of matrix; the solution to be separated is pushed through under high pressure.

In situ In position; in its original place.

In vitro Term used by biochemists to describe a process taking place in an isolated cell-free extract. Also used by cell biologists to refer to cells growing in culture (*in vitro*) as opposed to in an organism (*in vivo*).

Joule (J) The metric unit of energy and work. One joule is defined as the amount of energy exerted when a force of 1 newton is applied over a displacement of 1 metre.

Kilowatt (kW) A measure of electric power equal to 1,000 watts.

Kilowatt-hour (kWh) A unit of energy equivalent to 1,000 watts being used continuously for a period of 1 hour; the unit most commonly used to measure electrical energy, as opposed to kilowatt, which is simply a measure of available power.

Label Chemical group, radioactive atom or fluorescent dye added to a molecule in order to follow its progress through a biochemical reaction or to locate it specially. Also, as a verb, to add such a group or atom to a cell or molecule.

Lignin Along with cellulose, a component of plant cell walls. Lignin is largely responsible for the strength and rigidity of plants. (see **Lignocellulose**).

Lignocellulose A combination of lignin and cellulose that strengthens woody plant cells. (see **Lignin** and **Cellulose**)

Limited liability company (LLC) A business structure which is a hybrid between a corporation and a partnership. It provides certain tax advantages compared with incorporation, yet still provides protection from personal liability for its members or owners.

Liquefied petroleum gas (LP or LPG) A generic name for propane and butane. LPG occurs naturally in oil and gas fields. It is either separated during the extraction process from the oil or gas field or is produced as one of the by-products of oil refining.

Loan deficiency payment (LDP) A US farm programme payment, which equals the loan rate less the market price for a programme commodity times the producer's eligible production.

Lucerne A legume plant used for fodder. Also known as alfalfa.

Megajoule (MJ) 10^6 or one million joules (the metric unit of energy and work).

Megawatt (MW) A unit for measuring power corresponding to 10^6 or one million watts.

Metabolic pathway A series of chemical reactions within a cell, catalysed by enzymes, which either results in the removal of a molecule from the environment to be used/stored by the cell (metabolic sink), or to initiate another metabolic pathway.

Metabolism The sum total of the chemical processes that take place in living cells.

Metabolite A substance produced by metabolism.

Methyl tertiary-butyl ether (MTBE) A chemical compound that is manufactured by the chemical reaction of methanol and isobutylene. MTBE is almost exclusively used as a fuel additive in petrol. It is one of a group of chemicals commonly known as 'oxygenates' because they raise the oxygen content of petrol.

Metric ton A unit of weight equal to 1,000 kilograms, or 2,204.6 pounds.

Microarray Sets of miniaturized chemical reaction areas with molecules connected to a matrix or support in a specific arrangement relative to each other that may be used to test DNA fragments, antibodies, or proteins.

Micrometer (μm) One millionth (10^{-6}) of a metre.

Millimole (mM) One thousandth (10^{-3}) of a mole.

Moiety A component part of a complex molecule.

Mole The weight of a compound in grams that is numerically equal to its molecular weight. One mole equals 6.023×10^{23} molecules.

Molecular weight The sum of the atomic weights of the atoms in the molecules that form these compounds.

Messenger RNA (mRNA) RNA molecule that specifies the amino acid sequence of a protein. Produced by RNA splicing (in eukaryotes) from a larger RNA molecule made by RNA polymerase as a complementary copy of DNA. It is translated into protein in a process catalysed by ribosomes.

Monomer The basic chemical unit that can form polymers by combining identical or similar molecules. (see **Polymer** and **Oligomer**)

Nameplate capacity The maximum production capacity under ideal conditions as defined by the facility manufacturer.

Nanogram (ng) One billionth (10^{-9}) of a gram.

Nanometer (nm) One billionth (10^{-9}) of a metre.

Natural gas A combustible mixture of hydrocarbon gases, composed primarily of methane, but also including ethane, propane, butane and pentane. Natural gas is found primarily in porous geological formations beneath the earth.

Nuclear magnetic resonance (NMR) Resonant absorption of electromagnetic radiation at a specific frequency by atomic nuclei in a magnetic field, due to flipping of the orientation of their magnetic dipole moments. The NMR spectrum provides information about the chemical environment of the nuclei. Two-dimensional NMR is used widely to determine the three-dimensional structure of small proteins.

Nucleic acid RNA or DNA, a macromolecule consisting of a chain of nucleotides joined together by phosphodiester bonds.

Nucleoside A sugar-base compound that is a nucleotide precursor. Nucleotides are nucleoside phosphates, a nitrogen base linked to a sugar molecule.

Nucleotide Nucleoside with one or more phosphate groups joined in ester linkages to the sugar moiety. DNA and RNA are polymers of nucleotides.

Oilseed rape A plant with an oil-rich seed. (see **Canola**)

Oligomer Short polymer, usually consisting (in a cell) of amino acids (oligopeptides), sugars (oligosaccharides), or nucleotides (oligonucleotides). (see **Monomer** and **Polymer**)

Organization of the Petroleum Exporting Countries (OPEC) A cartel of 11 countries organized for the purpose of negotiating with oil companies on matters of oil production, price, and future concession rights. Members include Algeria, Indonesia, Iran, Iraq, Kuwait, Libya, Nigeria, Qatar, Saudi Arabia, the United Arab Emirates, and Venezuela. OPEC's members collectively supply about 40 per cent of the world's oil output, and possess more than three-quarters of the world's total proven crude oil reserves.

Oxygenate A substance which adds oxygen. Methyl tertiary-butyl ether (MTBE) and ethanol are common petrol oxygenates which are used to boost petrol's octane value, enhance combustion, and reduce exhaust emissions.

Pentose Any of a group of monosaccharides having a composition corresponding to the formula $C_5H_{10}O_5$ including xylose, ribose, arabinose, etc.

Peptide A molecule formed by two or more amino acids, bound by bonds called 'peptide bonds'. Proteins are made of multiple peptides.

Petrol A petroleum liquid mixture consisting primarily of hydrocarbons used as fuel in internal combustion engines. Also known as gasoline.

Picogram (pg) One trillionth (10^{-12}) of a gram.

Polymer A molecule composed of many repeated subunits. (see **Monomer** and **Oligomer**)

Polymerase chain reaction (PCR) Technique for amplifying specific regions of DNA by the use of sequence-specific primers and multiple cycles of DNA synthesis, each cycle being followed by a brief heat treatment to separate complementary strands.

Probe Defined fragment of RNA or DNA, radioactively or chemically labelled, used to locate specific nucleic acid sequences by hybridization.

Protein The major macromolecular constituent of cells. A linear polymer of amino acids linked together by peptide bonds in a specific sequence.

Proximate analysis The fuel composition as a percentage of ash, moisture, fixed carbon and volatile matter.

Pyrolysis A form of incineration that chemically decomposes organic materials by heat in the absence of oxygen. Pyrolysis typically occurs under pressure and at operating temperatures above 430°C (800°F).

Quad A unit of energy equal to 10^{15} (one quadrillion) Btu.

Quadrillion 10^{15} or 1,000,000,000,000,000. In European usage, one billiard.

Rack price In the petroleum industry, the published bulk rate paid at a city loading terminal by a tanker-load buyer without a long-term supply contract. (see **Wholesale price/markets**)

Recombinant DNA Any DNA molecule formed by jointing DNA segments from different sources. Recombinant DNAs are widely used in the cloning of genes, in the genetic modification of organisms, and in molecular biology generally.

Reformulated gasoline (RFG) In the USA, petrol formulated for use in motor vehicles to meet clean air standards in cities with air pollution problems. RFG has used methyl tertiary-butyl ether (MTBE) or ethanol as an additive to reduce air pollution while maintaining octane levels and fuel performance.

Renewable energy Energy obtained from sources that are essentially inexhaustible, such as conventional hydroelectric power, wood, waste, geothermal, wind, photovoltaic, and solar thermal energy.

Ribonucleic acid (RNA) Polymer formed from covalently linked ribonucleotide monomers.

Saccharification The process of converting (e.g., a starch) into sugar.

Soy diesel Biodiesel derived from soybean oil. It may also indicate a combination of filtered and clarified crude soybean oil and diesel fuel in a 20/80 blend. *Soy Biodiesel* is derived from virgin soybean oil. *B2* is a blend of 2% soy biodiesel with 98% petroleum diesel. *B20* is a blend of 20% soy biodiesel with 80% petroleum diesel. *B100* is 100% soy biodiesel. (see **Biodiesel**)

Spot price The current price for current delivery of product, usually quoted for a few key trading centres around the world. (see **Wholesale price/markets**)

Stoichiometric combustion The ideal combustion process in which a fuel is burned completely. All carbon is converted to carbon dioxide; all hydrogen is converted to water; and all sulphur is converted to sulphur dioxide.

Stover The stalks and leaves left after grain harvest.

Switchgrass (*Panicum virgatum*) A summer perennial grass that is native to North America.

Texas Panhandle The northernmost 26 counties of the US state of Texas.

Thermo-gravimetric analysis An indicator of the temperature at which the solid fuel will start releasing the volatile matter and also the rate at which the volatile matter is released from the fuel.

Ton In the USA, a unit of weight equal to 2,000 pounds avoirdupois (907.19 kilograms).

Transcription (DNA transcription) Copying of one strand of DNA into a complementary RNA sequence by the enzyme RNA polymerase.

Ultimate analysis The fuel elemental composition as a percentage of carbon, oxygen, nitrogen and sulphur.

Watt The unit of power; equal to 1 joule per second.

Wet-mill An ethanol production process where the grain is separated into starch, germ and fibre. The starch is separated and fermented. (see **Dry-mill**)

Wholesale price/markets Wholesale is a generic term for sales of goods to retailers for eventual sale to the consumer. For specific examples, see **Rack price**, **Contract price** and **Spot price**.

Xylose A five-carbon aldose (sugar).

Index

2002 farm bill *see Farm Security and Rural Investment Act of 2002*

Acid Rain Program 21
Aerobic digestion 13, 48–49, 64, 297–298
    benefits of 13
Africa 14
Agricultural Risk Protection Act of 2000 17
Agriculture as a Producer and Consumer of Energy conference (2004) viii, ix, 4, 23–26
Agriculture Resource Management Survey 74, 79
Alabama 243, 256
Alaska 2, 8, 22, 25, 120
Alberta 170–174, 177
Alternative Motor Fuels Act of 1988 20
American Jobs Creation Act of 2004 23
Anderson, David x
Animal wastes 182, 266–280
    as energy feedstock 267–280
Annamalai, Kalyan x
Arab oil embargo (1973) 1, 34, 112, 114, 116–117, 125
Arctic National Wildlife Refuge 22, 120, 128
Argentina 7, 24
Arkansas 77, 243
Asia financial crisis (1990s) 131
Australia 7, 24, 113
Austria 62

Bagasse 58, 186, 219
Ball, Eldon 93
Barley 174–175, 182, 195
    acreage changes from bioenergy production 55
    price effects from bioenergy production 56

Beef cattle
    farm business expenses 102–103, 105
    industry 266
    production and energy costs 80
Bergeron, Nancy x
Bermudagrass (*Cynodon dactylon*) 208
BIOCAP Canada 168
Biodiesel viii, 2, 9, 12, 17, 31, 43, 59–60, 63, 88–89, 182, 296–297
    advantages of 9–10, 43–44
    animal fats and 12, 43, 158
    as lubricity agent 21, 54
    B20 blend 12, 43–44, 297
    disadvantages of 44, 297
    feedstock supply 15, 51, 63, 157
    price and income effects 52, 165
    prices 45–46
    production 62, 64, 156–165
        costs 44–45
        studies 59–61, 156–165
    reduction of greenhouse gasses and 10
    reduction of pollutants and 9
    rendering industry and 12, 158
    sources of feedstock 12, 43–45, 157
Biodiesel Fuel Education Program 17
Bioenergy
    benefits of 13, 195, 300–301
    life cycle analysis 195–203, 317–330
    resource base 50–51, 59, 195–203
    rural electric cooperatives and 295–299
Biofuels
    carbon sequestration and 10, 300
    greenhouse gas emission offset and 303
Biogas 182
Biomass 112, 205
    co-firing with coal 13, 254–264
        ash sales 261–262
        costs 262–264
    co-milling with coal 255
    production rates 182

Biomass energy 30
　availability 121
　benefits of 182, 195
Biomass Research and Development Act of 2000 17, 89
Biorefining 42, 205–215, 317–330
　described 205–206, 317–320
　source of feedstock and 205
Boulding, Kenneth 113
Bovine spongiform encephalopathy 12
Boylan, Doug x
Brazil 7, 20, 24, 61–62, 64, 119
British Columbia 170–171
Brown grease 12, 157–158
Bush, George W. 8, 18, 22–23, 48

CAFE *see* Corporate Average Fuel Economy standard
California 11, 21, 47, 77, 81, 89, 116, 256
　banning of MTBE and 11
Canada 7, 15, 24, 61, 113, 116, 125, 168–179
Canola *see* Oilseed rape
Carbon 10, 303
　forests and 305, 314
　land use and 305, 314
　sequestration 10, 305
Carbon dioxide (CO_2) 34, 122, 126, 264, 267, 280, 301, 303–304, 306, 314, 320
　as ethanol production co-product 35–36, 141, 143, 154, 175, 185
　as source of global warming 301
Carbon monoxide (CO) 9, 34, 43, 54, 219
Caribbean Basin Initiative 20
Carter, Jimmy 3-4, 18
Cellulose 205
Cheese whey 182
Chicken *see* Poultry
China viii, 24, 61
Clark, R. Nolan x
Clean Air Act Amendments of 1990 11, 20–21, 34, 43, 53, 156
　Clear Skies Initiative and 302
　Oxygenated Fuels Program and 11
　Reformulated Gasoline Program and 11
Clinton, William J. 120
Coal 9, 39, 114–115, 120, 195, 258
　co-firing with animal waste 267–280
　　economics 268

　co-firing with biomass 120, 121, 254–264
Coal liquids 120
Collins, Keith ix–x, xv
Colorado 77
Commodity Credit Corporation Bioenergy Program 17–18, 153
Commodity Credit Corporation Charter Act 17
Confined animal feeding operations 266, 297
　locations 266
Connecticut 11
　banning of MTBE and 11
Conservation Reserve Program 14, 16, 54, 206
　biomass harvesting and 16, 54–55, 121, 206, 208, 215
Conservation Security Program 18
Conservation tillage 10, 83–84, 92
Conway, Roger xv, 24
Cornforth, Gerald x
Corporate Average Fuel Economy standard viii, 16, 20, 119–120, 128
Cotton
　acreage changes from bioenergy production 55
　acres irrigated 77
　farm business expenses 99–101
　price effects from bioenergy production 56
　production and energy use 73–75
Cottonseed 43
　oil 44
Crude oil
　demand 131, 134
　imports 8, 30
　prices viii, 2, 8–9, 131–133, 181, 218
　renewable energy and 15
　reserves 8
　sources of 15
Crude Oil Windfall Tax Act 19
Czarick, Michael x

Dairy
　farm business expenses 102–103, 105
　production and energy costs 80
　production and energy use 75
Dairyland Power Cooperative 297–298
Dale, Bruce x
De La Torre Ugarte, Daniel x
Denmark 62

Diesel 8–10, 21, 31, 43–44, 63, 73, 77, 80, 182, 205
  prices 32–33, 45, 63, 81–82, 107–108
  substitution for petrol in agriculture 70
  US consumption 32, 54, 182
    farm 69–71, 78–79, 96, 106
Diesel, Rudolf 1
Dinneen, Bob 25
Distillers dried grains and solubles 35–36, 63, 118, 140–141, 152, 154, 178, 185
Doering, Otto x, xv, 25
Dry-grind ethanol production process *see* Dry-mill ethanol production process
Dry-mill ethanol production process 11, 35–38, 139–140, 174–175, 177, 182
  considerations 139, 154
  economics 139–156, 182
  ethanol pricing strategies 146–147
  yield 36, 148–149
Dubman, Bob 93
Duffield, James ix–x, xv, 93
Duncan, Marvin xv

Eastis, Jack x
East River Electric Power Cooperative 296
Egypt 1
Eidman, Vernon x, xv, 24
Electricity 69, 77, 80, 90, 182, 195
  prices 33, 81–82, 85, 107–108
  US consumption 32
    farm 69, 72, 79, 96
El-Halwagi, Mahmoud x
Energy Conservation Reauthorization Act of 1998 20
Energy consumption
  farm 10–11, 68–93
    direct 68, 70–71, 73, 75, 84, 96, 106
    indirect 68, 70–71, 75–76, 84, 106
Energy costs
  direct farm 75–76
  indirect farm 75–76
Energy crisis (1970s) 1, 3, 8, 69, 76
Energy Policy Act of 1992 12, 18–20, 43, 296
Energy Policy Act of 2003 156–157
Energy Policy and Conservation Act of 1975 16
Energy Security Act of 1980 14, 19
Energy shock (1978) 2

Energy Tax Act of 1978 19
English, Burton x
Epplin, Francis x
Ernstes, David x, xv
Ethanol viii, 1–2, 9–10, 17, 20, 31, 34–42, 59–60, 63, 91, 112, 124–125, 139–165, 168–179, 181–193, 195–203, 205–215, 218–230, 317–330
  alternative fuelled vehicles and 20, 219
  as octane enhancer 11, 146, 181, 219
  as oxygenate 9, 34, 38, 52–53, 181
  benefits of 9, 34, 118
  blended with petrol 169, 219, 317, 319
    advantages 34, 219
    disadvantages 35
  Canada and 168–179
    public policy 169–172
  energy balance 10, 119, 185–186, 195
    lignocellulosic process 185–186
  feedstock supply 50–52, 195–203
  negative facets 118
  price and income effects 52, 165
  prices 37–39, 54, 140, 145–147
  production 59–61, 156–165, 169–170, 173, 219
    costs 36–37, 41, 177–178
    dry-mill ethanol production process *see* Dry-mill ethanol production process
    employment and 219, 229
    grain as feedstock 35–39, 63, 88, 174–175, 182
    incentives 19–23, 118–119, 169–171
    lignocellulose as feedstock *see* Lignocellulose as ethanol feedstock
    maize as feedstock 11, 15, 63, 139, 157, 169, 177, 185
    new technologies 42, 219
    sugar as feedstock 24, 61
    wet-mill ethanol production process *see* Wet-mill ethanol production process
  studies
    economic 36–42, 139–165, 181–193, 205–215, 218–230
    environmental impacts 195–203, 300–314, 317–330

increased production 51–52, 156–165
US production 32, 220
US public policy and 18–23
world production 61–62
Ethanol Expansion Program (Canada) 168–170
Europe 24, 62, 135
European Union 7, 61–62, 64, 161

Fairchild, Brian x
Farm credit crisis (1980s) 5
Farm Foundation xv, 4
Farm Security and Rural Investment Act of 2002 viii, 17–18, 88–89, 161, 209
  energy title and 17, 89, 299
Federal Biobased Product Procurement Preference Program 17
Federal Production Tax Credit 48, 64
Ferris, John (Jake) x
Fertilizer 8, 11, 25, 72, 80
  consumption in US agriculture 72, 97, 99, 106
  energy use and 72, 90
  input use and 72
  prices 3
  US industry 25, 90
Fischer, James xv
Flexible fuel vehicles 20
Florida 77, 89–90
Ford, Henry 1
Fossil fuels 9
France 61–62
Fruit farms
  business expenses 102–103, 105
  production and energy use 75

Gasifier combined cycle technology 60–61
Gasohol 169, 171
Gasoline *see* Petrol
Georgia 243, 256–257
Geothermal energy 30
Germany 61–62
Gillig, Dhazn x
Global warming 219, 300–301, 322
Gray, Boyden 24
Greenhouse gasses 9–10, 34, 168, 178, 300–302

emission offsets and forests 304
reasons for reducing 302–303
sources of 304

Halbrook, Steve x, xv
Hemicellulose 205
Herbst, Brian xi
High fructose corn syrup 91
Highway Trust Fund 22
Hogs *see* Swine
Holtzapple, Mark xi
Homestead Act of 1862 115
Hybrid electric vehicles 16
Hybrid poplar (*Populus* x) 13–14, 16, 54–55, 195
Hydroelectric energy 30
Hydrogen 182

Idaho 77
Illinois 16, 158, 167, 196, 221, 297, 318, 320
India 7, 24, 61
Indiana 114, 154, 167, 196, 221
Intergovernmental Panel on Climate Change 300
Iowa 16, 47–48, 59, 121, 196, 206, 221, 297
Iran 2
Iran-Iraq war (1980-1988) 2
Iraq 2, 15
Iraq war (2003–) 3
Irrigation 8, 77, 104
  US farm acres 77, 104
Israel 1
Italy 62

Joshi, Satish xi
Jumpstart Our Business Strength Act 299

Kansas 196, 221
Kim, Seungdo xi
Klein, K.K. xi
Klose, Steven xi
Kumarappan, Subramanian 93
Kuwait 2, 15
Kyoto Protocol 168, 301
  ratification and Canada 168
  United States and 302

Lard 43, 157

Latin America 14
Lau, Michael xi
Lee, Heng–Chi xi
Lignin 205, 318–320
Lignocellulose 205, 283–284
    as ethanol feedstock 12–13, 34, 38–42, 57, 60–64, 169, 174–175, 181–193, 205–215, 218–230, 283–284, 317–330
        described 183–185, 283–284
        economic modelling 181–193, 205–215, 218–230
        economic viability 206, 215
        environmental impacts 195–203, 317–330
        feedstock supply 59–61, 64, 186, 195–203, 206, 215, 221
        infrastructure 206
Lindemar, Kevin xi, 25
Liquefied petroleum (LP) gas 49, 64, 75, 114, 243
Liu, Z. Lewis xi
Louisiana 90, 132
Lovins, Amory 112, 114
LP gas
    broiler house use and 244
    US farm consumption 69, 78–79, 96, 106
Lucerne 61
    acreage changes from bioenergy production 55
    acres irrigated 77

Maize 15, 38, 43, 83, 91, 118, 120, 139, 156–165, 174–175, 181, 195–197, 206, 219, 296, 317, 320
    acreage changes from bioenergy production 55, 59–60
    acres 10, 196
        irrigated 77
    as ethanol production feedstock *see* Ethanol production maize as feedstock
    farm business expenses 99–101
    price effects from bioenergy production 14, 52, 56, 59
    prices 6, 38–39, 140–141, 153
    production and banning of MTBE 11
    production and energy use 73–75
    stover 15, 39–41, 58–59, 195–203, 206, 218–230, 317–330
        collection challenges 206
        collection cost estimates 200, 215
        prices 59
Maize oil 44, 158
Manitoba 170–174
Manure *see* Animal wastes
Mapemba, Lawrence xi
Maryland 296
McCarl, Bruce xi
Menard, R. Jamey xi
Mercury 9
Methane (CH_4) 9, 13, 34, 48, 88, 112, 267, 304, 306
Methyl tertiary-butyl ether (MTBE)
    as oxygenate 9, 34, 52–54, 64
    banning of 11, 21, 181
        effects 62
    groundwater contamination and 9, 34, 37
    prices 54
Mexico 2, 15, 24, 116
Michigan 297
Middle East viii, 1, 2, 15, 25
Minnesota 16, 47–48, 153–154, 196, 221, 297
Miranowski, John xi, xv, 24
Mississippi 77
Missouri 196, 221
Monchuk, Daniel 93
Montana 77
MTBE *see* Methyl tertiary-butyl ether
Mukhtar, Saqib xi

National Biodiesel Board 17
National Biomass Ethanol Program (Canada) 168
National Energy Act of 1978 18
National Energy Policy Development Group 8, 18, 22
National Rural Electric Cooperative Association 296
Natural gas 2, 9, 25, 64, 72, 75, 77, 90, 115–116, 119–120, 143, 195
    demand 132, 134
    importance as energy source 126
    imports 30

prices viii, 3, 8, 32–33, 63, 81–82, 107–108, 122, 126, 131–133, 147, 243
   reserves 3, 125, 132
   US consumption 32, 85, 89
      farm 69, 96, 106
Nebraska 77, 196, 221
Nelson, Richard xi
New Brunswick 172
New Mexico 266
New York 11, 16, 115
   banning of MTBE and 11
Nigeria 15
Nitrogen oxides (NO_x) 9, 10, 44, 254, 280
   emissions 10, 262, 264
Nitrous oxide (N_2O) 34, 304, 306
North America 131–133, 135, 172, 174, 177
Norway 2
Nuclear power 9
Nursery and greenhouse producers
   production and energy use 75

Oats 195
   acreage changes from bioenergy production 55
   price effects from bioenergy production 56
Ohio 196, 221
Oilseed rape 43–44, 157
Oklahoma 16, 90, 207–209, 215, 266
Olar, Maria xi
Old world bluestem (*Bothriochloa* sp.) 208
Ontario 171–173
Oregon 77
Organization of the Petroleum Exporting Countries 1, 112, 131, 133
Outlaw, Joe xi, xv
Oxygenated Fuels Program 11, 34

Peanut oil 1
Peanuts 43
   farm business expenses 99–101
   production and energy use 73, 75
Pennsylvania 16, 115
Persian Gulf War (1991) 2
Pesticides 72
   consumption in US agriculture 73, 97, 106
   energy use and 72

Petrol 1, 2, 8, 10, 15, 20, 31, 61, 70, 73, 77, 80, 119, 168, 182, 185, 205, 317
   fuel efficiency and 119
   prices 32–33, 37, 53–54, 63, 81–82, 107–108, 139, 181, 218
   reformulated 9, 34, 52–54
   US consumption 32, 53, 118, 182
      farm 69–71, 78–79, 96, 106
Philippines 24
Pigs *see* Swine
Polyhydroxyalkanoates 317
   wet-mill process and 319
Pork *see* Swine
Potatoes 182
Poultry
   farm business expenses 102–103, 105
   housing 243-252
      energy efficiency and 243-252
   production and energy costs 80
   production and energy use 75, 243–252
Poultry industry 92, 243–252, 266
Prince Edward Island 172
Priyadarsan, Soyuz xi
Public Utility Regulatory Policies Act of 1978 18

Qin, Xiaoyun xi
Quebec 171–173, 177

Raisins 182
Rapeseed 44, 62
Reauthorization Act of 1998 12
Reformulated Gasoline Program 11, 34
Renewable Energy Production Incentive 19
Renewable Energy Systems and Energy Efficiency Improvements Program 17
Renewable Fuels Standard 12, 156
Renewable Portfolio Standard 23
Rice 7, 195
   acreage changes from bioenergy production 55
   farm business expenses 99–101
   price effects from bioenergy production 56
   production and energy costs 80
   production and energy use 73–75
   straw 58, 219

Romain, Robert xi
Rural electric cooperatives 295–299
Russell, Kathy xi
Russia 2, 7, 25
Rye 195

Sant, Roger 114, 128
Saskatchewan 170–173, 175
Saudi Arabia 15
Shale oil 120
Shapouri, Hosein xv
Sheehan, John xi
Slininger, Patricia xi
Solar photovoltaic (solar-PV) cells 232–241, 254
Solar power 2, 13, 30, 112, 232–241
Sorghum
    acreage changes from bioenergy production 55
    price effects from bioenergy production 14, 52, 56
    production and energy use 73–74
    silage as ethanol feedstock 181–193
    stover 58
South Africa 24
South America 140
South Dakota 196, 221, 296
Southern Maryland Electric Cooperative 296–297
Soybean meal 140
    price effect from bioenergy production 52
Soybean oil 44
    biodiesel production and 12, 44, 63
    price effect from bioenergy production 14, 52, 59
    prices 158
Soybeans 12, 43, 62, 83, 140, 156–165, 296, 317
    acreage changes from bioenergy production 55, 59–60
    acres 7, 10
    acres irrigated 77
    farm business expenses 99–101
    price effect from bioenergy production 14, 52, 59
    prices 12
    production and energy costs 80
    production and energy use 73–75

Spain 61–62
Steel Trade Liberalization Program Implementation 20
Stochastic simulation 186
Strategic Petroleum Reserve 2
Sugar 85
Sugarcane *see* Bagasse
Sulphur (S) 9, 43, 121
Sulphur dioxide (SO_2) 9, 43, 54, 182, 264, 280, 320
Sunflowers 62, 157
Sweden 62
Sweeten, John xii
Swift, Gustavius 115
Swine 91
    farm business expenses 102–103, 105
    production and energy costs 80
    production and energy use 75
Switchgrass (*Panicum virgatum*) 13–14, 16, 39, 54–56, 120–121, 195, 206, 254
    co-firing in coal-fired power plants 254–264, 267
Syria 1

Taiwan 24
Tall fescue (*Festuca arundinacea*) 208
Tallow 43–44, 157
Tembo, Gelson xii
Terrorist attack, September 11, 2001 3
Texas 47, 54, 77, 90, 181–182, 186, 192–193, 232, 266
Thailand 61
Thien, Ben xii
Thompson, Mary xv
Tiffany, Douglas xii
Tobacco 296
    farm business expenses 99–101
    production and energy use 75
Trans Alaska Pipeline 2
Tree nut farms
    business expenses 102–103, 105
    production and energy use 75
Trinidad and Tobago 25

Ukraine 7, 25
Union of Soviet Socialist Republics (former) 5

United Kingdom 24, 61–62
United Nations Framework Convention on
    Climate Change 21, 301
University of Idaho 17
Uruguay 24
US Congress viii, 12, 15, 18, 20, 22–23, 48
US Department of Agriculture xv, 4, 17
    Bioenergy Program 12
    Office of Energy 4
    Office of Energy Policy and New Uses
        xv, 4
US Federal Energy Regulatory Commission
        124
Utah 77

Van Houten, Gretchen xv
Vegetable farms
    business expenses 102–103
Venezuela 15, 25
Vick, Brian xii
Volatile organic compounds 9, 34
Volumetric Ethanol Tax Credit 23

Walsh, Marie xii
Washington (US state) 77
Wet-mill ethanol production process 11,
        35–37, 139, 175, 177, 317–330
    described 318
    reduction of greenhouse gasses and 10,
        168–169, 219, 303
    yield 36, 318
Wheat 7, 61, 174–175, 195, 317–318
    acreage changes from bioenergy
        production 55
    farm business expenses 99–101
    price effects from bioenergy production
        56
    production and energy use 74–75
    straw 39
Whitman, Carol xii
Willow (*Salix* sp.) 13–14, 16, 54–55, 195
Wilson, Steve xii, 25
Wind power 13, 23, 31, 46, 63–64, 88, 112
    advantages of 46
    disadvantages of 46
    production costs 46–47
    US production 32
    wind energy developers and 48

Windmills
    solar photovoltaic cells and 232–241
    water pumping and 232–241
Wisconsin 196, 221, 297
Wood energy 49–50, 64, 115, 120
    US production 32, 50
Worley, John xii
Wyoming 77

Yellow grease 12, 43–44, 157–158
    prices 158
Young, Sandy xv

Zemo, Billy xii